I0044540

Petroleum Reservoir Engineering

Petroleum Reservoir Engineering

Editor: Michael Dedini

RCALLISTO REFERENCE

www.callistoreference.com

Callisto Reference,
118-35 Queens Blvd., Suite 400,
Forest Hills, NY 11375, USA

Visit us on the World Wide Web at:
www.callistoreference.com

© Callisto Reference, 2019

This book contains information obtained from authentic and highly regarded sources. Copyright for all individual chapters remain with the respective authors as indicated. All chapters are published with permission under the Creative Commons Attribution License or equivalent. A wide variety of references are listed. Permission and sources are indicated; for detailed attributions, please refer to the permissions page and list of contributors. Reasonable efforts have been made to publish reliable data and information, but the authors, editors and publisher cannot assume any responsibility for the validity of all materials or the consequences of their use.

ISBN: 978-1-64116-093-3 (Hardback)

Trademark Notice: Registered trademark of products or corporate names are used only for explanation and identification without intent to infringe.

Cataloging-in-Publication Data

Petroleum reservoir engineering / edited by Michael Dedini.
    p. cm.
Includes bibliographical references and index.
ISBN 978-1-64116-093-3
1. Petroleum engineering. 2. Oil reservoir engineering. I. Dedini, Michael.
TN870 .P48 2019
665.5--dc23

Table of Contents

Preface VII

Chapter 1   Laboratory studies of rice bran as a carbon source to stimulate indigenous
             microorganisms in oil reservoirs 1
             Chun-Mao Chen, Jin-Ling Wang, Jung Bong Kim, Qing-Hong Wang, Jing Wang,
             Brandon A. Yoza and Qing X. Li

Chapter 2   Selective dissolution of eodiagenesis cements and its impact on the quality
             evolution of reservoirs in the Xing'anling Group, Suderte Oil Field,
             Hailar Basin, China 13
             Zhen-Zhen Jia, Cheng-Yan Lin, Li-Hua Ren and Chun-Mei Dong

Chapter 3   HSE training matrices templates for grassroots posts in petroleum and
             petrochemical enterprises 29
             Shao-Lin Qiu, Lai-Bin Zhang and Mu Liu

Chapter 4   Establishment of a multi-cycle generalized Weng model and its application in
             forecasts of global oil supply 39
             Yi Jin, Xu Tang, Cui-Yang Feng, Jian-Liang Wang and Bao-Sheng Zhang

Chapter 5   Features and genesis of Paleogene high-quality reservoirs in lacustrine mixed
             siliciclastic–carbonate sediments, central Bohai Sea, China 45
             Zheng-Xiang Lü, Shun-Li Zhang, Chao Yin, Hai-Long Meng, Xiu-Zhang Song
             and Jian Zhang

Chapter 6   Understanding aqueous foam with novel CO_2- soluble surfactants for
             controlling CO_2 vertical sweep in sandstone reservoirs 56
             Guangwei Ren and Quoc P. Nguyen

Chapter 7   A new mathematical model for horizontal wells with variable density
             perforation completion in bottom water reservoirs 88
             Dian-Fa Du, Yan-Yan Wang, Yan-Wu Zhao, Pu-Sen Sui and Xi Xia

Chapter 8   Naturally fractured hydrocarbon reservoir simulation by elastic fracture
             modeling 100
             Mehrdad Soleimani

Chapter 9   Impact of formation water on the generation of H_2S in condensate reservoirs:
             a case study from the deep Ordovician in the Tazhong Uplift of the Tarim
             Basin, NW China 116
             Jin Su, Yu Wang, Xiao-Mei Wang, Kun He, Hai-Jun Yang, Hui-Tong Wang,
             Hua-Jian Wang, Bin Zhang, Ling Huang, Na Weng, Li-Na Bi and Zhi-Hua Xiao

Chapter 10   Fracture prediction in the tight-oil reservoirs of the Triassic Yanchang
             Formation in the Ordos Basin, northern China 129
             Wen-Tao Zhao and Gui-Ting Hou

Chapter 11   **Sensitivity-based upscaling for history matching of reservoir models** 152
Saad Mehmood and Abeeb A. Awotunde

Chapter 12   **Combination and distribution of reservoir space in complex carbonate rocks** 166
Lun Zhao, Shu-Qin Wang, Wen-Qi Zhao, Man Luo, Cheng-Gang Wang,
Hai-Li Cao and Ling He

Chapter 13   **Reservoir stress path and induced seismic anisotropy: results from linking
coupled fluid-flow/geomechanical simulation with seismic modelling** 179
D. A. Angus, Q. J. Fisher, J. M. Segura, J. P. Verdon, J.-M. Kendall, M. Dutko
and A. J. L. Crook

Chapter 14   **Hydrocarbon charge history of the Paleogene reservoir in the northern
Dongpu Depression, Bohai Bay Basin, China** 195
You-Lu Jiang, Lei Fang, Jing-Dong Liu, Hong-Jin Hu and Tian-Wu Xu

Permissions

List of Contributors

Index

Preface

Petroleum reservoir engineering is a branch of petroleum engineering that studies the application of scientific principles to maximize the economic recovery of crude oil from reservoirs. Numerous drainage problems arise in production processes. Petroleum reservoir engineering builds on the tools developed from subsurface geology, applied mathematics and basic physics and chemistry. It strives to understand phase behavior of crude oil and natural gas to develop working tools of reservoir engineering. The key functionalities of surveillance, analysis, production and simulation modeling are also explored in the domain of reservoir engineering. Reservoir engineering is also significant for field development planning and framing cost-effective reservoir depletion schemes in order to optimize recovery of petroleum from deposits. The various advancements in petroleum reservoir engineering are glanced at in this book and their applications as well as ramifications are looked at in detail. Different approaches, evaluations, methodologies and advanced studies in this domain have been included in this book. It is aimed at engineers, geologists, students and other professionals involved in this field.

Various studies have approached the subject by analyzing it with a single perspective, but the present book provides diverse methodologies and techniques to address this field. This book contains theories and applications needed for understanding the subject from different perspectives. The aim is to keep the readers informed about the progresses in the field; therefore, the contributions were carefully examined to compile novel researches by specialists from across the globe.

Indeed, the job of the editor is the most crucial and challenging in compiling all chapters into a single book. In the end, I would extend my sincere thanks to the chapter authors for their profound work. I am also thankful for the support provided by my family and colleagues during the compilation of this book.

Editor

Laboratory studies of rice bran as a carbon source to stimulate indigenous microorganisms in oil reservoirs

Chun-Mao Chen[1,2] · Jin-Ling Wang[1] · Jung Bong Kim[3] · Qing-Hong Wang[1] ·
Jing Wang[1] · Brandon A. Yoza[4] · Qing X. Li[2]

Abstract There is a great interest in developing cost-effi-
cient nutrients to stimulate microorganisms in indigenous
microbial enhanced oil recovery (IMEOR) processes. In
the present study, the potential of rice bran as a carbon
source for promoting IMEOR was investigated on a laboratory scale. The co-applications of rice bran, K_2HPO_4 and
urea under optimized bio-stimulation conditions significantly increased the production of gases, acids and emulsifiers. The structure and diversity of microbial community
greatly changed during the IMEOR process, in which
Clostridium sp., *Acidobacteria* sp., *Bacillus* sp., and
Pseudomonas sp. were dominant. Pressurization, acidifi-
cation and emulsification due to microbial activities and
interactions markedly improved the IMEOR processes.
This study indicated that rice bran is a potential carbon
source for IMEOR.

Keywords Rice bran · Bio-stimulation · Petroleum ·
Microbial diversity · Indigenous microbial enhanced oil
recovery

1 Introduction

Increasing demand for crude oil is promoting the development of oil extraction technologies. Among these, microbial
enhanced oil recovery (MEOR) is a promising tertiary oil
recovery technology for depleted oil fields (Brown 2010; Sen
2008). Indigenous microorganisms, which naturally inhabit
oil reservoirs, show a greater metabolic activity than
exogenous ones due to their long-term adaptation (Castoren-
Cortés et al. 2012; Lazar et al. 2007). Therefore, indigenous
microorganisms activated MEOR (IMEOR) has greater
efficiency than using exogenous ones (Yao et al. 2012;
Zhang et al. 2012). Stimulation, growth and propagation of
beneficial microorganisms that can contribute to producing
effective metabolites are critical to the application of
IMEOR (Zhang et al. 2010). The metabolites mainly include
gases, acids and emulsifiers (Gao and Zerki 2011). Gases can
pressurize the oil reservoir and reduce the viscosity of crude
oil (Kobayashi et al. 2012; Spirov et al. 2014). Acids may
increase carbonate rock porosity and permeability, thereby
promote the exudation of remained oil (Sen 2008). Emulsifiers can emulsify crude oil, lower its viscosity and improve
its fluidity (Banat et al. 2010; Dastgheib et al. 2008; Kitamoto
et al. 2009; Sarafzadeh et al. 2013; She et al. 2011). The
multiple effects of various metabolites improve oil flooding
and thus enhance crude oil recovery.

Efficient production of beneficial metabolites is needed
during an IMEOR process. However, nutrients in an oil
reservoir are often insufficient and unbalanced, thus cannot
provide adequate bio-stimulation (Wang et al. 2012). The

✉ Chun-Mao Chen
chunmaochan@163.com

✉ Qing X. Li
qingl@hawaii.edu

[1] State Key Laboratory of Heavy Oil Processing, China
University of Petroleum, Beijing 102249, China

[2] Department of Molecular Biosciences and Bioengineering,
University of Hawaii at Manoa, Honolulu, HI 96822, USA

[3] Department of Agro-Food Resources, National Institute of
Agricultural Sciences, Rural Development Administration,
Jeonju 55365, Republic of Korea

[4] Hawaii Natural Energy Institute, University of Hawaii at
Manoa, Honolulu, HI 96822, USA

Edited by Xiu-Qin Zhu

injection of nutrients can optimize bio-stimulation conditions in an oil reservoir (Gao et al. 2013). The injected nutrients should stimulate beneficial microorganisms but restrain harmful ones, while they should not cause formation damage or contamination (da Silva et al. 2014). More importantly, the nutrients should be economical. Of these nutrients, carbon sources have an overwhelming influence on bio-stimulation effects and application costs. Carbohydrates such as molasses, corn syrup, malt dextrin sucrose and starch have been explored for use in IMEOR in past decades (Bao et al. 2009; Joshi et al. 2008). As market prices of carbohydrates are rising, finding an economical carbon source has been a significant motivation. Rice bran, as an abundant agricultural by-product, has been used to prepare culture media for enzymatic solid-state fermentations (Ng et al. 2010; Noike and Mizuno 2000; Tanaka et al. 2006). However, it has not been investigated as a bio-simulator to promote IMEOR.

The objective of this study was to investigate the industrial potential of rice bran to promote IMEOR. Additionally, the bio-stimulation mechanism of rice bran can be acquired by the microbial diversity analysis during an IMEOR process. The results could be hopefully beneficial to reduce IMEOR cost and would generate value from agricultural by-products.

2 Materials and methods

2.1 Materials

The formation water and crude oil were both sampled from the Qixi block of Karamay oil field, which is located in Xinjiang Uygur Autonomous Region, Northwest China. Water flooding has been implemented by recycling production water for 40 years. The sampling and storage methods were as previously described (Tang et al. 2012). The initial temperature, pH and salinity of the formation water were 33.5 °C, 7.85 and 5054 mg/L, respectively. The kinematic viscosity (50 °C) and density (20 °C) of the crude oil were 62.33 mm^2/s and 0.912 g/cm^3, respectively.

The nutrients that were used to stimulate microorganisms in an IMEOR process commonly included carbon, nitrogen and phosphorus sources. The carbon nutrients were selected from several agricultural by-products including rice bran (reducing sugar at 390 mg/g and total nitrogen at 2.21 mg/g), wheat bran (reducing sugar at 280 mg/g and total nitrogen at 0.45 mg/L), corn residue, rice husk, glycerol residue and molasses. Rice bran and wheat bran were pulverized to 100 mesh particles prior to the experiment. The nitrogen sources were urea, NH_4Cl, KNO_3 and $NaNO_3$. The phosphorus sources were K_2HPO_4, $NH_4H_2PO_4$, NaH_2PO_4, KH_2PO_4, $(NH_4)_2HPO_4$ and

Na_2HPO_4. The agricultural by-products were purchased from a local market. All nitrogen and phosphorus sources were from Beijing Chemical Reagents Co., China.

2.2 Optimization experiments of bio-stimulation conditions

Preferred nutrients and their basic concentrations were selected by single factor optimization (SFO) experiments. Concentrations of preferred nutrients were further optimized by orthogonal design experiments (Chen et al. 2010). The optimized bio-stimulation conditions were quantified with response surface methodology (RSM) experiments based on a Box–Behnken design (Chen et al. 2007). Response surface regression analysis and analysis of variance (ANOVA) were performed by Minitab software (Version 16, Minitab Inc, State College, PA, USA). The experiments were carried out with 250 mL SIBATA fermentation bottles. Various nutrients at different concentrations and 50 mL distilled water were introduced into the bottles, followed by pH adjustment to 7.85. After sterilization at 115 °C for 30 min, the bottles were inoculated with 50 mL of formation water and 5 g of crude oil, and then incubated at 33.5 °C and 110 rpm for 7 days. Gas production and surface tension were indicators of bio-stimulation effects. All experiments were done in triplicate.

2.3 Laboratory-scale experiments of IMEOR

The potential of rice bran was investigated in 250 mL SIBATA fermentation bottles to simulate an IMEOR process. The initial broth contained 50 mL of formation water, 5 g of crude oil and optimized nutrient components, and was diluted to a final volume of 100 mL with distilled water. Sterilization followed the protocols described in the above-mentioned optimization experiments, as well as the incubation conditions with an extended 10 day incubation period. An aliquot of 0.05 mL Resazurin solution (0.1 wt%) was added to each bottle as a visual anaerobic status indicator. Simulated IMEOR experiments were carried out in triplicate.

The bio-stimulation effects were measured during experiments. The produced gases which mainly reflected the re-pressurization of an oil reservoir were collected in a Devex gas collecting bag (0.3 L), and the gas volume was recorded by a graduated syringe. The gas composition was analyzed on a HP 6890 gas chromatograph (GC) (Agilent, Wilmington, DE, USA) equipped with a TCD detector (80 °C) and a Porapak Q packing column (60 °C). Helium was used as carrier gas. The pH values which indicated the production of acids were measured on a PHSJ-4 meter (Leici, Shanghai, China). Two mL of broth was extracted by a graduated syringe and injected to a 5-mL centrifuge tube for pH measurement. Bacterial numbers were counted by a cell counting method (Zhou and Wang 2004). Surface tension and emulsification degree both characterized the production of emulsifiers such as biosurfactants, biopolymer, acids and solvents. Surface tension was measured with a JK99B tensiometer (Powereach, Shanghai, China), and emulsification degree was obtained with the emulsification index (%, EI 24) method (Reddy et al. 2010).

2.4 Analysis of microbial diversity

Describing the relationship between the microbial community structures and their functions will contribute to understanding of bio-stimulation mechanisms (Fuhrman 2009). Polymerase chain reaction-denaturing gradient gel electrophoresis (PCR-DGGE) has been widely used in microbial community analysis (Wang et al. 2008a, b). Aliquots (2 mL) of fermentation broth were collected from the bottles. After centrifugation at 10,000 rpm for 5 min, the total DNA was extracted with a genomic DNA extraction kit (TaKaRa, Dalian, China) according to the supplier's instructions. PCR-DGGE was performed as previously described in literature (Ji et al. 2009). DGGE profiles, including the presence, intensity and abundance of the bands, were analyzed with Quantity One software (Version 4.4, Bio-Rad, Hercules, CA, USA). Microbial diversity was calculated by Shannon-Wiener's indexes (H) (Andreoni et al. 2004). Dominant bands were excised from the gels and re-amplified, and the fragments were recovered and cloned again. The positive clones were selected and sequenced (Sangon, Shanghai, China). Typical sequences were analyzed using the NCBI BLAST database to identify the closest relatives. A phylogenetic tree was constructed with MEGA software (Version 5.0).

3 Results

3.1 Optimized bio-stimulation conditions

Rice bran, K_2HPO_4 and urea were selected as preferred carbon, phosphorus and nitrogen sources, respectively (Fig. 1). An optimal bio-stimulation condition was preliminarily estimated to be 2.0 g/100 mL of rice bran, 0.05 g/100 mL of K_2HPO_4 and 0.05 g/100 mL of urea under SFO experiments (Fig. 2). The concentrations of rice bran, K_2HPO_4 and urea were further optimized to be 3.0, 0.07 and 0.07 g/100 mL, respectively, under 3^3 orthogonal design experiments (Table 1). According to the Box–Behnken experiments (Table 2) and subsequent response surface regression analysis (Fig. 3), the optimal bio-stimulation conditions were estimated as 3.36 g/100 mL of rice bran, 0.075 g/100 mL of K_2HPO_4 and 0.076 g/100 mL of urea.

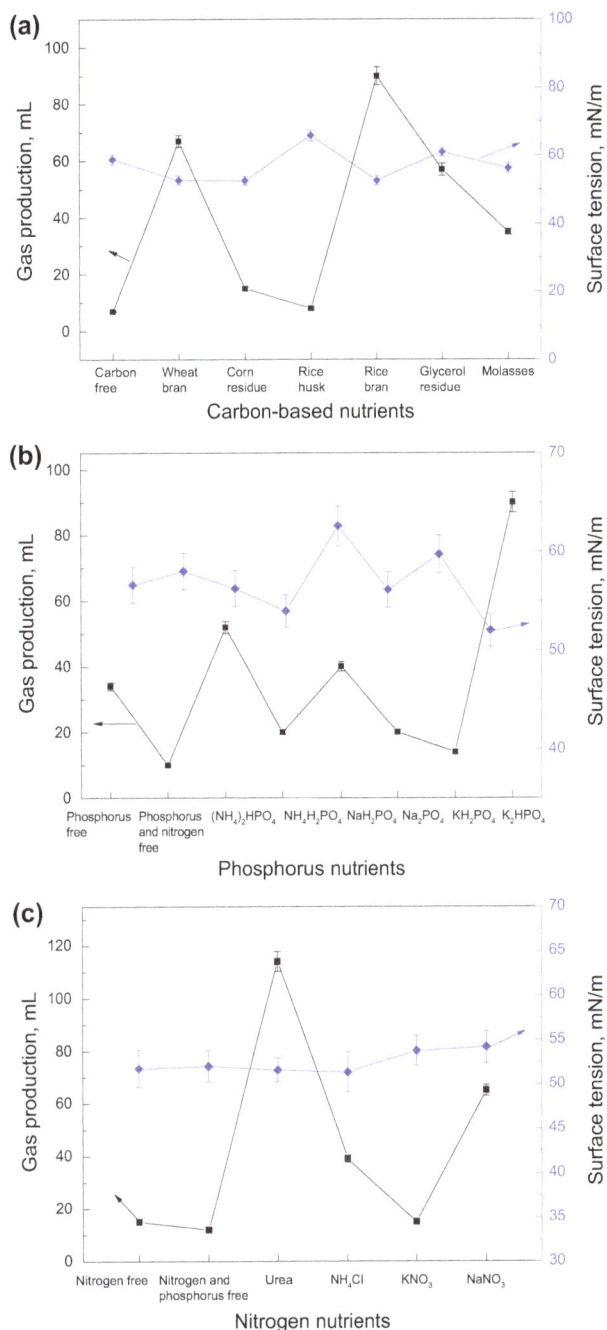

Fig. 1 Screening of preferred carbon nutrient (1.0 g/100 mL carbon sources, 0.1 g/100 mL urea and 0.1 g/100 mL $(NH_4)_2HPO_4$) (a), phosphorus nutrient (0.1 g/100 mL phosphorus sources, 1.0 g/100 mL rice bran and 0.1 g/100 mL urea) (b), and nitrogen nutrient (0.1 g/100 mL nitrogen sources, 1.0 g/100 mL rice bran and 0.1 g/100 mL K_2HPO_4) (c)

3.2 Bio-stimulation effects of rice bran

Color changes of the fermentation broth from red to pink and then colorless indicated the gradual conversion from aerobic (0–1st day), facultative (2nd–3rd days) to anaerobic (4–10th days) metabolic stages in the simulated

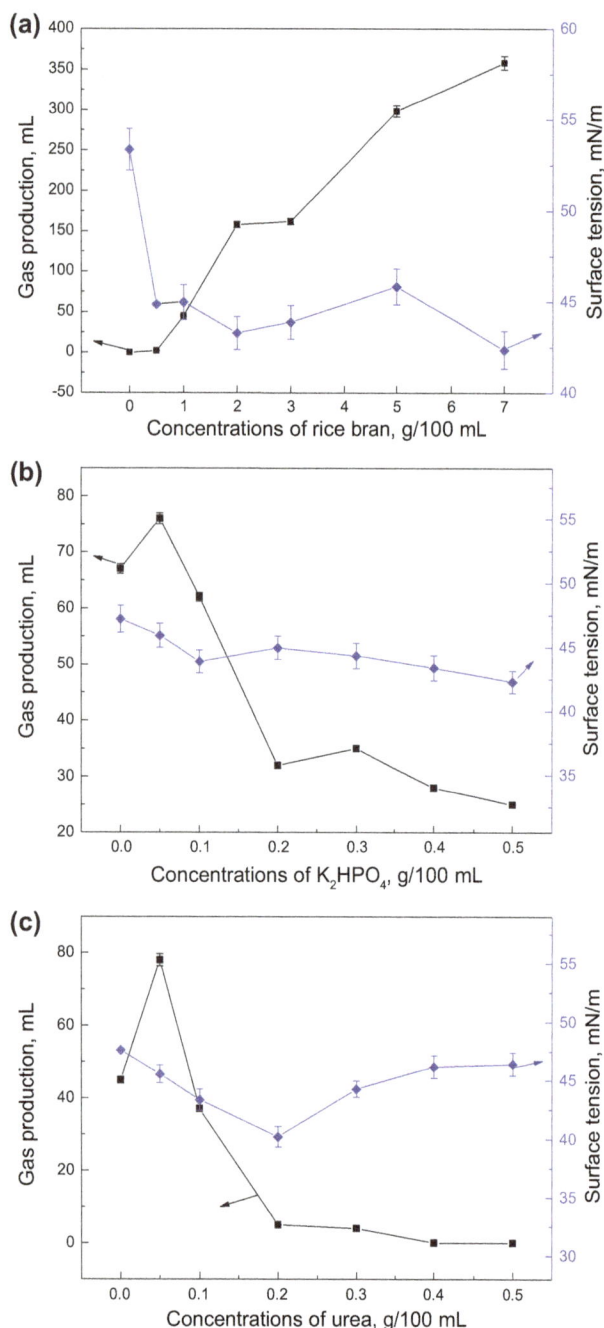

Fig. 2 Preliminary screening of concentrations of rice bran (0.1 g/ 100 mL urea and 0.1 g/100 mL K_2HPO_4,) (**a**), K_2HPO_4 (1.0 g/ 100 mL rice bran and 0.05 g/100 mL urea) (**b**), and urea (1.0 g/ 100 mL rice bran and 0.1 g/100 mL K_2HPO_4) (**c**)

IMEOR process. The numbers of bacteria rapidly increased from the initial 0.07×10^9 cell/mL to 2.4×10^9 cell/mL in the aerobic stage and reached a peak value of 4.9×10^9 cell/mL in the anaerobic stage (5th day), followed by a slight drop to 4.5×10^9 cell/mL at the end of bio-stimulation (Fig. 4a). The pH values significantly decreased from the initial 7.85 to 5.26 in the facultative stage (3rd day) and then finally decreased to 4.94 (Fig. 4a).

Gas production lasted for 8 days; a total 322 mL of gases was collected mainly in the facultative stage where the highest gas production rate was observed (Fig. 4b). Gas components varied markedly among stages (Table 3). CO_2 concentration promptly increased from 0.03 to 42.7 vol% along with sharp depletion of O_2 in the aerobic and facultative stages; CO_2 and H_2 were the main gases in the anaerobic stage. Surface tension dropped from 62.5 to 37.6 mN/m in the early anaerobic stage (5th day) when the EI 24 increased rapidly from zero to 61.1 % (Fig. 4c).

3.3 Microbial community

There were 17 major bands in the DGGE profiles (Fig. 5a) with obvious variations in the intensities (Fig. 5b), H-index, numbers of bands and microbial abundances among lanes (Table 4). All 17 bands were isolated from the gel, re-amplified, and sequenced. Except for bands 1, 2 and 3, all of the identified genera had sequence similarities of 98 % or higher (Table 5). The phylogenetic tree (Fig. 6) shows that the identified sequences could be divided into three classifications, *Proteobacteria*, *Firmicutes* and *Acidobacteria*. Sequences from bands 3, 4, 5 and 13 were not classified with the phylogenetic tree, which was presumed to be not reported.

4 Discussion

4.1 Optimized bio-stimulation conditions

Nutrient-rich polysaccharides stimulate indigenous microorganisms in oil reservoirs (Cheng et al. 2010; Feng et al. 2012). Rice bran and wheat bran showed higher emulsification and gas production than the other carbon sources. The market prices of rice bran and wheat bran are similar at approximately $180 per ton. Rice bran contained more reducing sugar, total nitrogen and vitamins than wheat bran under previous composition analysis (Wang 2013); therefore, it was identified as a preferred carbon source. Urea and K_2HPO_4 contributed largely to the gas production and were determined as the preferred nitrogen and phosphorus nutrients. The concentration of rice bran displayed a positive correlation with gas production, while an excess of rice bran would increase the costs. The concentrations of K_2HPO_4 and urea showed negligible influence on bio-stimulation. The concentrations of rice bran, K_2HPO_4 and urea at, respectively, 2.0 0.05 and 0.05 g/ 100 mL basically were the balanced nutrient ratios.

The surface tension that directly accounted for oil-flooding effects was selected as the decisive indictor to optimize bio-stimulation conditions. The significance order of nutrient concentrations were rice bran > urea > K_2HPO_4

Table 1 Experimental results of 3^3 orthogonal designs

Runs	Rice bran, g/100 mL	Urea, g/100 mL	K_2HPO_4, g/100 mL	Gas production, mL	Surface tension, mN/m
1	1 (−1)	0.03 (−1)	0.03 (−1)	67	48.8
2	1	0.05 (0)	0.05 (0)	41	51.0
3	1	0.07 (+1)	0.07 (+1)	61	46.5
4	2 (0)	0.03	0.05	92	46.2
5	2	0.05	0.07	64	35.4
6	2	0.07	0.03	38	34.7
7	3 (+1)	0.03	0.07	191	35.8
8	3	0.05	0.03	120	39.9
9	3	0.07	0.05	125	33.9

(−1) for low level, (0) for medium level and (+1) for high level

Table 2 Experimental results of Box–Behnken designs

Runs	Rice bran, g/100 mL	Urea, g/100 mL	K_2HPO_4, g/100 mL	Gas production, mL	Surface tension, mN/m
1	2 (−1)	0.04 (−1)	0.07 (0)	93	48.5
2	3 (0)	0.07 (0)	0.07	78	38.0
3	3	0.04	0.1 (+1)	153	46.2
4	4 (+1)	0.07	0.1	250	41.8
5	3	0.1 (+1)	0.1	219	42.3
6	4	0.1	0.07	310	42.6
7	2	0.07	0.1	160	45.8
8	3	0.07	0.07	189	40.5
9	3	0.1	0.04 (−1)	154	41.4
10	2	0.07	0.04	100	47.8
11	2	0.1	0.07	101	48.2
12	4	0.07	0.04	38	44.2
13	3	0.04	0.04	85	48.1
14	3	0.07	0.07	22	39.0
15	4	0.04	0.07	210	42.1

(−1) for low level, (0) for medium level and (+1) for high level

according to the analysis of orthogonal design experiments. By applying response surface regression analysis to the surface tension data from the Box–Behnken experiments, the following second-order equation was established:

$$Y = 114.276 - 22.4558C - 568.595N - 444.793P$$
$$+ 3.30217C^2 + 3229.07N^2 + 2696.85P^2 \quad (1)$$
$$+ 6.2500CN - 3.46667CP + 786.667NP$$

where Y is the surface tension, mN/m; C, N and P are the concentrations (g/100 mL) of rice bran, urea and K_2HPO_4, respectively. C^2, N^2, C, and N show significant influences on surface tension with P values of 0.028, 0.043 and 0.029, respectively. ANOVA results indicated a good consistency between experimental and predicted values with a P value of 0.017 and a R^2 value of 0.902. The experimental surface tension (40.6 mN/m) under the optimized bio-stimulation conditions agreed well with the predicted value, which the predicted minimum surface tension (38.6 mN/m) was also obtained under the optimal bio-stimulation conditions.

4.2 Bio-stimulation effects of rice bran for promoting IMEOR

Indigenous microorganisms became active and reached a logarithmic growth phase shortly after addition of rice bran supplemented with urea and K_2HPO_4. A stationary growth phase appeared in the later anaerobic stage due to depletion of nutrients. Acids were rapidly produced in the aerobic and facultative stages, and their production and utilization reached an approximate balance in the anaerobic stage. Acid production favors flooding, whereas a lower pH value is harmful to some indigenous microorganism species.

Gas-producing bacteria were sensitive to rice bran, and gas production mainly occurred in the aerobic and facultative stages. As O_2 was consumed, CO_2 was generated

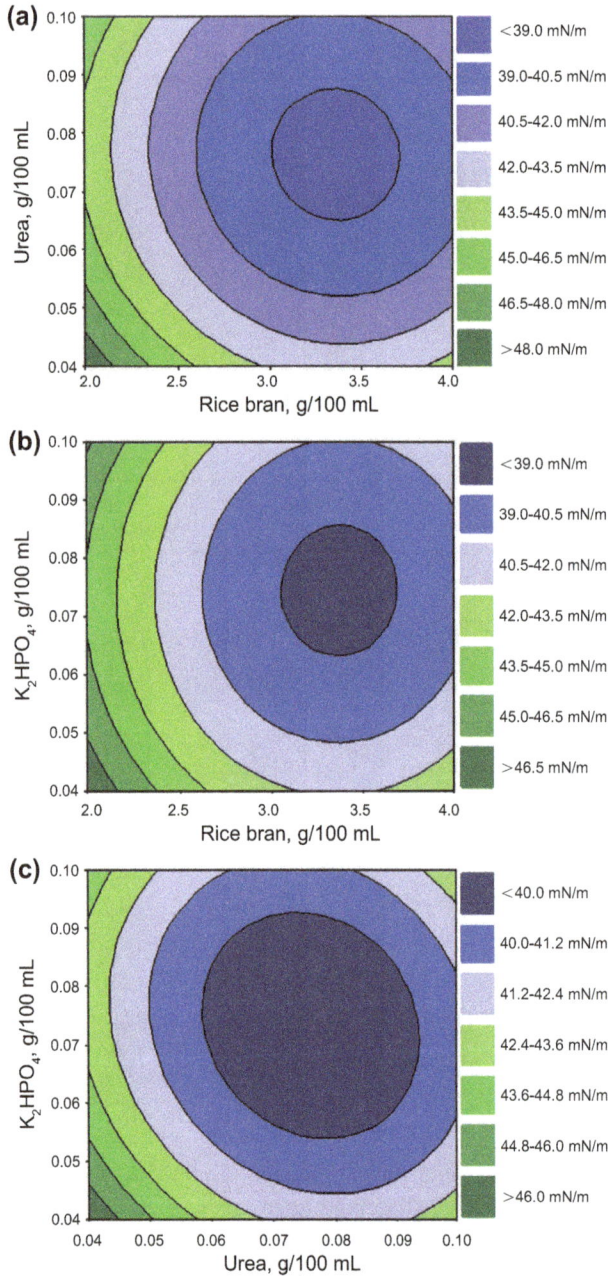

Fig. 3 Interaction effects of rice bran and urea with K_2HPO_4 concentration of 0.07 g/100 mL (**a**), rice bran and K_2HPO_4 with urea concentration of 0.07 g/100 mL (**b**), and urea and K_2HPO_4 with rice bran at a concentration of 3.0 g/100 mL (**c**) on surface tension

Fig. 4 pH values and numbers of bacteria (**a**), gas production and production rate (**b**), surface tension and EI 24 (**c**) of fermentation broth in a simulated IMEOR process

rapidly with the growth of aerobic bacteria. The emergence of H_2 signified that anaerobic bacteria became active, while aerobic bacteria were inhibited in the anaerobic stage. Injecting an extra amount of air or O_2 can extend the aerobic and facultative stages, and contribute to re-pressurization and CO_2 flooding.

Aerobic bacteria rapidly produce emulsifiers utilizing sufficient nutrients and O_2 in the aerobic stage. After that, facultative and anaerobic bacteria became dominant and

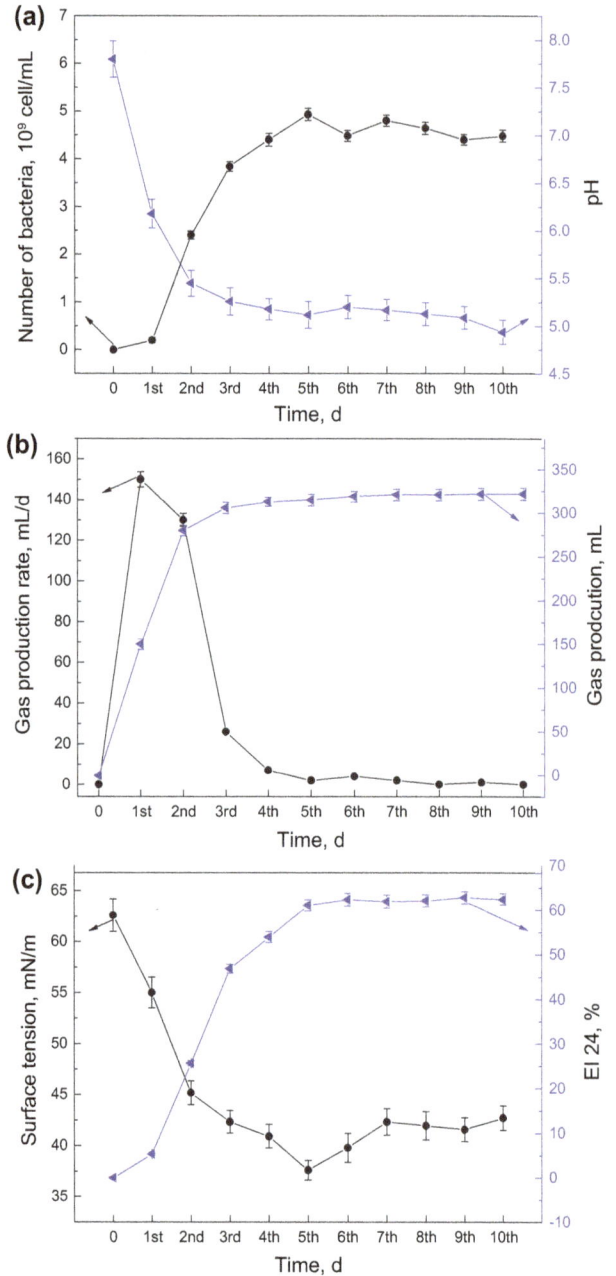

emulsifier production decreased; therefore, no further significant decrease of surface tension was observed. The EI 24 demonstrated the high emulsification of crude oil at the end of bio-stimulation. Emulsifiers play a critical role in MEOR processes. Continuous supply of nutrients and O_2 to an oil reservoir would contribute to the production of a greater amount of emulsifiers.

The rice bran, as a carbon nutrient stimulator, successfully stimulated beneficial indigenous microorganisms and

Table 3 Composition of produced gases at different days during a simulated IMEOR process

Components	0 day		2nd day		3rd day		8th day		10th day	
	vol%	mL	vol%	mL	vol%	mL	vol%	mL	vol%	mL
N_2	78	117	39.5	118.5	33.9	101.7	0	0	0	0
O_2	21	31.5	2.3	6.9	1.9	5.7	0	0	0	0
H_2	0	0	12.8	38.4	30.7	92.1	6.3	18.9	20.4	61.2
CO_2	0.03	0.05	42.7	128.1	32.1	96.3	89.8	269.4	76.1	228.3
Others (C_2H_6, C_3H_8, etc.)	0.97	0.15	2.7	8.1	1.4	4.2	3.9	11.7	3.5	10.5

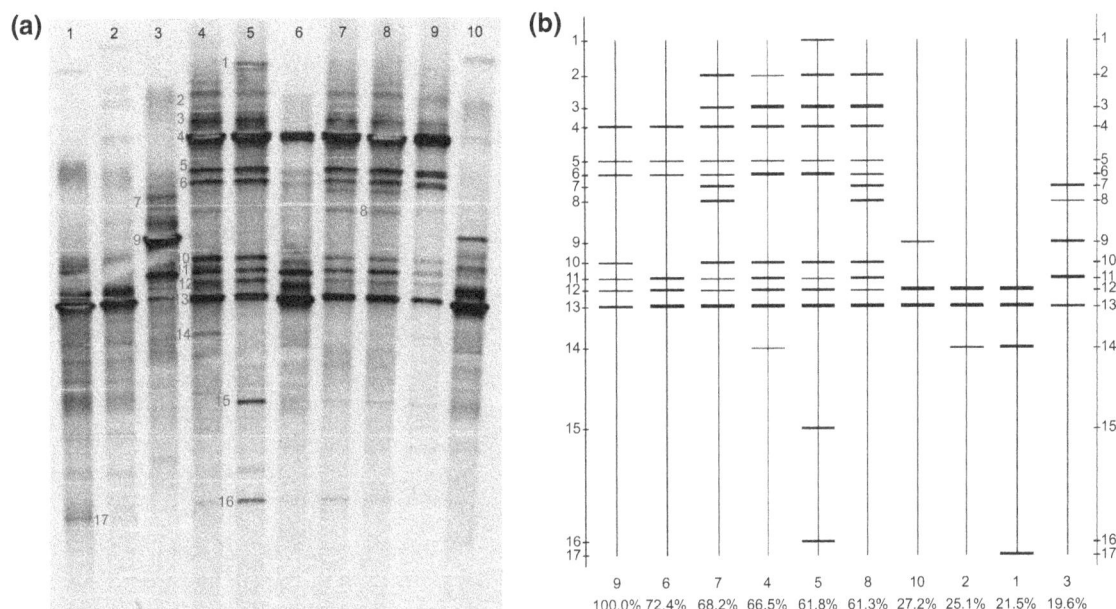

Fig. 5 DGGE profile of bacterial samples from different days (**a**), schematic diagram of relative band intensities in DGGE profiles (**b**). *Note* Lane number represented the bacterial samples collected in different days. Lane 1—the 1st day; Lane 2—the 2nd day; and so on; Lane 10—the 10th day

Table 4 Diversity index (H), numbers of bands and abundances of DGGE profiles

Lane no.	1	2	3	4	5	6	7	8	9	10
H-index	2.45	2.40	2.05	2.77	2.61	1.93	2.61	2.43	1.98	1.99
Numbers of bands	12	14	8	21	15	9	14	13	10	10
Microbial abundance	1665	157	1150	476	368	282	1908	358	239	226

enhanced the production of acids, gases and emulsifiers, and hence effectively promoted the IMEOR process.

4.3 Diversity and functions analysis of microbial community

Changes in microbial community structure occurred in the simulated IMEOR process (Table 6). Microbial diversity (H-index) was significant on the 1st, 4th, 5th and 7th days and microbial abundance peaked markedly on the 1st, 3rd

and 7th days. *Clostridium* sp., *Acidobacteria* sp., *Pseudomonas* sp. and *Bacillus* sp. were the dominant bacteria. Of these bacteria, *Bacillus* sp. and *Pseudomonas* sp. dominated the aerobic stage and *Acidobacteria* sp. dominated the facultative and anaerobic stages. *Clostridium* sp. was highly distributed throughout the simulated IMEOR process.

Aerobic bacteria were stimulated and caused a higher microbial diversity and abundance in the aerobic stage (1st day). *Clostridium* sp., uncultured *Acidobacteria* bacterium,

Table 5 Comparisons of nucleotide sequences of sequenced DGGE bands

Band no.	Closest relatives	Accession No.	Identity, %
1	*Clostridium* sp.	HM801879.1	95
	Uncultured *Clostridium* sp.	JX273758.1	95
2	*Clostridium sulfidigenes* strain	HM163534.1	96
	Uncultured *Enterococcus* sp.	DQ232854.1	96
3	*Achromobacter* sp.	AB772984.1	96
	Uncultured bacterium clone	KC465632.1	97
4	*Bacillus cereus* strain	KC683782.1	98
	Uncultured bacterium clone	GU002857.1	98
5	*Pseudomonas aeruginosa* strain	KC570343.1	100
	Uncultured *Pseudomonas* sp.	KC470004.1	100
6	*Clostridium indolis* strain	JX960755.1	100
	Uncultured bacterium clone	KC000040.1	100
7	*Brevundimonas* sp.	HF571531.1	99
	Uncultured *Brevundimonas* sp. clone	JQ701321.1	99
8	*Pantoea agglomerans* strain	KC009691.1	100
	Uncultured bacterium clone	KC299005.1	100
9	*Ochrobactrum* sp.	KC493414.1	100
	Uncultured *Ochrobactrum* sp.	KC502956.1	100
10	*Arcobacter* sp.	FN397894.1	99
	Uncultured *Arcobacter* sp.	HM245616.1	99
11	*Agrobacterium vitis* strain	KC196472.1	99
	Uncultured bacterium clone	JX872344.1	99
12	Uncultured *Acidobacteria* bacterium clone	DQ829628.1	99
13	*Clostridium* sp.	KC331196.1	100
	Uncultured *Clostridium* sp.	KC110473.1	100
14	*Bacillus aerius* strain	KC469617.1	100
	Uncultured bacterium clone	KC414651.1	100
15	*Sulfurospirillum carboxydovorans* strain	AY740528.1	99
	Uncultured *epsilon proteobacterium*	AJ576003.1	99
16	*Paenibacillus* sp.	KC134361.1	99
	Uncultured bacterium clone	JX223186.1	99
17	*Pseudomonas* sp.	KC433644.1	100
	Uncultured *Pseudomonas* sp. clone	KC253432.1	100

Bacillus aerius, and *Pseudomonas* sp. were stimulated and resulted in the increase of bacteria numbers. *B. aerius* and *Pseudomonas* sp. began to produce acids and emulsifiers (e.g., biosurfactants) by utilizing O_2, nutrients and even crude oil (Abdel-Mawgoud et al. 2009; Banat 1995), then both pH and surface tension decreased.

Aerobic bacteria (e.g., *Pseudomonas* sp.) were gradually inhibited with consumption of O_2, and facultative bacteria became dominant in the facultative stage. Microbial abundance rebounded in the 3rd day, while the genera of bacteria were not rich. *Clostridium* sp. began to produce gases by utilizing nutrients (Taguchi et al. 1996). *Brevundimonas* sp., *Pantoea agglomerans*, *Ochrobactrum* sp., and *Agrobacterium vitis* were stimulated. Among these, *Brevundimonas* sp. and *Ochrobactrum* sp. can decompose

crude oil to produce biosurfactants and acids (Arulazhagan and Vasudevan 2011; Ghosal et al. 2010; Ruggeri et al. 2009; Takahashi et al. 1999); *Agrobacterium vitis* can promote the metabolism of petroleum-degrading bacteria (Chalneau et al. 1995). The functions of *P. agglomerans* in an IMEOR process have not been reported. The co-effects of microbial community increased the bacteria numbers and EI 24, but decreased both the pH and surface tension.

As anaerobic bacteria were stimulated in the anaerobic stage, the number of bacterial genera significantly increased on the 4th day and the microbial diversity reached its maximum level. *Clostridium* sp., *Clostridium sulfidigenes*, *Clostridium indolis*, *Achromobacter* sp., *Bacillus cereus*, *B. aerius*, *Pseudomonas aeruginosa*, *Arcobacter* sp., *A. vitis*, uncultured *Acidobacteria* bacteria

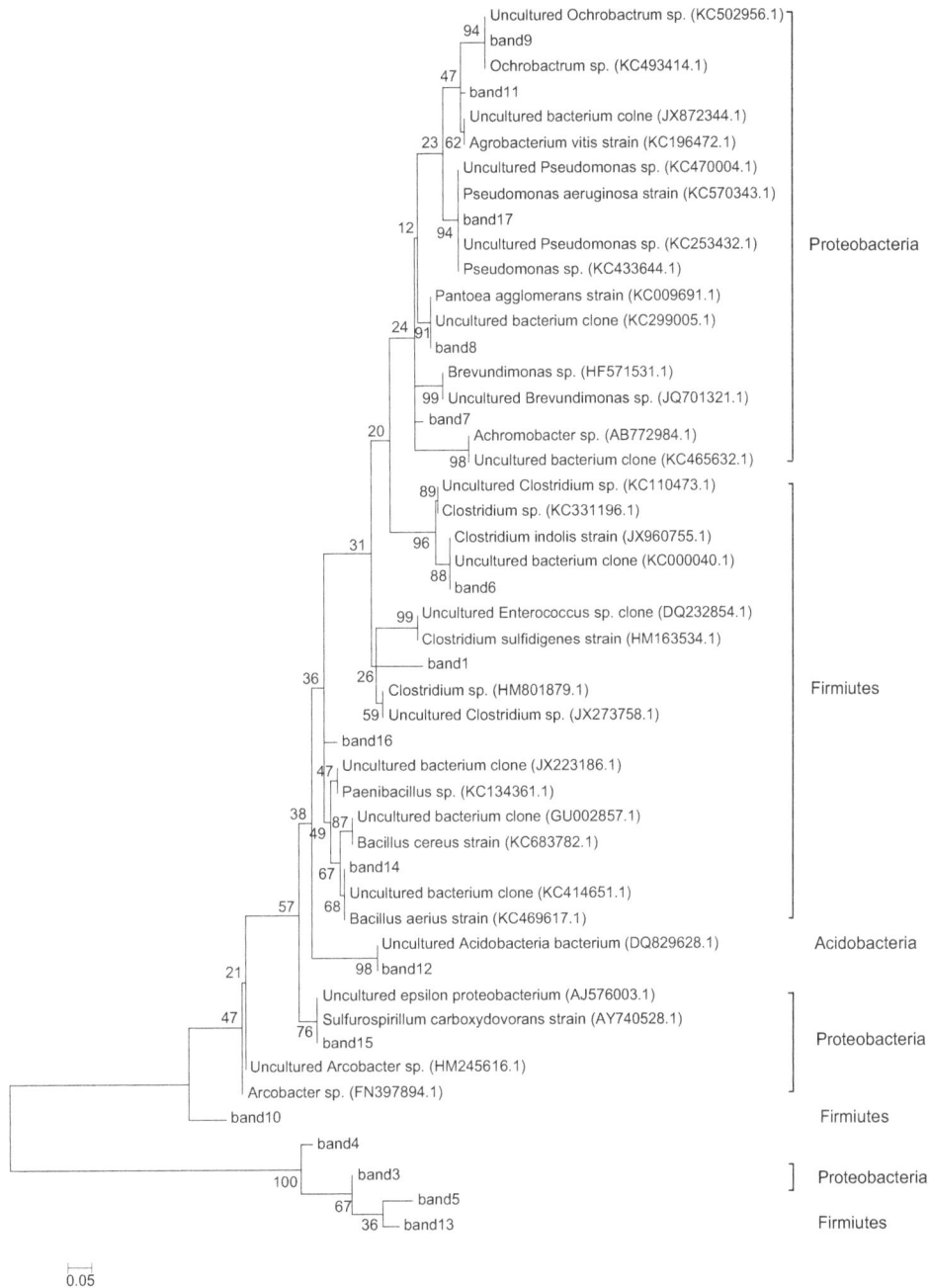

Fig. 6 Phylogenetic tree of 16S rDNA sequences from DGGE profiles

and the other uncultured bacteria co-built up the most complex structure of microbial community. *B. cereus* can produce acids but not gases in anaerobic conditions (Nazina et al. 2003). *P. aeruginosa* can produce biosurfactants by degrading crude oil (Das and Mukherjee 2007). *Sulfurospirillum carboxydovorans* and *Paenibacillus* sp. appeared on the 5th day. The former can inhibit sulfate-reducing bacteria (Hubert and Voordouw 2007), and the latter may produce biopolymers (Wang et al. 2008a, b). The bacteria abundance reached a peak value and surface

tension at a minimum on the 5th day when the pH and EI 24 almost attained their lowest and highest values, respectively; meanwhile, the gas production declined significantly. The number of bacterial genera decreased, but the microbial abundance increased, and *B. cereus*, *A. vitis*, uncultured *Acidobacteria* and *Clostridium* sp. became dominant on the 6th day. As anaerobic bacteria becoming adapted, microbial abundance recovered again and reached its peak value on the 7th day when the microbial diversity remained high. After that, microbial diversity and

Table 6 Dominant bacteria and the community succession during a simulated IMEOR process

Band no.	Genus	Dominance during a simulated IMEOR process									
		1st day	2nd day	3rd day	4th day	5th day	6th day	7th day	8th day	9th day	10th day
1	*Clostridium* sp.	−	−	−	−	+	−	−	−	−	−
	Uncultured *Clostridium* sp.										
2	*Clostridium sulfidigenes* strain	−	−	−	+	+	−	+	+	−	−
	Uncultured *Enterococcus* sp.										
3	*Achromobacter* sp.	−	−	−	+	+	−	+	+	−	−
	Uncultured bacterium clone										
4	*Bacillus cereus* strain	−	−	−	+	+	+	+	+	+	−
	Uncultured bacterium clone										
5	*Pseudomonas aeruginosa* strain	−	−	−	+	+	+	+	+	+	−
	Uncultured *Pseudomonas* sp.										
6	*Clostridium indolis* strain	−	−	−	+	+	+	+	+	+	−
	Uncultured bacterium clone										
7	*Brevundimonas* sp.	−	−	+	−	−	−	+	+	−	−
	Uncultured *Brevundimonas* sp. Clone										
8	*Pantoea agglomerans* strain	−	−	+	−	−	−	+	+	−	−
	Uncultured bacterium clone										
9	*Ochrobactrum* sp.	−	−	+	−	−	−	−	−	−	+
	Uncultured *Ochrobactrum* sp.										
10	*Arcobacter* sp.	−	−	−	+	+	−	+	+	+	−
	Uncultured *Arcobacter* sp.										
11	*Agrobacterium vitis* strain	−	−	+	+	+	+	+	+	+	−
	Uncultured bacterium clone										
12	Uncultured *Acidobacteria* bacterium clone	+	+	−	+	+	+	+	+	+	+
13	*Clostridium* sp.	+	+	+	+	+	+	+	+	+	+
	Uncultured *Clostridium* sp.										
14	*Bacillus aerius* strain	+	+	−	+	−	−	−	−	−	−
	Uncultured bacterium clone										
15	*Sulfurospirillum carboxydovorans* strain	−	−	−	−	+	−	−	−	−	−
	Uncultured *epsilon proteobacterium*										
16	*Paenibacillus* sp.	−	−	−	−	+	−	−	−	−	−
	Uncultured bacterium clone										
17	*Pseudomonas* sp.	+	−	−	−	−	−	−	−	−	−
	Uncultured *Pseudomonas* sp. Clone										

(+) represent the dominant bacteria, (−) represent the non-dominant bacteria or absence

abundance both reduced to their minimum values at the end of bio-stimulation with the depletion of nutrients.

The structure and relative abundance of the microbial community, especially the interactions between different microorganisms, have great influence on promoting IMEOR. Rice bran, with the participation of the other preferred nutrients, could enhance the microbial diversity as well as strengthen the multiple functions of producing gases, acids and emulsifiers. The results indicated that rice bran can effectively promote IMEOR.

5 Conclusion

The potential of rice bran as a carbon nutrient for promoting IMEOR was investigated. Rice bran showed great bio-stimulation effects on producing gases, acids and emulsifiers with the supplements of K_2HPO_4 and urea. The oil reservoirs were re-pressurized and pH and surface tension decreased, which contributed to the IMEOR process. The indigenous microbial community showed remarkable successions, and the beneficial functioning bacteria were stimulated in the simulated IMEOR process. *Clostridium* sp., *Acidobacteria* sp., *Bacillus* sp., and *Pseudomonas* sp. dominated the IMEOR process. Microbial activities and interactions among indigenous microorganisms enhanced the IMEOR process. The results indicated the potential of rice bran for promoting IMEOR due to its effectiveness and low cost.

Acknowledgments This project was supported in part by the National Natural Science Foundation of China (Nos. 51209216 and 21306229) and the Korean RDA Grant (No. PJ009472). C. Chen was a China Scholarship Council scholarship recipient.

References

Abdel-Mawgoud AM, Aboulwafa MM, Hassouna NAH. Characterization of rhamnolipid produced by *Pseudomonas aeruginosa* isolate Bs20. Appl Biochem Biotechnol. 2009;157(2):329–45.

Andreoni V, Gavalca L, Rao MA. Bacteria communities and enzyme activities of PAHs polluted soils. Chemosphere. 2004;57:401–12.

Arulazhagan P, Vasudevan N. Biodegradation of polycyclic aromatic hydrocarbons by a halotolerant bacterial strain *Ochrobactrum* sp. VA1. Mar Pollut Bull. 2011;62(2):388–94.

Banat IM. Biosurfactants production and possible uses in microbial enhanced oil recovery and oil pollution remediation: a review. Bioresour Technol. 1995;51(1):1–12.

Banat IM, Franzetti A, Gandolfi I, et al. Microbial biosurfactants production, applications and future potential. Appl Microbiol Biotechnol. 2010;87(2):427–44.

Bao M, Kong X, Jiang G, Wang X, Li X. Laboratory study on activating indigenous microorganisms to enhance oil recovery in Shengli Oilfield. J Pet Sci Eng. 2009;66(1–2):42–6.

Brown LR. Microbial enhanced oil recovery (MEOR). Curr Opin Microbiol. 2010;13(3):316–20.

Castoren-Cortés G, Zapata-Peñasco I, Roldán-Carrillo T, et al. Evaluation of indigenous anaerobic microorganisms from Mexican carbonate reservoirs with potential MEOR application. J Pet Sci Eng. 2012;81:86–93.

Chalneau CH, Morel JL, Oudot J. Microbial degradation in soil microcosms of fuel oil hydrocarbons from drilling cuttings. Environ Sci Technol. 1995;29(6):1615–21.

Chen X, Wang J, Li D. Optimization of solid-state medium for the production of inulinase by Kluyveromyces S120 using response surface methodology. Biochem Eng J. 2007;34(2):179–84.

Chen JL, Au KC, Wong YS, et al. Using orthogonal design to determine optimal conditions for biodegradation of phenanthrene in mangrove sediment slurry. J Hazard Mater. 2010;176(1–3):666–71.

Cheng H, Liu M, Hu J, et al. Study on novel nutrient system based on starch-cellulose for microbial flooding. Acta Pet Sin. 2010;31(1):105–9 (in Chinese).

da Silva MLB, Soares HM, Furigo A Jr, et al. Effects of nitrate injection on microbial enhanced oil recovery and oilfield reservoir souring. Appl Biochem Biotechnol. 2014;174(5):1810–21.

Das K, Mukherjee AK. Crude petroleum-oil biodegradation efficiency of *Bacillus subtilis* and *Pseudomonas aeruginosa* strains isolated from a petroleum-oil contaminated soil from North-East India. Bioresour Technol. 2007;98(7):1339–45.

Dastgheib SM, Amoozegar MA, Elahi E, Asad S, Banat IM. Bioemulsifier production by a halothermophilic Bacillus strain with potential applications in microbially enhanced oil recovery. Biotechnol Lett. 2008;30(2):263–70.

Feng Q-X, Ma X-P, Cheng H-Y, et al. Application of a novel amylum-cellulose nutrient system for microbial flooding in Dagang oilfield, China. In: SPE improved oil recovery symposium, Vol. 1. Tulsa: Society of Petroleum Engineers; 2012. p. 25–34.

Fuhrman JA. Microbial community structure and its functional implications. Nature. 2009;459(7244):193–9.

Gao CH, Zerki A. Applications of microbial-enhanced oil recovery technology in the past decade. Energy Sour Part A. 2011;33(10):972–89.

Gao P, Li G, Dai X, et al. Nutrients and oxygen alter reservoir biochemical characters and enhance oil recovery during biostimulation. World J Microbiol Biotechnol. 2013;29(11):2045–54.

Ghosal D, Chakraborty J, Khara P, et al. Degradation of phenanthrene via meta-cleavage of 2-hydroxy-1-naphthoic acid by *Ochrobactrum* strain PWTJD. FEMS Microbiol Lett. 2010;313(2):103–10.

Hubert C, Voordouw G. Oil field souring control by nitrate-reducing *Sulfurospirillum* spp. that outcompete sulfate-reducing bacteria for organic electron donors. Appl Environ Microbiol. 2007;73(8):2644–52.

Ji G, Liao B, Tao H, et al. Analysis of bacteria communities in an upflow fixed-bed (UFB) bioreactor for treating sulfide in hydrocarbon wastewater. Bioresour Technol. 2009;100:5056–62.

Joshi S, Bharucha C, Jha S, et al. Biosurfactant production using molasses and whey under thermophilic conditions. Bioresour Technol. 2008;99(1):195–9.

Kitamoto D, Morita T, Fukuoka T. Self-assembling properties of glycolipid biosurfactants and their potential applications. Curr Opin Colloid Interface Sci. 2009;14(5):315–28.

Kobayashi H, Kawaguchi H, Endo K, et al. Analysis of methane production by microorganisms indigenous to a depleted oil reservoir for application in microbial enhanced oil recovery. J Biosci Bioeng. 2012;113(1):84–7.

Lazar I, Petrisor IG, Yen TF. Microbial enhanced oil recovery (MEOR). Pet Sci Technol. 2007;25(11):1353–66.

Nazina TN, Sokolova DS, Grigoryan AA, et al. Production of oil-releasing compounds by microorganisms from the Daqing Oil Field, China. Microbiology. 2003;72(2):173–8.

Ng IS, Li CW, Chan SP, et al. High-level production of a thermoacidophilic β-glucosidase from *Penicillium citrinum* YS40-5 by solid-state fermentation with rice bran. Bioresour Technol. 2010;101(4):1310–7.

Noike T, Mizuno O. Hydrogen fermentation of organic municipal wastes. Water Sci Technol. 2000;42(12):155–62.

Reddy MS, Naresh B, Leela T, et al. Biodegradation of phenanthrene with biosurfactant production by a new strain of *Brevibacillus* sp. Bioresour Technol. 2010;101(20):7980–3.

Ruggeri C, Franzetti A, Bestetti G, et al. Isolation and characterisation of surface active compound-producing bacteria from hydrocarbon-contaminated environments. Int Biodeterior Biodegrad. 2009;63(7):936–42.

Sarafzadeh P, Hezave AZ, Ravanbakhsh M, et al. Enterobacter cloacae as biosurfactant producing bacterium: differentiating its effects on interfacial tension and wettability alteration mechanisms for oil recovery during MEOR process. Colloid Surface B. 2013;105(1):223–9.

Sen R. Biotechnology in petroleum recovery: the microbial EOR. Prog Energy Combust. 2008;34(6):714–24.

She YH, Zhang F, Xia JJ, et al. Investigation of biosurfactant-producing indigenous microorganisms that enhance residue oil recovery in an oil reservoir after polymer flooding. Appl Microbiol Biotechnol. 2011;163(2):223–34.

Spirov P, Ivanova Y, Rudyk S. Modelling of microbial enhanced oil recovery application using anaerobic gas-producing bacteria. Pet Sci. 2014;11(2):272–8.

Taguchi F, Yamada K, Hasegawa K, et al. Continuous hydrogen production by *Clostridium* sp. strain no. 2 from cellulose hydrolysate in an aqueous two-phase system. J Ferment Bioeng. 1996;82:80–3.

Takahashi R, Yamayoshi K, Fujimoto N, et al. Production of (*S*,*S*)-ethylenediamine-*N*,*N*-disuccinic acid from ethylenediamine and fumaric acid by bacteria. Biosci Biotechnol Biochem. 1999;63(7):1269–73.

Tanaka T, Hoshina M, Tanabe S, et al. Production of d-lactic acid from defatted rice bran by simultaneous saccharification and fermentation. Bioresour Technol. 2006;97(2):211–7.

Tang YQ, Li Y, Zhao JY, et al. Microbial communities in long-term, water flooded petroleum reservoirs with different in situ temperatures in the Huabei Oilfield, China. PLoS One. 2012;7(3):e33535.

Wang J. Study of activator screening and its impact on indigenous microbial communities in the Xinjiang Oilfield. Master Dissertation. China University of Petroleum, Beijing. 2013. p. 26–28 (in Chinese).

Wang J, Ma T, Zhao L, et al. Monitoring exogenous and indigenous bacteria by PCR-DGGE technology during the process of microbial enhanced oil recovery. J Ind Microbiol Biotechnol. 2008a;35(6):619–28.

Wang J, Yan G, An M, et al. Study of a plugging microbial consortium using crude oil as sole carbon source. Pet Sci. 2008b;5(4):367–74.

Wang X, Li D, Hendry P, et al. Effect of nutrient addition on an oil reservoir microbial population: implications for enhanced oil recovery. J Pet Environ Biotechnol. 2012;3(2):118.

Yao C, Lei G, Ma J, et al. Laboratory experiment, modeling and field application of indigenous microbial flooding. J Pet Sci Eng. 2012;90–91:39–47.

Zhang F, She YH, Ma SS, et al. Response of microbial community structure to microbial plugging in a mesothermic petroleum reservoir in China. Appl Microbiol Biotechnol. 2010;88(6):1413–22.

Zhang F, She YH, Li HM, et al. Impact of an indigenous microbial enhanced oil recovery field trial on microbial community structure in a high pour-point oil reservoir. Appl Microbiol Biotechnol. 2012;95(3):811–21.

Zhou QY, Wang SF. Microbiology of environmental engineering. Beijing: Higher Education Press; 2004. p. 292–3 (in Chinese).

Selective dissolution of eodiagenesis cements and its impact on the quality evolution of reservoirs in the Xing'anling Group, Suderte Oil Field, Hailar Basin, China

Zhen-Zhen Jia[1,2] · Cheng-Yan Lin[1,2] · Li-Hua Ren[1,2] · Chun-Mei Dong[1,2]

Abstract Reservoirs in the Xing'anling Group in the Suderte Oil Field, Hailar Basin exhibit ultra-low to low permeability and high tuffaceous material content. This study comprehensively analyzed diagenesis and quality evolution of these low-permeability reservoirs using thin sections, SEM samples, rock physical properties, pore water data, as well as geochemical numerical simulations. Calcite and analcite are the two main types of cements precipitated in the eodiagenetic stage at shallow burial depths in the reservoirs. These two cements occupied significant primary intergranular pores and effectively retarded deep burial compaction. Petrography textures suggest selective dissolution of massive analcite and little dissolution of calcite in the mesodiagenetic stage. Chemical calculations utilizing the Geochemist's Workbench 9.0 indicated that the equilibrium constant of the calcite leaching reaction is significantly smaller than that of the analcite leaching reaction, resulting in extensive dissolution of analcite rather than calcite in the geochemical system with both minerals present. Numerical simulations with constraints of kinetics and pore water chemistry demonstrated that the pore water in the Xing'anling group is saturated with respect to calcite, but undersaturated with analcite, leading to dissolution of large amounts of analcite and no dissolution of calcite. Significant secondary intergranular pores have formed in analcite-cemented reservoirs from selective dissolution of analcite in the mesodiagenetic stage; the analcite dissolution formed preferential flow paths in the reservoirs, which promoted feldspar dissolution; and dissolution of such minerals led to the present reservoirs with medium porosity and low permeability. Calcite-cemented tight reservoirs have not experienced extensive dissolution of cements, so they exhibit ultra-low porosity and permeability.

Keywords Eodiagenetic cements · Calcite · Analcite · Selective dissolution · Secondary porosity · Hailar Basin

✉ Cheng-Yan Lin
  ycdzycms@126.com

[1] School of Geosciences, China University of Petroleum, Qingdao 266580, Shandong, China

[2] Key Laboratory of Reservoir Geology in Shandong Province, Qingdao 266580, Shandong, China

Edited by Jie Hao

1 Introduction

Early cementation (e.g., carbonate cementation, zeolite cementation) in clastic reservoirs has considerable impact on reservoir quality evolution, as it may slow subsequent burial compaction and provides significant potential minerals for burial dissolution (Yuan et al. 2015a; Dutton and Loucks 2010; Fu et al. 2010; Yu and Lai 2006; Schmidt and McDonald 1979; Zhu et al. 2012). Recent studies have suggested that early zeolite cements (e.g., analcite, laumontite) can be dissolved extensively at the mesodiagenetic stage to form secondary pores, enhancing reservoir porosity and permeability (Tang et al. 1997; Zhu 1985; Meng et al. 2013; Sun et al. 2014; Zhu et al. 2011). However, there is significant disagreement among various authors regarding burial dissolution of carbonate cements in clastic reservoirs. In broad terms, there are two major schools of thought (Bjørlykke and Jahren 2012; Giles 1987; Giles and Marshall 1986; Yuan et al. 2015a, b). One group of authors considers that carbonate cements can be leached during burial to generate significant secondary porosity and improve reservoir quality (Schmidt and McDonald 1979;

Zhong et al. 2003; Yu and Lai 2006). The other group of authors suggests that carbonate cement cannot be dissolved extensively at the deep burial stage, but that carbonate cementation degrades reservoir porosity and permeability (Bjørlykke and Jahren 2012; Giles and Marshall 1986; Taylor et al. 2010; Yuan et al. 2013, 2015a, b, c; Shou 2005). The reservoirs in the Xing'anling Group in the Suderte Oilfield of the Hailar Basin, which are characterized by abundant volcanic materials, are mostly low to ultra-low permeability reservoirs (Wang et al. 2012). Calcite and analcite are two main types of early cements in these reservoirs, and thin sections and SEM samples demonstrate extensive selective dissolution of analcite cement. However, there is not much evidence supporting the dissolution of associated calcite cement in the Xing'anling Group. Reservoirs with extensive dissolution of analcite are characterized by significant secondary pores, and the core porosity of such reservoirs can reach up to 20 %–25 %, while the core porosity of poor-quality reservoirs with massive calcite cement is generally lower than 10 %–15 % due to limited burial dissolution. To this effect, understanding the origin and processes at work in the selective dissolution of these two cements benefits reservoir quality prediction (Yuan et al. 2015a).

The selective dissolution phenomena of different minerals in sediments have garnered quite a bit of attention from geologists in recent years (Cao et al. 2014; Macquaker et al. 2014; Turchyn and Depaolo 2011; Yuan et al. 2015a). Macquaker et al. (2014) and Turchyn and Depaolo (2011), for example, reported similar phenomena in fine-grained sediments (Macquaker et al. 2014; Turchyn and Depaolo 2011) and Yuan et al. (2015a) investigated the selective dissolution between feldspars and calcite in buried sandstones. Other geologists have studied genetic mechanisms of the selective dissolution phenomena between zeolites and calcite via static plots, including Gibbs free energy versus temperature (burial depth) profiles (Meng et al. 2013, 2014; Qi 2013; Xiu 2008; Zhao 2005) and equilibrium constant versus temperature profiles (Qi 2013; Zhao 2005) of leaching reactions of different minerals. Studies of water–rock interactions in geochemical systems considering both minerals over extended periods of time and kinetics-related constraints, however, are relatively few. In effort to remedy this, the present study was conducted with the following main objectives: (1) to investigate sandstone diagenesis using cores, thin sections, and SEM sample analysis, (2) to analyze various impacts of selective dissolution of analcite and calcite on reservoir quality evolution by testing physical properties, and (3) to decipher the genetic mechanism of selective dissolution between analcite and calcite using the Geochemist's Workbench (GWB) 9.0 with constraints of pore water chemistry and kinetic data.

2 Geological settings

Suderte Oilfield, with an exploration area of about 200 km^2, is located near the middle of the Suderte tectonic zone in the Beier Sag of the Hailar Basin. The tectonic zone is cut by NE-trend and WE-trend faults, splitting the oilfield into several NEE-trend fault blocks. The sediments in the oilfield contain the Triassic Budate Formation, Cretaceous Tongbomiao, Nantun, Damoguaihe, Yimin, and Qingyuangang Groups, and Cenozoic Formation, from base to top. The Tongbomiao Group and the first member of the Nantun Group have been further divided into six oil group members marked X0–XV (Fig. 1), which constitute the Xing'anling Oil Group, the major oil-bearing sequence in the area. The sedimentary strata in the study area feature large structural altitude differences and significant thickness variations caused by fault impacts. Reservoirs in oil group members XI and XII, which show stable spatial distribution, are the main focus of this study.

During the depositional period of the first member of the Nantun Group, the Suderte tectonic zone was a small rift basin controlled by multiple fault terraces. The steep slope zone of the rift basin was characterized by large altitude differences and, of course, steep slopes. The alluvial fan depositional systems, with mainly southwest and southeast source supplies, entered the lakes quickly to form fan deltas; multisource fan deltas were interconnected to form fan delta aprons distributed along the fault margins (Wang et al. 2012). Studies on zircon U–Pb chronology have demonstrated that volcanic activity in the Hailar Basin continued during the late Jurassic to the early Cretaceous period, with stronger activity in 128–117 and 116–113 Ma (Chen et al. 2015; Zhao et al. 2013). Impacted by simultaneous volcanic eruptions during the depositional period, rocks in the XI–XIV oil group members generally contain volcanic debris (Xiao et al. 2011), consisting mainly of tuff conglomerate, tuff sandstones, and some sedimentary tuff.

Burial and thermal histories of the Suderte tectonic zone show that the Nantun Formation experienced four stages: rapid subsidence, uplift, slow subsidence, and stability (Fig. 2) (Song 2013; Shen et al. 2013). The paleo Formation temperature gradient in the Cretaceous was about 4.2–5.6 °C/100 m, higher than the present temperature gradient (3.30 °C/100 m) (Cui et al. 2011). Burial and thermal histories of well Bei-30 show that the Nantun Formation experienced its deepest burial depth and highest formation temperature by the end of the early Cretaceous, after which formation temperature decreased continually with uplift and decreasing temperature gradient. The present burial depth and temperature are below maximum burial depth and temperature.

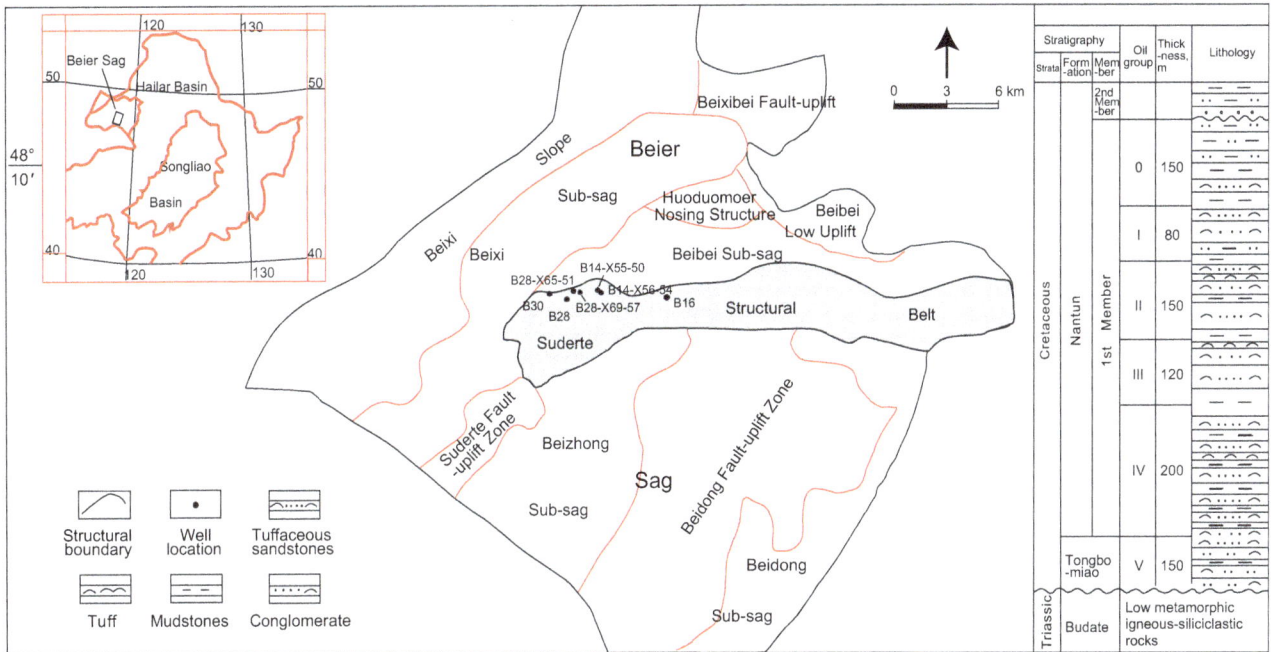

Fig. 1 Location map of the Beier Sag, Hailar Basin, subunits in the Beier Sag, and sedimentary strata in the Suderte Oilfield. Modified from Wang et al. (2012)

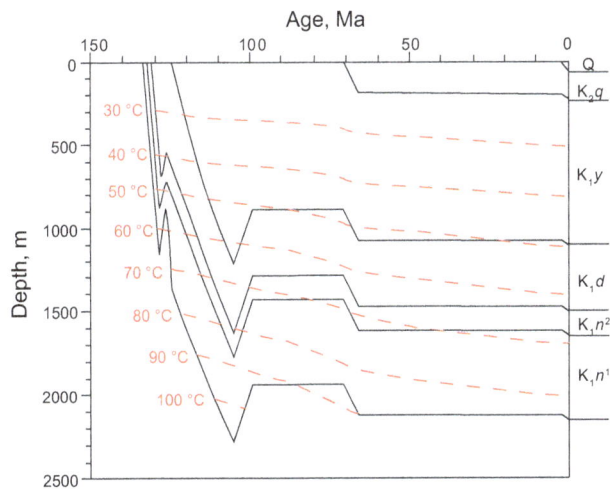

Fig. 2 Burial and thermal history of well Bei-30 in the Suderte Oilfield. Modified from Song (2013)

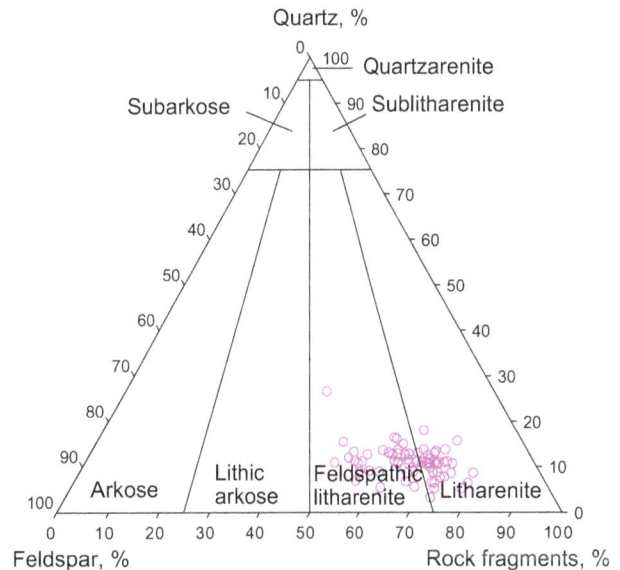

Fig. 3 Ternary plots showing grain composition of sandstones in the Xing'anling Group, Suderte Oilfield

3 Petrography

Members XI–XII in the Xing'anling Group in the Suderte Oilfield are mainly composed of fine sandstones, medium- to coarse-grained sandstones, fine-grained conglomerates, and siltstones, as evidenced by core observation and thin section identification. The siltstone and conglomerates are rich in tuffaceous matrix, and the sandstones are composed of feldspathic litharenite and litharenite (Fig. 3) that are texturally and compositionally immature. The sandstones contain an average of 10 % detrital quartz grains, 20 % feldspar, and 70 % rock fragments which consist mainly of andesitic and rhyolitic tuff (up to 80 %), followed by sedimentary rock fragments. The amount of intergranular fillings ranges from 10 % to 25 %, including 8 %–20 % authigenic cements and less than 5 % matrix (detrital clays and volcanic ash). Authigenic minerals consist mainly of calcite, analcite, and kaolinite. On the whole, detrital grains are moderately to poorly sorted, detrital rock fragments are present in subangular shapes and detrital quartz grains in subangular or subrounded shapes. Grain contacts are mainly line–line and concavo–convex.

4 Diagenesis

4.1 Types and characteristics of diagenesis

4.1.1 Compaction

Compaction is a major factor affecting porosity and permeability reduction during reservoir burial (Lu et al. 2015; Zhang et al. 2014; Zhu et al. 2010; Xu et al. 2008; Cao et al. 2012, 2014; Pittman and Larese 1991; Xi et al. 2015).

Due to their high content of ductile tuff debris, Xing'anling Group rocks have low compaction resistance and are generally strongly compacted (Mousavi and Bryant 2013; Pittman and Larese 1991) (Fig. 4a). The development of early cementation can effectively retard compaction. In sandstones with little cement, detrital grain contacts are mainly line–line and concavo–convex as opposed to point–line. In sandstones with abundant cements, conversely, grain contacts are mainly line–point and point–line. Quantitative statistics show that in reservoirs with high cement (calcite) content, reservoirs exhibit porosity loss from compaction of about 10 %–20 %. In reservoirs with relatively low cement content but well-developed intergranular secondary pores formed during analcite dissolution (analcite cementation occurred in the eodiagenetic stage), porosity loss due to compaction is nearly 20 %–

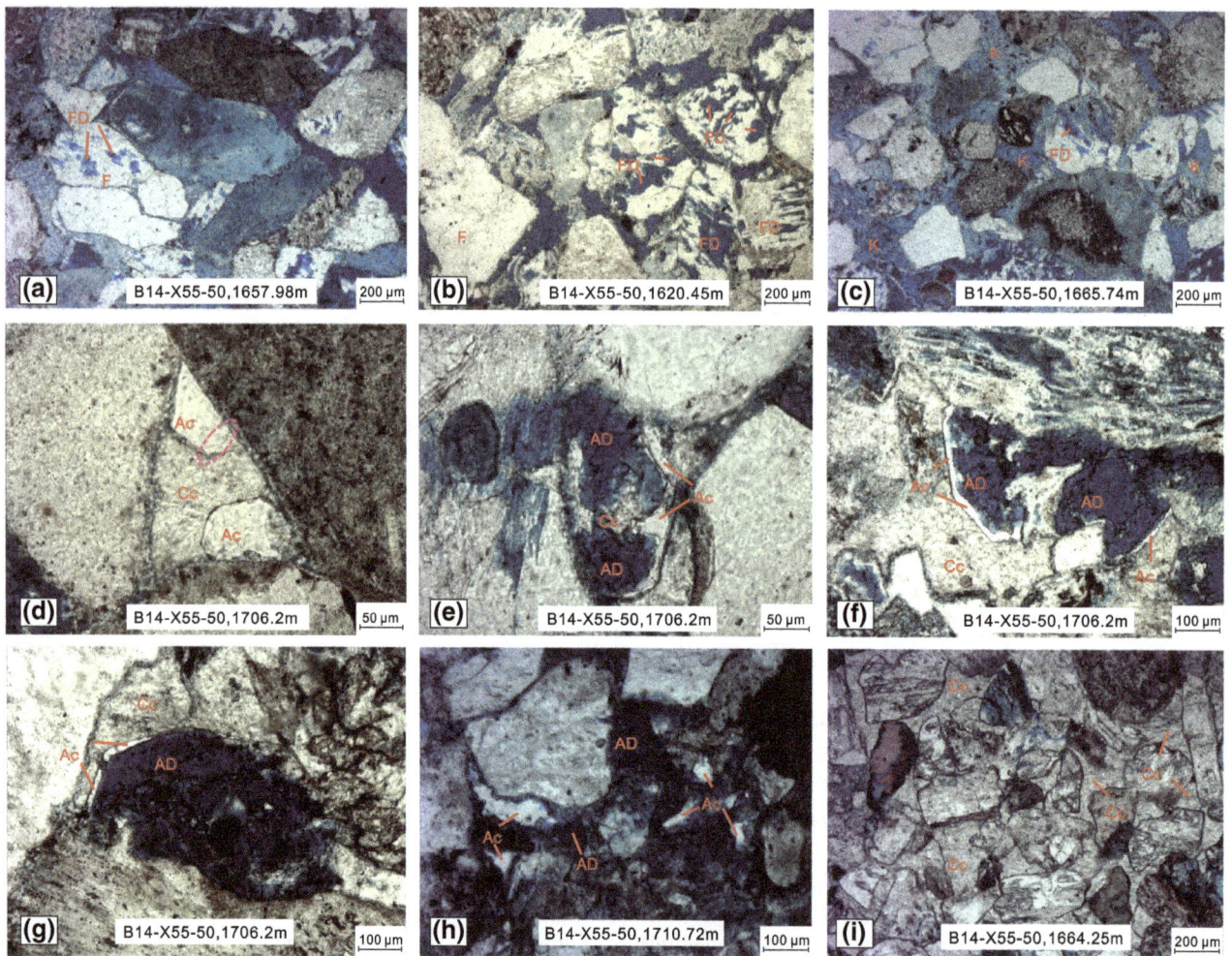

Fig. 4 Microscope photos of thin sections of sandstones in the Xing'anling Group. **a** Mechanical compaction, grains in concavo–convex-linear contact; **b** dissolved feldspars and intergranular secondary pores; **c** dissolved feldspar and authigenic kaolinite; **d** analcite dissolution remnants and early calcite cement, where analcite was partially replaced by calcite; **e** late carbonate cement in secondary pores formed by analcite dissolution; **f**, **g** selective dissolution of analcite in the presence of early calcite cement; **h** analcite remnants after dissolution; **i** tight sandstones cemented by large amounts of early calcite. *FD* secondary pores formed by feldspar dissolution, *K* kaolinite, *Cc* carbonate cement, *Ac* analcite

25 %. While in reservoirs with relatively low-cement content and poorly developed secondary pores (weak cementation occurred in the eodiagenetic stage), porosity loss due to compaction is approximately 30 %–35 %.

4.1.2 Cementation

Cementation also has significant impact on reservoir quality evolution during burial. Various cements in the Xing'anling Group reservoirs include calcite, analcite, kaolinite, and a small amount of quartz.

1. Carbonate cementation.

   Carbonate is the most important cement component in the Xing'anling sandstones. The mineral texture of carbonate cement and the relationship between carbonate cement and secondary pores indicate the existence of abundant stage-I calcite cement (early calcite) and a small amount of stage-II calcite cement (late calcite). In early calcite-cemented tight sandstones, cements occupy almost all primary pores and can account for 25 %–30 % of the total sandstone volume. Generally, early calcite-cemented sandstones are supported by detrital grains with point–point contacts or with floating texture (Fig. 4i), implying there was little compaction when cementation occurred (Gluyas and Coleman 1992; Yuan et al. 2015b, c). The content of late calcite cement is generally less than 1 % and this has only slight impact on reservoir quality. Late calcite cement occurred mainly in secondary pores in the study reservoirs, forming after the dissolution of feldspars and analcite.

2. Analcite cementation.

   The abundance of volcanic materials in the reservoirs of the Xing'anling Group and the early alkaline environment following deposition facilitated cementation of zeolite (Tang et al. 1997; Xiao et al. 2011; Zhu et al. 2011). Textures in thin sections indicate a competitive relationship between analcite and calcite cements (Fig. 4d–g), which plays a key role in controlling reservoir quality evolution. In the reservoirs, this competitive relationship is indicated by the negative relationship between the amount of carbonate cement and the amount of analcite and analcite secondary pores (Fig. 5). The analcite cement was precipitated in the form of pyritohedron single crystals and blocky aggregates in intergranular pores (Fig. 6b), and currently occurs primarily as dissolution remnants (Fig. 4e–h). Petrography texture of the replacement of analcite by calcite (Fig. 4d) indicates that the formation of analcite cement was just prior to or synchronous with that of calcite cement.

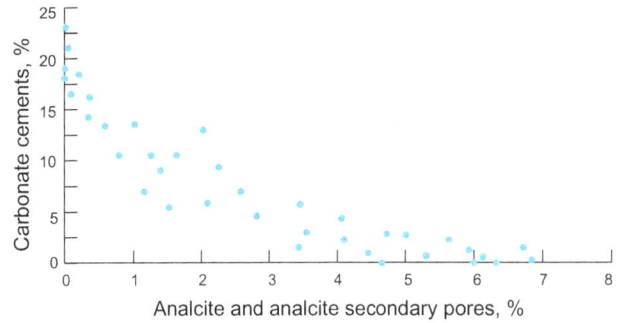

Fig. 5 Negative relationship between the amount of carbonate cement and the amount of analcite and analcite secondary pores in the reservoirs in the Xing'anling Group

3. Authigenic clays.

   Volcanic rock fragments in reservoirs of the Xing'anling Formation consist mainly of acidic andesite and rhyolitic tuff, exhibiting weak dissolution. Some secondary micropores occurred in fragments through partial hydrolysis of these grains. Tuffaceous matrix was abundant in the reservoirs and this has experienced complex diagenesis. In the eodiagenetic stage, the tuffaceous matrix was transformed to chlorite and illite/smectite (I/S) in the presence of alkaline water. Chlorite was precipitated mainly as grain rims (Fig. 4e), which may have retarded quartz cementation in the mesodiagenetic stage. Kaolinite is the most important authigenic clay mineral in the reservoirs, at relative content up to 70 %–80 % of the total clay minerals in the sandstones (Fig. 7). Kaolinite occurs mainly in the sandstones with abundant secondary pores, and is scarce in calcite-cemented, tight rocks. This textural relationship indicates that the precipitation of kaolinite occurred after early carbonate cementation, probably in the mesodiagenetic stage, and as a byproduct of dissolution of feldspars and analcite. Single kaolinite crystals generally show hypidiomorphic pseudohexagonal structure, and the size of kaolinite platelets is usually less than 5–8 μm in width with thickness less than 0.5 μm. Kaolinite aggregates, mainly in short vermiform shape and booklet-like shape, are generally less than 20 μm in length. Kaolinite is distributed in a relatively dispersed pattern in the study area, and contains abundant intercrystal micropores (Figs. 4c, 6c).

4. Quartz cementation.

   Petrography reveals that quartz cementation is relatively weak in the sandstones. Quartz cement occur mainly as small quartz crystals (<5–10 μm) (Fig. 6d), and quartz overgrowth cannot be identified in thin section or SEM samples. The impact of quartz cementation on physical properties of reservoirs is

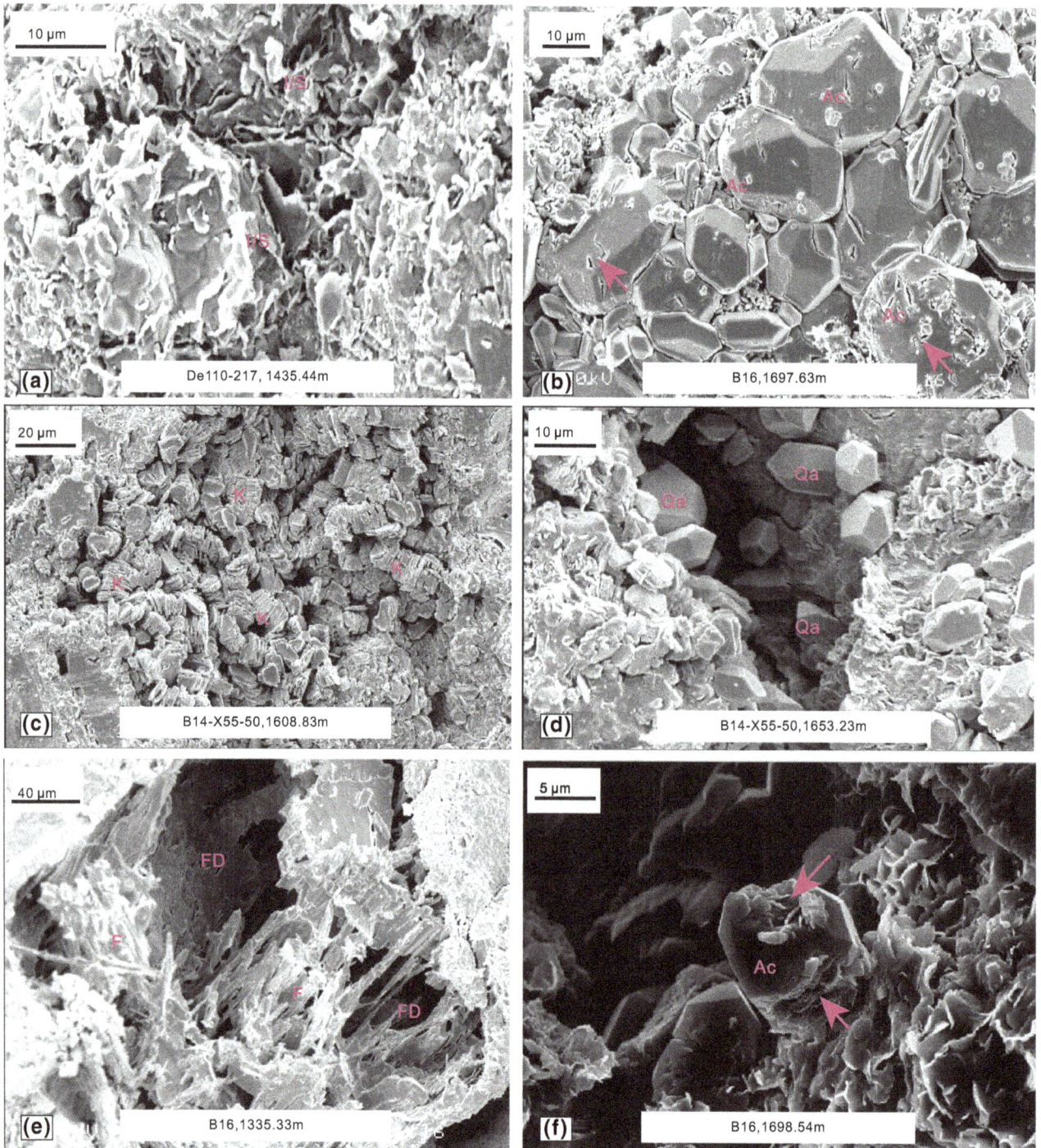

Fig. 6 SEM micrographs of sandstones in the Xing'anling Group, Suderte Oilfield. **a** Mixed layer illite/smectite (I/S) in sandstones; **b** analcite (Ac) in intergranular pores; **c** kaolinite (K) in pores; **d** quartz crystals (Qa) in intergranular pores; **e** extensively dissolved feldspar (F); **f** analcite (Ac) and corroded notches

insignificant. Authigenic quartz occurs mainly in the reservoirs with abundant secondary pores, and is scarce in calcite-cemented tight rocks, indicating that quartz cementation occurred after early carbonate cementation, probably in the mesodiagenetic stage, and also as a byproduct of dissolution of analcite and feldspars.

4.1.3 Selective dissolution of minerals

Mineral dissolution generally improves reservoir quality (Schmidt and McDonald 1979; Surdam et al. 1984; Yuan et al. 2015c; Cao et al. 2012; Zhu et al. 2007; Han et al. 2007). Vitrinite reflectance (R_0 %) of organic matter in

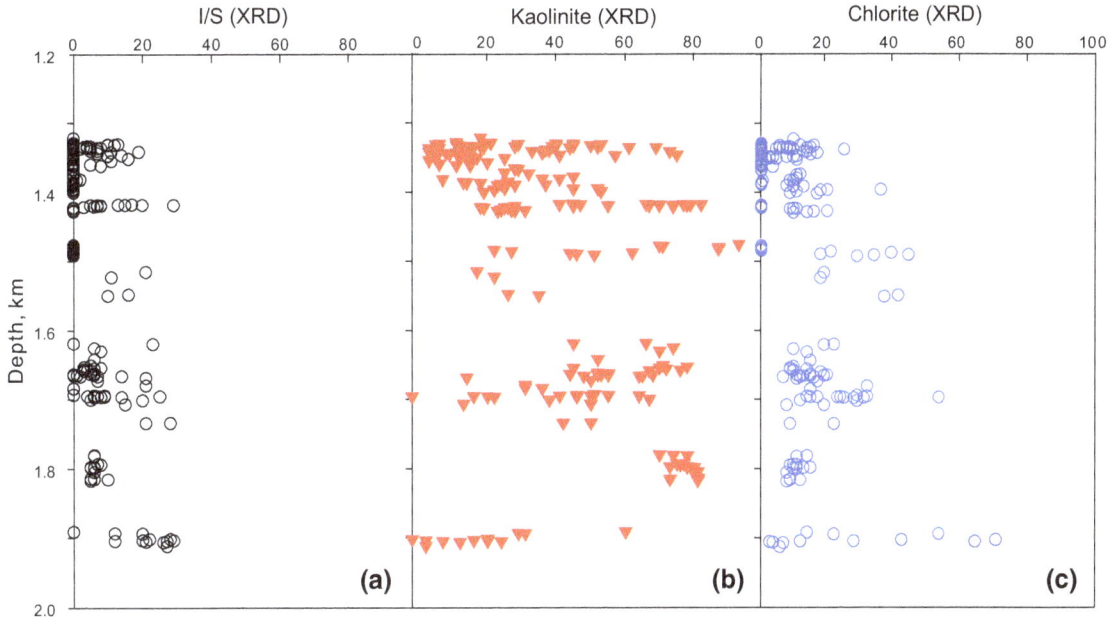

Fig. 7 Relative contents of illite–smectite, kaolinite, and chlorite in sandstones in the Xing'anling Group, Suderte Oilfield

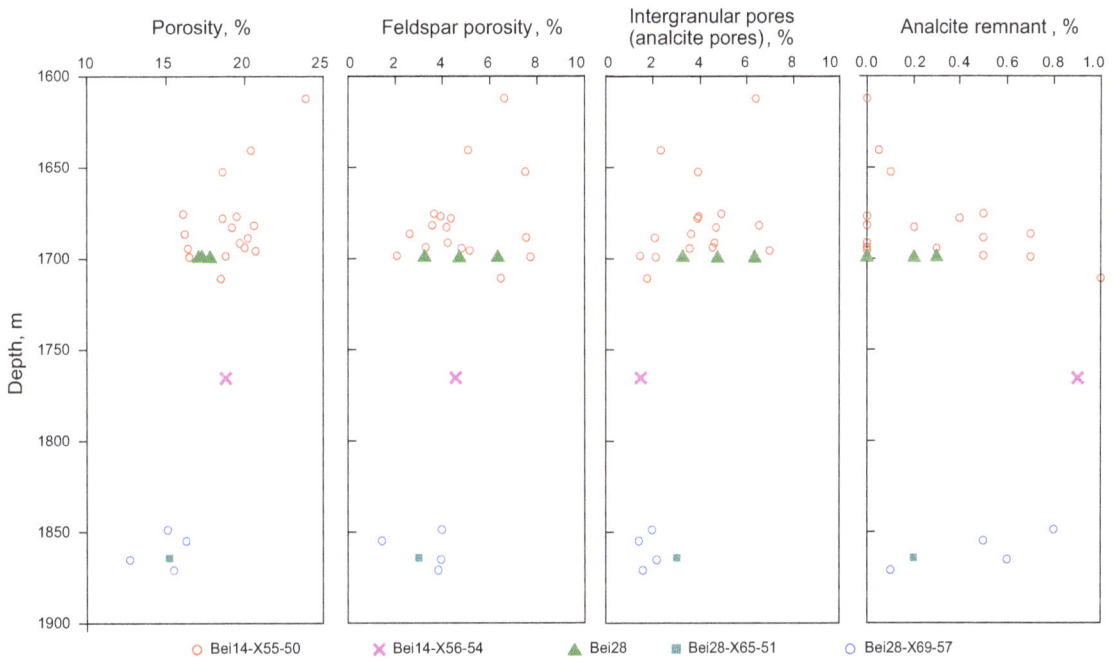

Fig. 8 Plots of porosity, feldspar porosity, intergranular pores (analcite pores), and remnant analcite in high-quality reservoirs in the Xing'anling Formation

interbedded mudstones in the Xing'anling Group is up to 0.8 %–1.0 %, and thermal evolution of such organic matter has produced large amounts of CO_2 and organic acids, which have probably leached unstable minerals (Schmidt and McDonald 1979; Surdam et al. 1984; Yuan et al. 2015a). Petrography textures show that in sandstones with extensive cementation of early calcite, the calcite cement was apparently not leached. As the calcite cement clogged fluid flow paths, feldspar in the sandstones were also not dissolved extensively. In sandstones with weak calcite cementation but extensive analcite cementation, extensive dissolution of analcite and feldspars occurred and formed a large amount of secondary porosity (Figs. 4b, c, 4e–h, 6e, 8). There is no petrographic evidence, however, supporting dissolution of the associated calcite cements in such porous sandstones (Fig. 4d–g). Differing from past studies which

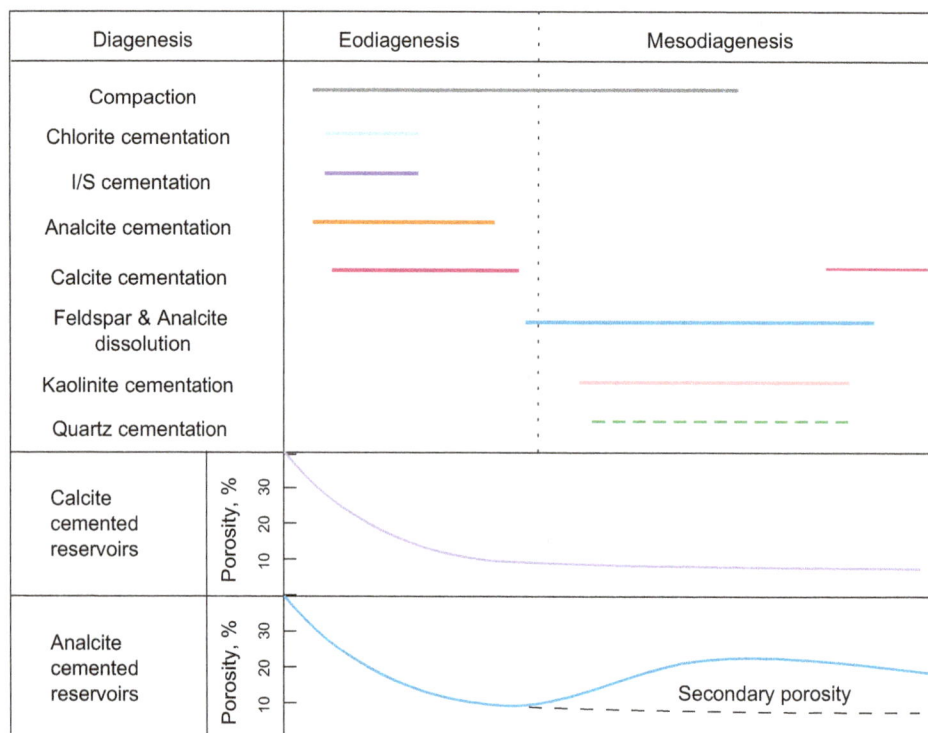

Fig. 9 Diagenetic sequences and porosity evolution model of different reservoirs in the Xing'anling Formation

suggested that carbonate minerals could be easily dissolved by acidic fluids, this selective dissolution of analcite in the presence of calcite in the studied reservoirs is quite interesting (Macquaker et al. 2014; Yuan et al. 2015a).

4.2 Diagenetic sequences

The relative timing of the major diagenetic sequence of the sandstones in the Xing'anling Group, which has been determined from thin sections and SEM examination, is based on texture relationship of cementation, dissolution, and replacement of various minerals (Figs. 4, 6). In summary, the integrated diagenetic sequences consist of compaction/chlorite cementation/early analcite cementation/early calcite cementation–analcite dissolution/feldspar dissolution/authigenic kaolinite precipitation/quartz cementation–late carbonate cementation (Fig. 9).

5 Early cements and reservoir properties

The selective dissolution of early analcite cement and calcite cement at the mesodiagenetic stage led to significant differences in properties among various reservoirs.

5.1 Reservoirs with eodiagenetic cementation and mesodiagenetic nondissolution of calcite

Thin sections and SEM samples demonstrate that in reservoirs with extensive calcite cementation, primary intergranular pores are occupied almost entirely by calcite cement. Reservoir spaces consist of a few residual micropores and secondary pores formed by dissolution of feldspars and analcite, and the content of pores in thin sections is commonly lower than 0.1 %. Physical properties of such sandstones show that the calcite-cemented tight sandstones are typically found in low-porosity and ultra-low permeability reservoirs (Fig. 10a1, b1, c), with porosity lower than 15 % and permeability lower than 0.1 mD. These sandstones are characterized by poor pore structures, generally with micropores and microthroats. High-pressure mercury injection tests matched with thin section examination demonstrate that mercury injection curves are characterized by high initial replacement pressure and low injection saturation. Initial replacement pressure is generally higher than 5 MPa, maximum pore-throat radius is lower than 0.2 μm, average pore-throat radius is lower than 0.05 μm, mercury injection saturation is below 50 %, and mercury withdrawal efficiency is below 30 %.

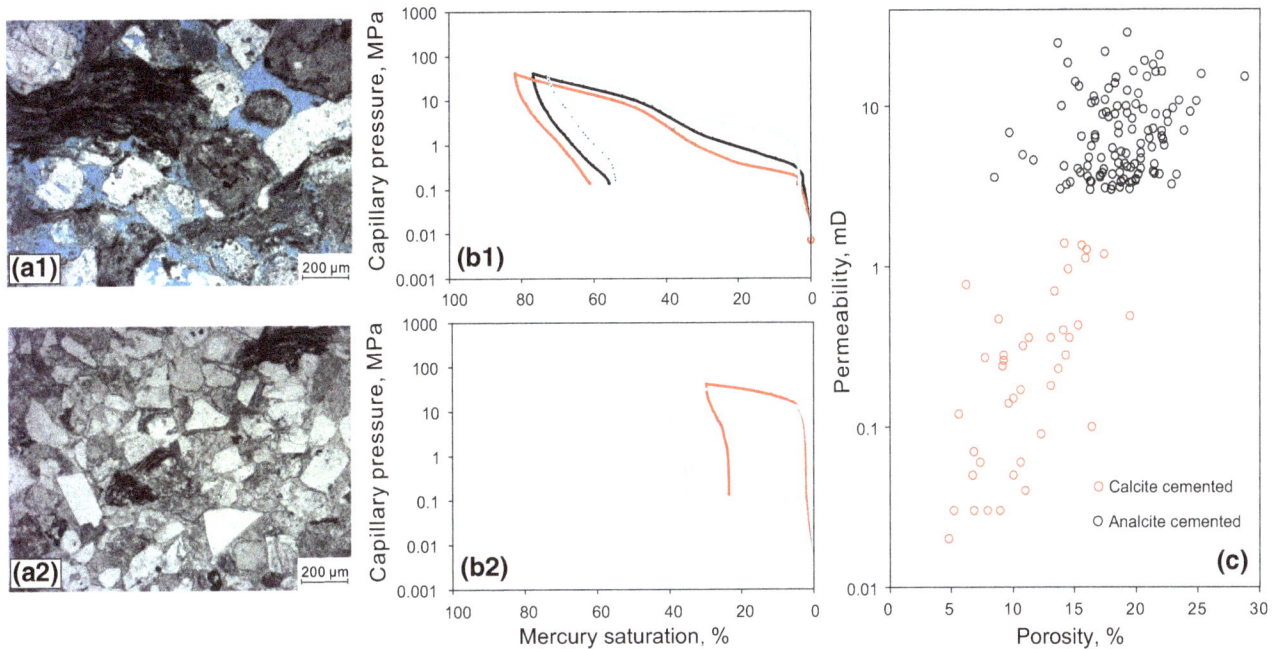

Fig. 10 Comparison of reservoir pores, pore structures, and physical properties of two different types of reservoirs. **a1–b1** Pores and mercury injection curves of reservoirs with extensive analcite dissolution; **a2–b2** pores and mercury injection curves of reservoirs with extensive early calcite cementation; **c** porosity versus permeability plots of the two different types of reservoirs

5.2 Reservoirs with eodiagenetic cementation, mesodiagenetic dissolution of analcite

Thin sections and SEM samples indicate that in reservoirs with strong early cementation and extensive late dissolution of analcite, spaces are mainly secondary pores formed by dissolution of analcite and feldspars while micropores in kaolinite aggregates, and primary pores are undeveloped. Quantitative analysis of different pores in thin sections shows that secondary pores formed by analcite dissolution range from 2 % to 6 % with an average of 4 %, and secondary pores formed by feldspar dissolution range from 2 % to 7 % with an average of 5 %. Physical properties of these sandstones are much more favorable than those of calcite-cemented tight sandstones, with porosity higher than 15 % and permeability higher than 1 mD; these reservoirs typically show medium porosity and low permeability (Fig. 10a2, b2, c), and sandstones are characterized by moderate pore structures with generally moderate pores and fine- to micro-throats. High-pressure mercury injection tests matched with the thin sections demonstrate that mercury injection curves are characterized by low initial replacement pressure and high injection saturation. The initial replacement pressure is generally lower than 0.1–1 MPa, maximum pore-throat radius is lower than 1 μm, and average pore-throat radius is lower than 0.2 μm.

5.3 Evolution models of reservoir physical properties

For reservoirs with extensive carbonate cementation, burial dissolution is especially weak in reservoirs with abundant calcite, and chemical reactions nearly cease after extensive calcite cementation. The porosity of such sandstones decreased significantly due to strong compaction and extensive calcite cementation, and without significant burial dissolution, the reservoirs still show low porosity now (Fig. 9).

For reservoirs with strong analcite cementation, analcite and feldspars were dissolved extensively, and such reservoir experienced a relatively complete diagenetic sequence. The porosity of these sandstones also decreased significantly with intensive compaction and cementation at the eodiagenetic stage. Extensive dissolution of analcite and feldspars at the mesodiagenetic stage, however, formed large-volume secondary pores, and the current porosity recovered to a relatively high level (Fig. 9).

6 Genetic mechanism of selective dissolution reaction

Laboratory experiments indicated that carbonate minerals can be dissolved much faster than aluminosilicate minerals (Arvidson et al. 2003; Yuan et al. 2015a). However, recent studies suggested that feldspars can be selectively dissolved in the presence of carbonate minerals in buried sandstones (Yuan et al. 2015a). The selective dissolution of analcite in the presence of calcite is also interesting as a departure from the results of traditional laboratory experiments (Savage et al. 1999). This section reports the numerical calculations and simulations we conducted using Geochemist's Workbench 9.0 to investigate the chemical reactions in the analcite–calcite–acid (CO_2 acid or other acids) H_2O system, and proposes a genetic mechanism of the selective dissolution reaction.

6.1 Chemical reactions

The leaching reactions of analcite and calcite by CO_2 can be expressed by the following equations (Giles and Marshall 1986; Yuan et al. 2015a; Zhang et al. 2011):

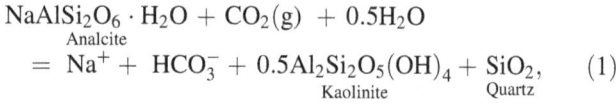

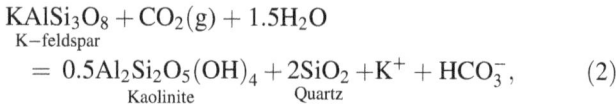

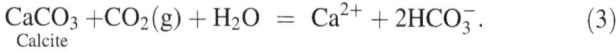

$$NaAlSi_2O_6 \cdot H_2O + CO_2(g) + 0.5H_2O$$
$$\underset{\text{Analcite}}{} = Na^+ + HCO_3^- + 0.5\underset{\text{Kaolinite}}{Al_2Si_2O_5(OH)_4} + \underset{\text{Quartz}}{SiO_2}, \quad (1)$$

$$\underset{\text{K--feldspar}}{KAlSi_3O_8} + CO_2(g) + 1.5H_2O$$
$$= 0.5\underset{\text{Kaolinite}}{Al_2Si_2O_5(OH)_4} + 2\underset{\text{Quartz}}{SiO_2} + K^+ + HCO_3^-, \quad (2)$$

$$\underset{\text{Calcite}}{CaCO_3} + CO_2(g) + H_2O = Ca^{2+} + 2HCO_3^-. \quad (3)$$

The log equilibrium constant of the above three reactions can be expressed as

$$logK_1 = -logf[CO_2(g)] - loga[H_2O] + loga[N^+] + loga[HCO_3^-],$$

$$logK_2 = loga[K^+] + loga[HCO_3^-] - logf[CO_2(g)] - 1.5loga[H_2O],$$

$$logK_3 = loga[Ca^{2+}] + 2loga[HCO_3^-] - logf[CO_2(g)] - loga[H_2O].$$

Instead of CO_2, when other acids are used to leach analcite and calcite, the chemical reactions can be expressed by as follows:

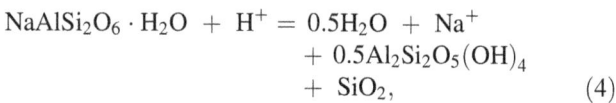

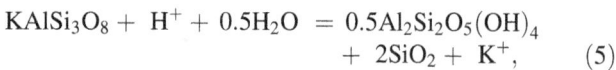

$$NaAlSi_2O_6 \cdot H_2O + H^+ = 0.5H_2O + Na^+ + 0.5Al_2Si_2O_5(OH)_4 + SiO_2, \quad (4)$$

$$KAlSi_3O_8 + H^+ + 0.5H_2O = 0.5Al_2Si_2O_5(OH)_4 + 2SiO_2 + K^+, \quad (5)$$

$$CaCO_3 + H^+ = Ca^{2+} + HCO_3^-. \quad (6)$$

The log equilibrium constant of the above reactions can be expressed as

$$logK_4 = -logf[CO_2(g)] - loga[H_2O] + loga[N^+] + loga[HCO_3^-],$$

$$logK_5 = loga[K^+] - loga[H^+] - 0.5loga[H_2O],$$

$$logK_6 = loga[Ca^{2+}] + 2loga[HCO_3^-] - logf[CO_2(g)] - loga[H_2O].$$

The values of log equilibrium constant of the four reactions are shown in Fig. 11, where $logK_1$ is higher than $logK_2$ and $logK_3$ is higher than $logK_4$, indicating that the equilibrium constant of analcite leaching reactions are much higher than that of calcite leaching reactions.

6.2 Kinetic data

For kinetically controlled mineral dissolution and precipitation, the following simple rate law was applied (Yuan et al. 2015a):

$$r_m = k_m A_m (1 - Q/K), \quad (7)$$

where m is the mineral index, r_m is the reaction rate (mol/s, positive for dissolution and negative for precipitation), k_m is the rate constant (in $mol/cm^2/s$), A_m is the mineral's surface area (in cm^2), and Q and K are the activity product and equilibrium constants for the dissolution reaction, respectively. The temperature dependence of the reaction rate constant can be expressed reasonably well via the Arrhenius equation (Lasaga 1984; Steefel and Lasaga 1994). Because the rate constants for K-feldspar, calcite, and secondary minerals are generally reported at around 25 °C, it is reasonable to approximate the rate constant dependency as a function of temperature (Xu et al. 2005):

$$k = k_{25} \exp\left[\frac{-Ea}{R}\left(\frac{1}{T} - \frac{1}{298.15}\right)\right], \quad (8)$$

where Ea is the activation energy, k_{25} is the rate constant at 25 °C, R is the gas constant, and T is the absolute temperature.

Mineral dissolution and precipitation rates are a product of the kinetic rate constant and the reactive surface area described in Eq. (7). Parameters used for the kinetic rate expression of calcite and K-feldspar are provided in Table 1. Temperature-dependent kinetic rate constants were calculated using Eq. (8) and the precipitation of possible secondary minerals is represented utilizing the same kinetic expression as that used for dissolution. Nucleation was also considered in the current simulations for mineral precipitation. Scientific publications were referenced for kinetic parameters and the specific surface areas of calcite, analcite, quartz, and kaolinite with specific grain sizes.

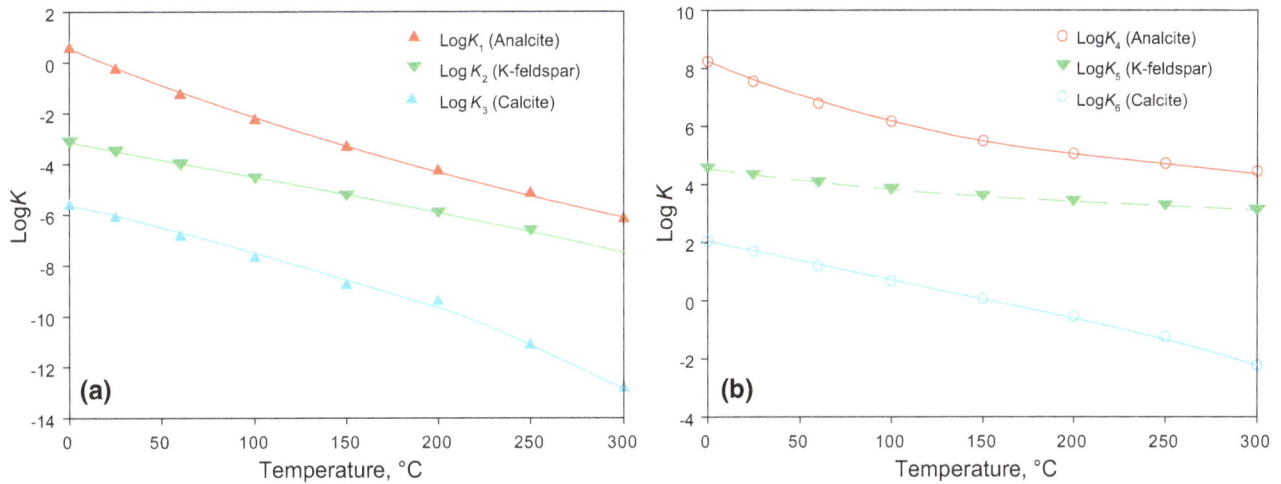

Fig. 11 Relations of temperature versus equilibrium constant in acid solution with analcite and calcite

Table 1 Kinetic data for different minerals used in numerical simulations

Minerals	K_m (25 °C), mol/cm^2/s	Ea, kJ/mol	K_m (80 °C), mol/cm^2/s	Specific surface area, cm^2/g	Nucleus, cm^2/cm^3	References
Calcite	1×10^{-9}	48.2	2.06×10^{-8}	2000	–	Pokrovsky et al. (2009); Sanz et al. (2011)
K-feldspar	1×10^{-17}	57.78	1.08×10^{-15}	1000	–	Kampman et al. (2009); Xu et al. (2005)
Analcite	–	–	3.16×10^{-14}	1800	–	Savage et al. (1999)
Quartz	1.26×10^{-18}	87.5	3.07×10^{-16}	1000	500	Harouiya and Oelkers (2004); Xu et al. (2005)
Kaolinite	1.26×10^{-17}	62.76	6.50×10^{-16}	10×10^4	500	Xu et al. (2005); Yang and Steefel (2008)

The evolution of surface area in natural geologic media is very complex and reported that specific surface areas vary based on mineral size and even literature reference. Specific surface areas of clean calcite grains measuring 125–250 and 25–53 μm by BET method were reported as 1700 and 2100 cm^2/g (Sanz et al. 2011), respectively. Also, the surface areas of calcite grains measuring 100–200 μm by BET method were reported as 662 cm^2/g according to Pokrovsky et al. (2005). The specific surface areas of clean K-feldspar and quartz grains measuring 50–100 μm by BET method were reported as 955 and 945 cm^2/g, respectively (Harouiya and Oelkers 2004), that for K-feldspar grains measuring 50–100 μm was 1400 cm^2/g (Alekseyev et al. 1997). Specific surface areas of clean analcite grains measuring 63–75, 45–63, and 32–45 μm by BET method were reported as 1700 and 2100 cm^2/g, respectively (Savage et al. 1999). The specific surface areas of clean K-feldspar and quartz grains measuring 50–100 μm were reported as 1030, 1450, and 830 cm^2/g, respectively, and the specific surface area of clean quartz grains measuring 50–100 μm was reported as 1000 cm^2/g (Harouiya and Oelkers 2004). For this study, the specific surface area of 2000 cm^2/g for calcite and 1800 cm^2/g for analcite are used. Kaolinite has a much larger specific surface area, up to approximately 10×10^4 cm^2/g (Yang and Steefel 2008) (Table 1).

6.3 Pore water

Data of 40 pore water samples from sandstone reservoirs in the Xing'anling Group in the Suderte Oilfield show that the pore water is characterized by NaHCO$_3$-water. The salinity of pore water is very low, ranging from 1737 to 10,813 mg/L, with an average of 5689 mg/L. Primary ions consist mainly of Na$^+$, Cl$^-$, and HCO$_3^-$ (Table 2). We employed one water sample with fairly average composition for numerical simulations.

6.4 Simulation results

6.4.1 Analcite–calcite–CO$_2$–H$_2$O system

Based on the diagenetic environment of the studied sandstones, 80 °C was employed in short-term (100 s) (Fig. 12) and long-term (1000 years) (Fig. 13) simulations, and 1.176 bar was set for partial pressure of CO$_2$ according to the equation $\log p_{CO_2} = -1.45 + 0.019T$ (Smith and Ehrenberg 1989). Our simulation results showed that the

Table 2 Composition of current pore water in the Xing'anling Group, Suderte Oilfield

	Salinity, mg/L	$Na^+ + K^+$, mg/L	Cl^-, mg/L	Ca^{2+}, mg/L	Mg^{2+}, mg/L	HCO_3^-, mg/L	SO_4^{2-}, mg/L
Maximum	10,813	6279	5021	114	50	3031	1575
Minimum	1737	436	272	1	1	0	66
Average	5689	1855	1356	29	17	1728	567
Sample no.	6548	4990	2184	24	10	1308	784

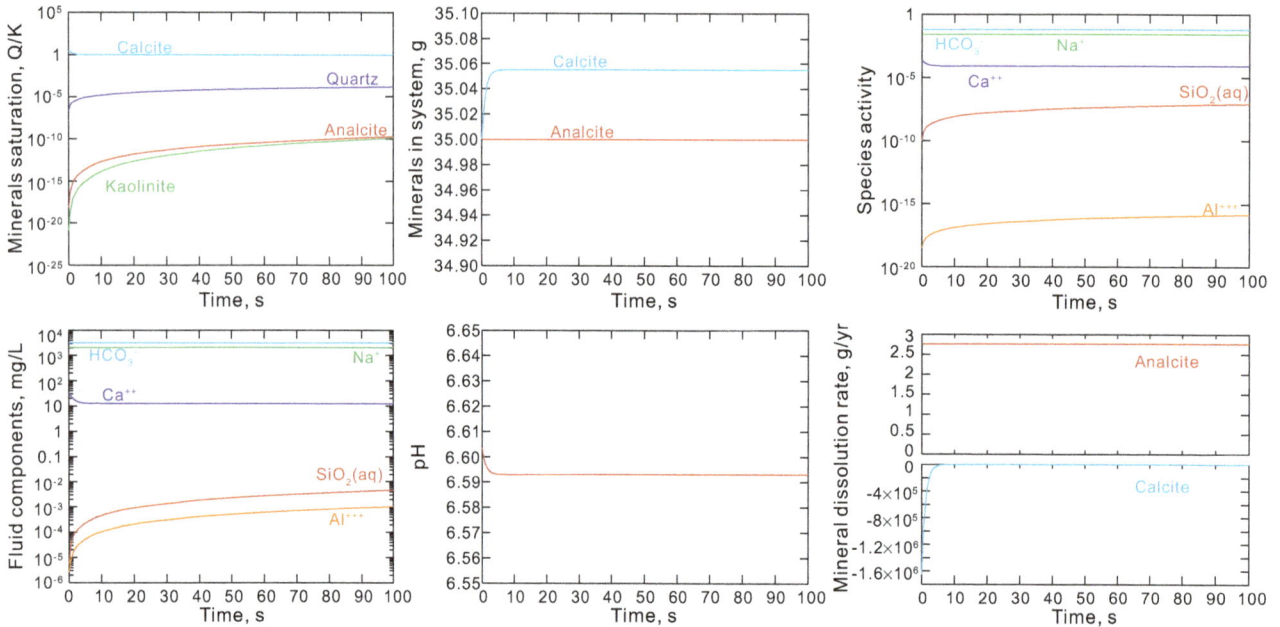

Fig. 12 Numerical simulation results of fluid-rock reactions in the calcite–analcite–CO_2–H_2O system for a short time (100 s)

chemical reaction processes at work in the reservoirs can be roughly divided into two stages.

Stage 1: Fast calcite precipitation—slow analcite dissolution (Fig. 12). At the beginning of the simulation, the mineral saturation index showed that pore water in the Xing'anling Formation was oversaturated with respect to calcite and undersaturated with respect to analcite, causing rapid calcite precipitation. Because the calcite reaction rate is high, this stage lasts a very short time (10 s). Only a little analcite can be dissolved in such a brief time, and the concentrations of Al^{3+} and SiO_2 (aq) in the fluids are low, so no precipitation of kaolinite or quartz occurs.

Stage 2: Slow calcite precipitation—slow analcite dissolution (Fig. 13). After stage 1, the mineral saturation index showed that pore water reached equilibrium with calcite while pore water was still undersaturated with analcite, leading to slow dissolution of analcite. As the concentrations of Al^{3+} and SiO_2 (aq) increased, the pore water became saturated with respect to kaolinite and quartz, leading to precipitation of secondary minerals. As

the dissolution rate of analcite is very low, this stage could last a long time.

6.4.2 Analcite–K-feldspar–CO_2–H_2O system

Based on the diagenetic environment of the studied sandstones, 80 °C was used in the short-term (100 s) (Fig. 9) and long-term (1000 years) (Fig. 10) simulations, and 1.176 bar was set for partial pressure of CO_2 according to the equation $\log p_{CO_2} = -1.45 + 0.019T$ (Smith and Ehrenberg 1989). Simulation results showed that in the analcite–K-feldspar–CO_2–H_2O system, dissolution of large volumes of analcite occurred more easily than that of K-feldspar and at a much faster rate (Fig. 14). In effect, in the geochemical system consisting both of analcite and K-feldspar, extensive feldspar dissolution probably occurred later than analcite dissolution, which is consistent with the petrography texture of the few analcite remnants in the reservoirs in the Xing'anling Formation. As the concentrations of Al^{3+} and SiO_2 (aq) increased, the pore water became saturated with respect to kaolinite and quartz,

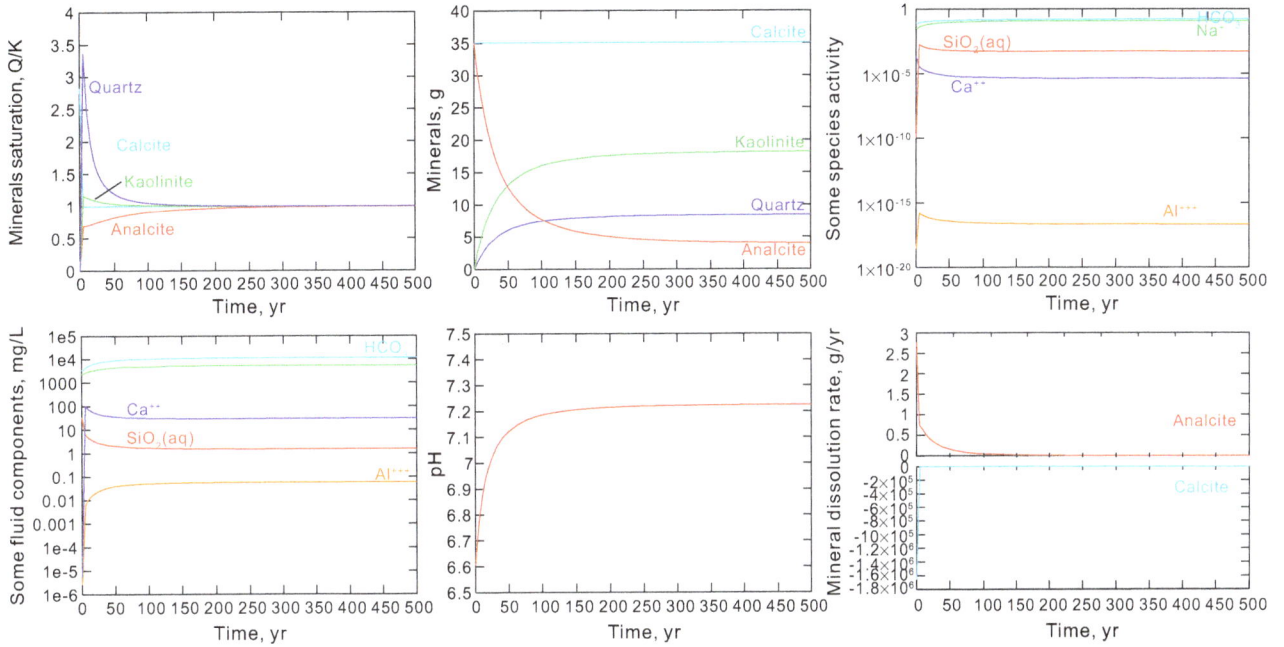

Fig. 13 Numerical simulation results of fluid-rock reactions in the calcite–analcite–CO_2–H_2O system for extended periods of time (500 years)

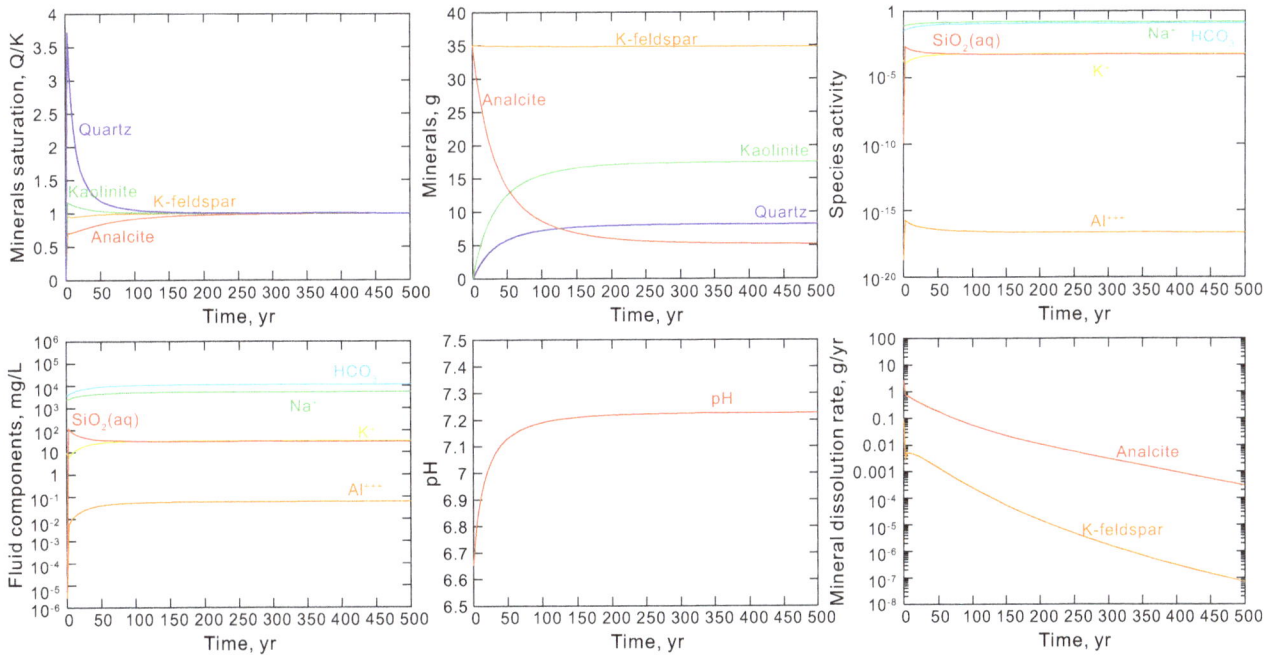

Fig. 14 Numerical simulation results of fluid-rock reactions in analcite–K-feldspar–CO_2–H_2O system for extended periods of time (500 years)

leading to precipitation of secondary minerals. Because the rate of analcite dissolution is very low, this stage lasts a long time.

A study by Yuan et al. (2015a) showed that feldspar dissolution occurs more easily than calcite dissolution in a geochemical system with both minerals. Constrained by pore water, equilibrium constants of different reactions, dissolution/precipitation rate, and saturation state of the pore water to minerals, analcite dissolution occurred more easily than calcite dissolution in the geochemical system with these two minerals, and analcite dissolution occurred more easily than feldspar dissolution in the geochemical system with these two minerals. Thus, we concluded that in geochemical systems with analcite, K-feldspar, and calcite, the dissolution trend is analcite > K-feldspar > calcite. The selective dissolution of analcite and K-feldspar in the presence of calcite is an inevitable natural result, indicating that different early cements can develop different dissolution features during the mesodiagenetic period in the presence of acidic fluids. Understanding these processes substantially assists high-quality reservoir prediction.

7 Conclusions

The most notable conclusions and implications of this study can be summarized as follows.

1. The Xing'anling Group reservoirs with abundant volcanic materials in the Suderte Oilfield are low-permeability and ultra-low permeability reservoirs, texturally and compositionally immature. The reservoirs consist mainly of litharenite and feldspathic litharenite and have generally experienced compaction/early analcite cementation/early calcite cementation–feldspar dissolution/analcite dissolution/authigenic kaolinite precipitation/quartz cementation–late carbonate cementation.

2. The main early cements in the study area are calcite and analcite. In the mesodiagenetic stage, abundant analcite was selectively dissolved and calcite was left intact. The equilibrium constant of the calcite leaching reaction by acidic fluids is significantly lower than that of the analcite leaching reaction, indicating that in the analcite–calcite–CO_2–H_2O system, calcite is prone to reach the precipitation–dissolution equilibrium stage. Simulations with constraints of pore water and kinetics demonstrated that the pore water is supersaturated with respect to calcite, thus calcite cannot be dissolved, and that the water is undersaturated with respect to analcite, leading to extensive dissolution of analcite.

3. Selective dissolution of different early cements resulted in differing impacts on reservoir quality evolution. Reservoirs with abundant analcite exhibit favorable physical properties through burial dissolution of analcite cement at the mesodiagenetic stage, while reservoirs with abundant early calcite exhibit poor physical properties with no dissolution of calcite cement.

Acknowledgments This study is financially supported by the National Science and Technology Special Grant (No. 2011ZX05009-003), China Postdoctoral Science Fund (2015M580617), Shandong Postdoctoral Innovation Fund (201502028), and 2014 Innovation Project of China University of Petroleum (YCX2014002). We thank anonymous reviewers for their constructive comments.

References

Alekseyev VA, Medvedeva LS, Prisyagina NI, et al. Change in the dissolution rates of alkali feldspars as a result of secondary mineral precipitation and approach to equilibrium. Geochim Cosmochim Acta. 1997;61(6):1125–42.

Arvidson RS, Ertan IE, Amonette JE, et al. Variation in calcite dissolution rates: a fundamental problem? Geochim Cosmochim Acta. 2003;67(9):1623–34.

Bjørlykke K, Jahren J. Open or closed geochemical systems during diagenesis in sedimentary basins: constraints on mass transfer during diagenesis and the prediction of porosity in sandstone and carbonate reservoirs. AAPG Bull. 2012;96(12):2193–214.

Cao YC, Yuan GH, Wang YZ, et al. Genetic mechanisms of low permeability reservoirs of Qingshuihe Formation in Beisantai area, Junggar Basin. Acta Pet Sin. 2012;33(5):758–71 (**in Chinese**).

Cao YC, Yuan GH, Li XY, et al. Characteristics and origin of abnormally high porosity zones in buried Paleogene clastic reservoirs in the Shengtuo area, Dongying Sag, East China. Pet Sci. 2014;11(3):346–62.

Chen CY, Gao YF, Wu HB, et al. Zircon U–Pb chronology and its stratigraphy implications for volcanic rocks from the Hailaer Basin. J Jilin Univ (Earth Sci Ed). 2015;45(S1):1503–19 (**in Chinese**).

Cui JP, Ren ZL, Chen YL. Study on the relations between geothermal history and oil–gas generation in Beier depression of Hailaer Basin. Acta Sedimentol Sin. 2011;29(2):388–94 (**in Chinese**).

Dutton SP, Loucks RG. Reprint of: diagenetic controls on evolution of porosity and permeability in lower Tertiary Wilcox sandstones from shallow to ultradeep (200–6700 m) burial, Gulf of Mexico Basin, USA. Mar Pet Geol. 2010;27(1):69–81.

Fu GM, Dong MC, Zhang ZS, et al. Formation process and distribution of laumontite in Yanchang 3 reservoir of Fuxian exploration area in north Shaanxi province and the controls of the high quality reservoirs. Earth Sci (J China Univ Geosci). 2010;35(1):107–14 (**in Chinese**).

Giles MR. Mass transfer and problems of secondary porosity creation in deeply buried hydrocarbon reservoirs. Mar Pet Geol. 1987;4(3):188–204.

Giles MR, Marshall JD. Constraints on the development of secondary porosity in the subsurface: re-evaluation of processes. Mar Pet Geol. 1986;3(3):243–55.

Gluyas J, Coleman M. Material flux and porosity changes during sediment diagenesis. Nature. 1992;356(6364):52–4.

Han S, Yu H, Si C, et al. Corrosion of analcite in reservoir of Junggar Basin. Acta Pet Sin. 2007;28(3):51–4 (**in Chinese**).

Harouiya N, Oelkers EH. An experimental study of the effect of aqueous fluoride on quartz and alkali-feldspar dissolution rates. Chem Geol. 2004;205(1–2):155–67.

Kampman N, Bickle M, Becker J, et al. Feldspar dissolution kinetics and Gibbs free energy dependence in a CO_2-enriched groundwater system, Green River, Utah. Earth Planet Sci Lett. 2009;284(3–4):473–88.

Lasaga AC. Chemical kinetics of water–rock interactions. J Geophy Res Solid Earth (1978–2012). 1984;89(B6):4009–25.

Lu ZX, Ye SJ, Yang X, et al. Quantification and timing of porosity evolution in tight sand gas reservoirs: an example from the Middle Jurassic Shaximiao Formation, western Sichuan, China. Pet Sci. 2015;12(2):207–17.

Macquaker JHS, Taylor KG, Keller M, et al. Compositional controls on early diagenetic pathways in fine-grained sedimentary rocks: implications for predicting unconventional reservoir attributes of mudstones. AAPG Bull. 2014;98(3):587–603.

Meng YL, Liang HT, Wei W, et al. Thermodynamic calculations of the laumontite dissolution and prediction of secondary porosity zones: a case study of horizon of Xujiaweizi fault depression. Acta Sedimentol Sin. 2013;31(3):509–15 (**in Chinese**).

Meng YL, Zhu HD, Li XN, et al. Thermodynamic analyses of dolomite dissolution and prediction of the secondary porosity zones: a case study of tight tuffaceous dolomites of the second member, Permian Lucaogou Formation, Santanghu Basin, NW China. Pet Explor Dev. 2014;41(6):754–60.

Mousavi MA, Bryant SL. Geometric models of porosity reduction by ductile grain compaction and cementation. AAPG Bull. 2013;97(12):2129–48.

Pittman ED, Larese R. Compaction of lithic sands: experimental results and applications. AAPG Bull. 1991;75(8):1279–99.

Pokrovsky OS, Golubev SV, Schott J. Dissolution kinetics of calcite, dolomite and magnesite at 25 °C and 0 to 50 atm pCO_2. Chem Geol. 2005;217(3–4):239–55.

Pokrovsky OS, Golubev SV, Schott J, et al. Calcite, dolomite and magnesite dissolution kinetics in aqueous solutions at acid to circumneutral pH, 25 to 150 °C and 1 to 55 atm $pCO2$: new constraints on CO_2 sequestration in sedimentary basins. Chem Geol. 2009;265(1–2):20–32.

Qi SC. The diagenesis of the sandstone in Chang 8–10 Layer, Yanchang Formation, Longdong Region, late Triassic of Ordos Basin and the thermodynamic behavior of laumontite. Master Thesis, Chengdu University of Technology; 2013 (**in Chinese**).

Sanz E, Ayora C, Carrera J. Calcite dissolution by mixing waters: geochemical modeling and flow-through experiments. Geol Acta. 2011;9(1):67–77.

Savage D, Rochelle C, Mihara M, et al. Dissolution of analcite under conditions of alkaline pH. Ninth annual VM Goldschmidt conference, August 22–27, 1999.

Schmidt V, McDonald DA. The role of secondary porosity in the course of sandstone diagenesis. SPEM Spec Publ. 1979;26:175–207.

Shen JN, Song T, Zhu J. Burial and subsidence history of Beier depression in Hailaer Basin. J Heilongjiang Inst Sci Technol. 2013;23(2):176–80 (**in Chinese**).

Shou JF. Kinetics of Sandstone Diagenesis. Beijing: Petroleum Industry Press; 2005. p. 10–5 (**in Chinese**).

Smith JT, Ehrenberg SN. Correlation of carbon dioxide abundance with temperature in clastic hydrocarbon reservoirs: relationship to inorganic chemical equilibrium. Mar Pet Geol. 1989;6(2):129–35.

Song T. The study of burial and subsidence history of Beier Depression in Hailar Basin. Ph.D. Dissertation. Daqing: Northeast Petroleum University; 2013 (**in Chinese**).

Steefel CI, Lasaga AC. A coupled model for transport of multiple chemical species and kinetic precipitation/dissolution reactions with application to reactive flow in single phase hydrothermal systems. Am J Sci. 1994;294(5):529–92.

Sun YS, Liu XN, Zhang YQ, et al. Analcite cementation facies and forming mechanism of high-quality secondary clastic rock reservoirs in western China. J Palaeogeogr. 2014;16(4):517–26 (**in Chinese**).

Surdam RC, Boese SW, Crossey LJ. The chemistry of secondary porosity: Part 2. Aspects of porosity modification. Paper in AAPG special volumes: clastic diagenesis; 1984. pp. 127–49.

Tang Z, Parnell J, Longstaffe FJ. Diagenesis and reservoir potential of Permian–Triassic fluvial/lacustrine sandstones in the southern Junggar Basin, northwestern China. AAPG Bull. 1997;81(11):1843–65.

Taylor TR, Giles MR, Hathon LA, et al. Sandstone diagenesis and reservoir quality prediction: models, myths, and reality. AAPG Bull. 2010;94(8):1093–132.

Turchyn AV, Depaolo DJ. Calcium isotope evidence for suppression of carbonate dissolution in carbonate-bearing organic-rich sediments. Geochim Cosmochim Acta. 2011;75(22):7081–98.

Wang JP, Fan TL, Wang HY, et al. Reservoir heterogeneity characteristics in the framework of multi-grade base level cycle of the oil layers of Tongbomiao and Nantun Formations in the Suderte Oil Field. Earth Sci Front. 2012;19(2):141–50 (**in Chinese**).

Xi KL, Cao YC, Wang YZ, et al. Factors influencing physical property evolution in sandstone mechanical compaction: the evidence from diagenetic simulation experiments. Pet Sci. 2015;12(3):391–405.

Xiao YY, Fan TL, Wang HY. Characteristics and diagenesis of the volcaniclastic rock reservoirs from the Nantun Formation within the Suderte structural zone in the Buir depression. Sedim Geol Tethyan Geol. 2011;31(2):91–8 (**in Chinese**).

Xiu HW. Diagenesis research and reservoir evaluation in Quan3 and Quan4 Members of northern Songliao Basin. Ph.D. Dissertation, Northeast Petroleum University; 2008 (**in Chinese**).

Xu T, Apps JA, Pruess K. Mineral sequestration of carbon dioxide in a sandstone–shale system. Chem Geol. 2005;217(3–4): 295–318.

Xu ZY, Zhang XY, Wu SH, et al. Genesis of the low-permeability reservoir bed of upper Triassic Xujiahe Formation in Xinchang gas field, western Sichuan Depression. Pet Sci. 2008;5(3):230–7.

Yang L, Steefel CI. Kaolinite dissolution and precipitation kinetics at 22 °C and pH 4. Geochim Cosmochim Acta. 2008;72(1):99–116.

Yu BS, Lai XY. Carbonic acid system of groundwater and the solubility of calcite during diagenesis. Acta Sedimentol Sin. 2006;24(5):627–35 (**in Chinese**).

Yuan GH, Cao YC, Yang T, et al. Porosity enhancement potential through mineral dissolution by organic acids in the diagenetic process of clastic reservoir. Earth Sci Front. 2013;20(5):207–19 (**in Chinese**).

Yuan GH, Cao YC, Gluyas J, et al. Feldspar dissolution, authigenic clays, and quartz cements in open and closed sandstone geochemical systems during diagenesis: typical examples from two sags in Bohai Bay Basin, East China. AAPG Bull. 2015a;99(11):2121–54.

Yuan GH, Cao YC, Jia ZZ, et al. Selective dissolution of feldspars in the presence of carbonates: the way to generate secondary pores in buried sandstones by organic CO_2. Mar Pet Geol. 2015b;60:105–19.

Yuan GH, Gluyas J, Cao YC, et al. Diagenesis and reservoir quality evolution of the Eocene sandstones in the northern Dongying Sag, Bohai Bay Basin, East China. Mar Pet Geol. 2015c;62:77–89.

Zhang XH, Huang SJ, Lan YF, et al. Thermodynamic calculation of laumontite dissolution and its geologic significance. Lithol Reserv. 2011;23(2):64–9 (**in Chinese**).

Zhang Q, Zhu XM, Steel RJ, et al. Variation and mechanisms of clastic reservoir quality in the Paleogene Shahejie Formation of the Dongying Sag, Bohai Bay Basin, China. Pet Sci. 2014;11(2):200–10.

Zhao GQ. Study of petrological characteristics and thermodynamic mechanism of secondary pores in deep reservoirs, Songliao Basin. Ph.D. Dissertation, China University of Geosciences (Beijing); 2005 (**in Chinese**).

Zhao L, Gao FH, Zhang YL, et al. Zircon U–Pb chronology and its geological implications of Mesozoic volcanic rocks from the Hailaer Basin. Acta Pet Sin. 2013;29(3):864–74 (**in Chinese**).

Zhong D, Zhu X, Zhang Z, et al. Origin of secondary porosity of Paleogene sandstone in the Dongying Sag. Pet Explor Dev. 2003;30(6):51–3 (**in Chinese**).

Zhu GH. Formation of lomonitic sand bodies with secondary porosity and their relationship with hydrocarbons. Acta Pet Sin. 1985;6(1):1–8 (**in Chinese**).

Zhu XM, Wang YG, Zhong DK, et al. Pore types and secondary pore evolution of Paleogene reservoirs in the Jiyang Sag. Acta Geol Sin. 2007;81(2):197–204 **(in Chinese)**.

Zhu XM, Zhu SF, Xian BZ, et al. Reservoir differences and formation mechanisms in the Ke-Bai overthrust belt, northwestern margin of the Junggar Basin, China. Pet Sci. 2010;7(1):40–8.

Zhu SF, Zhu XM, Wang XL, et al. Zeolite diagenesis and its control on petroleum reservoir quality of Permian in northwestern margin of Junggar Basin. Sci China Earth Sci. 2011;41(11):1602–12 **(in Chinese)**.

Zhu SF, Zhu XM, Wang XL, et al. Zeolite diagenesis and its control on petroleum reservoir quality of Permian in northwestern margin of Junggar Basin, China. Sci China Earth Sci. 2012;55(3):386–96.

3

HSE training matrices templates for grassroots posts in petroleum and petrochemical enterprises

Shao-Lin Qiu[1] · Lai-Bin Zhang[1] · Mu Liu[2]

Handling editor: Jian Shuai

Abstract This paper aims to standardize health, safety and environment (HSE) training matrices for effectively identifying training requirements and enhancing job competency of grassroots posts in petroleum and petrochemical enterprises. After an investigation into HSE training performance in petroleum and petrochemical enterprises, HSE training matrices templates for 239 grassroots posts involving 22 primary petroleum and petrochemical disciplines were developed based on the technical requirements of HSE training matrices templates and the studies of template development processes and technical methods at grassroots level. Applications of these templates in 12 companies demonstrate their effectiveness in improving HSE training and management for grassroots posts, strengthening risk control and promoting the application of HSE system and tools in grassroots units.

Keywords Petroleum and petrochemical enterprise · Grassroots posts · HSE · Training matrices · Template

✉ Mu Liu
  paulday@163.com

[1] China University of Petroleum (Beijing), Beijing 102249, China

[2] CNPC Research Institute of Safety and Environment Technology, Beijing 102206, China

Edited by Yan-Hua Sun

1 Introduction

Statistical data in recent years indicate that more than 80% of incidents and accidents affecting production safety, environment pollution and staff health in petroleum and petrochemical enterprises were caused by illegal operations by grassroots staff with weak HSE awareness and competency (Bush and Ingram 1996; Gitomer 2012; Graduate School of Chinese Academy of Social Sciences 2002; Hu 2012; Jing 2012; Shafiababy et al. 2007; Wang et al. 2008; Wen et al. 2012; Yao 2011). HSE training is an important means to publicize corporate HSE policies and improve employees' HSE knowledge and skills and their abilities in job-related risk identification and control. It is also an integral part of HSE management system and a significant measure to ensure job under control, build HSE culture and continuously improve HSE performance. HSE training for grassroots posts in domestic and foreign petroleum and petrochemical industries mainly has the following four problems: (1) Training content and post demands are not consistent; (2) Training methods are simple and boring, making trainees reluctant to accept them; (3) Positive initiatives for attending training are not strong, without direct relation to staff performance assessment; and (4) Responsibilities of training management are not clear, even none is fixed be in charge of training management in some corporation, so it is difficult to implement linear responsibility. However, so far, no training matrices based on HSE technical requirements of grassroots posts are available in China to specific professional classifications (He 2013; Kernberg 2014; Li 2010; Shevchenko and Kudryavtsev 2012; Weldy et al. 2014; Yao et al. 2011, 2012; Zhang et al. 2016). Since 2007, Chinese oil enterprises began to introduce and explore HSE training matrices management tools, promote HSE training in grassroots units with HSE training matrices as the basic carrier and implement "training-on-demand" models to meet job HSE needs (Carr et al. 2000; Finke 2004; Graff and Karsten 2012; Yang et al. 2013). Through effective identification, HSE training matrices can be an effective tool to link HSE training needs and posts and acquire clear instructions on training contents, grasp level, training frequency, etc. (Ahmed and Kolachi 2013; Falkenstein et al, 2016; Puhakainen and Siponen 2010; Yan 2011; Zia et al. 2014). It spells out HSE training requirements in an adequate, vivid, clear and concise manner. However, more applications of HSE training matrices demonstrate deficiencies such as unreasonable design, inaccurate identification of training need and lack of consistency of posts with competence, which restricts the role of HSE training matrices (Hogan et al 2014). Facing the fact that training status is not completely based on real needs of enterprises, it is necessary to research and develop HSE training matrices templates which should be operative, general and suitable for the competency and requirements of grassroots posts and bring a positive effect on the application of HSE training matrices. This paper studies HSE training matrices templates based on technical requirements and following the professional classification of grassroots posts and designs training content, period and methods. Using these templates, hierarchical responsibilities could be implemented and training requirements and technical support measures are defined. Applications of these templates have proved that they can effectively solve training problems of grassroots posts and significantly improve training performance.

2 Technical requirements for template development

Application of HSE training matrices to strengthen HSE training for grassroots posts is very different in training mode, form design, requirement analysis and application effect in Chinese oil enterprises. To identify basic training requirements and develop general HSE training matrices templates for key posts in different disciplines (MacDuffie 2007; Montreuil 1996; Nor 2009; Glushenkov 2009; Shadlovskiy and Kovaliov 2013; Shafiabady et al. 2007), the following requirements should be considered:

2.1 Scope

HSE training matrices templates for grassroots posts should cover primary business activities in petroleum and petrochemical enterprises, such as exploration and development, refining and petrochemistry, marketing, natural gas and pipelines, engineering technology, surface construction and equipment manufacturing. The specific HSE training matrices template for every business type should cover primary

jobs, such as oil production, oil gathering and transportation, refining, petrochemistry, gas stations, oil storage, drilling, downhole operations, geophysical exploration, well logging, surface construction and equipment manufacturing.

2.2 Operability

HSE training matrices templates for grassroots posts shall have adequate operability. They should be specific to target posts and specify procedures required for every workplace. Training content should be detailed down to the smallest unit of an operation.

2.3 Generality

The development process of HSE training matrices templates for grassroots posts should be standardized by defining business types, breaking down operation, minimizing operating unit and specifying training period, form, trainer and mastery of training content, etc.

2.4 Suitability

HSE training matrices templates for grassroots posts should include necessary training content for meeting specific job competency, design appropriate training methods and provide training courses and training materials consistent with grassroots posts. After being trained, all trainees should grasp job skills, understand and be able to control job risks.

3 Process and technical methods of template development

After analyzing the HSE training performance in petroleum and petrochemical enterprises, and combing the technical requirements for developing HSE training matrices, the following development process of HSE training matrices for grassroots units was proposed (Fig. 1).

3.1 Technical preparation

These data should be prepared as the base for HSE training matrices development, including applicable laws and regulations for related disciplines, grassroots posts, job descriptions and responsibilities, staff structure, available equipment and facilities, processes, procedures.

3.2 Division of management units

Aiming at production and management characteristics, job management units should be divided in consideration of related equipment and facilities, processes and (or)

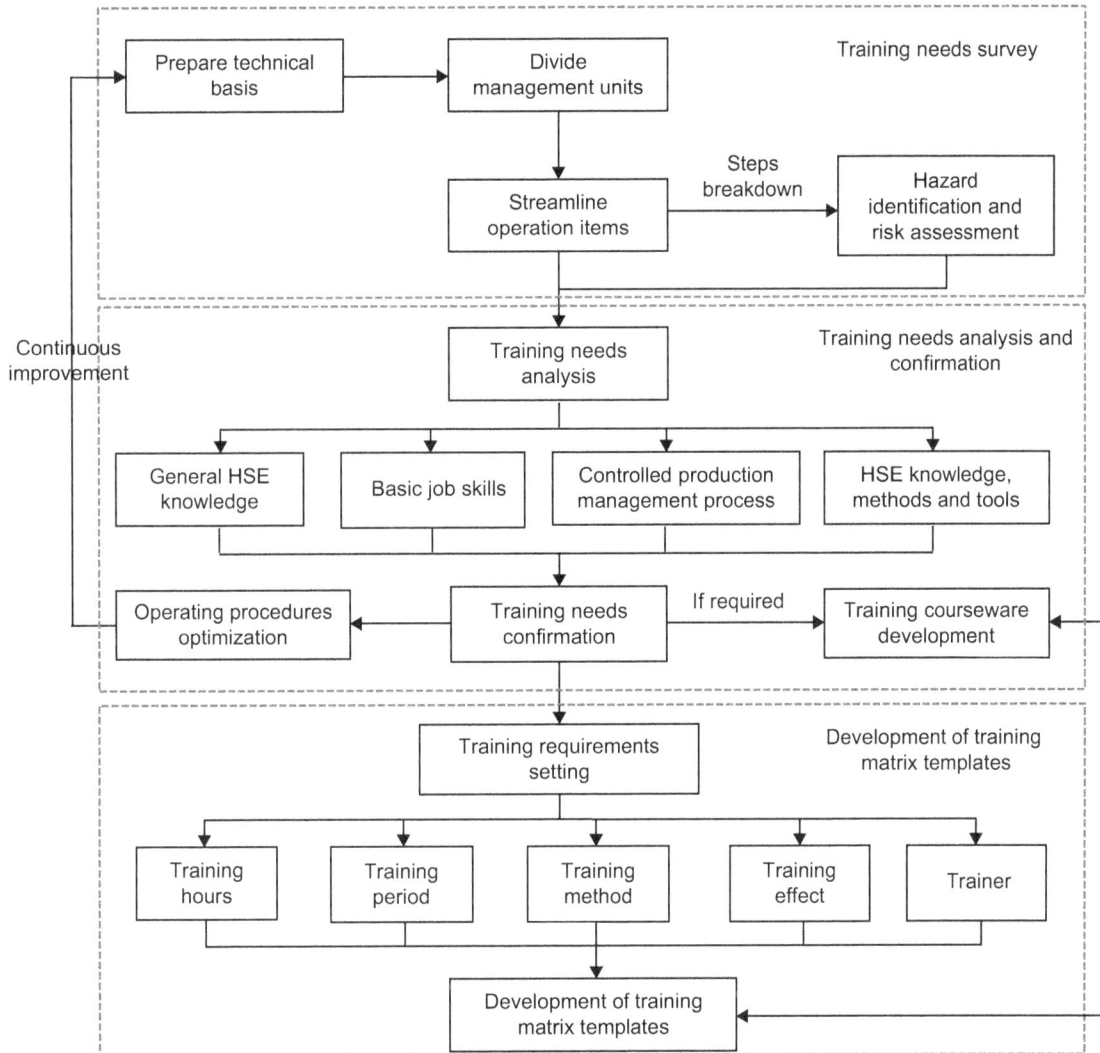

Fig. 1 Development process of HSE training matrices for grassroots units

worksites. For example, oil production activity can be divided into 30 management units in accordance with specific scope of work, job description, workflow, equipment and facilities. Table 1 shows the management unit list.

3.3 Analysis of operations

To meet the requirements for keeping every operation relatively independent and complete, operational risks can be identified and control measures can be implemented, and management units should be detailed to the most basic activities. In Table 2, 30 management units in oil production can be broken down into 245 activities.

3.4 Hazard identification and risk assessment

For each activity, it is recommended to break down operating procedures or how to dismantle key components of

equipment and facilities, conduct hazard identification and risk assessment and develop risk control measures using the job safety analysis (JSA) method, to ensure that the whole process risk management, risk control and emergency response measures are brought into the training needs, laying the foundation for job training matrices application.

3.5 Training requirement analysis

According to job responsibilities, operating procedures, risk control and emergency response measures, job-related training requirements are analyzed and determined. Training requirements mainly involve the following four aspects.

3.5.1 General HSE knowledge

The general part of HSE training matrices, i.e., general HSE knowledge closely related to staff's activities, daily

Table 1 Job management unit list of oil production

S/N	Management unit	S/N	Management unit
1	Pumping unit installation, uninstallation, transportation, operation, monitoring, adjustment and maintenance	16	Water buffer tank operation and maintenance
2	Screw pump operation, adjustment and maintenance	17	Water tank maintenance
3	Oil well operation, monitoring and maintenance	18	Centrifugal pump installation, operation and maintenance
4	Christmas tree maintenance	19	Processes in oil gathering and transportation room, and accessories maintenance
5	Well test	20	Operation and maintenance of gas /oil vacuum/phase change furnace
6	Metering separator operation, monitoring and maintenance	21	Processes in furnace operating room and accessories maintenance
7	Maintenance of processes and fittings in the production allocation room	22	Pump operation and maintenance in deep water well
8	High-pressure injection pump (piston pump) operation and maintenance	23	Test and analysis of produced oil
9	Injection well operation, monitoring and maintenance	24	Electrical equipment and facilities maintenance
10	Maintenance of processes and fittings in the water distribution room	25	Maintenance equipment inspection, operation and maintenance
11	Two-phase separator operation and maintenance	26	Inspection tour
12	Three-phase separator operation and maintenance	27	Upgrading and fixing of oil/gas/water pipelines
13	Oil skimmer operation and maintenance	28	General accessories maintenance
14	Dryer operation and maintenance	29	Single-well tank installation, operation, maintenance and adjustment
15	Air purifier operation and maintenance	30	Logistics support and management

Table 2 List of basic activities in oil production (partial)

S/N	Basic activity	S/N	Basic activity
2.1	Installation of pumping unit	2.3	Oil well operation, monitoring and maintenance
2.1.1	Overall installation of pumping unit	2.3.1	Measurement of tubing pressure
2.1.2	Assembly and disassembly of pumping unit	2.3.2	Measurement of casing pressure
...	...	2.3.3	Well sampling
2.2	Screw pump operation, adjustment and maintenance	2.3.4	Adjustment of collision avoidance distance for pumping well
2.2.1	Screw pump start-up	...	...
2.2.2	Screw pump shutdown	2.4	Well test
...	...	...	...

life and jobs and that should be mastered, includes job-related laws and regulations, job HSE risks, equipment and material used, emergency response and prevention measures and typical incidents as shown in Table 3.

3.5.2 Basic job skills

Professional skills of HSE training matrices refer to the training contents set specifically to a job based on job requirement analysis. It mainly includes training in operation procedures, such as technical skills for equipment management, operational activities and processes (Table 4). Through training, employees can grasp job-related skills, hazard identification and risk control methods, emergency response procedures, etc.

3.5.3 Controlled production management process

According to the training requirements set for controlled production management, all employees should understand or master related controlled management requirements applicable to their jobs and apply them to HSE management, which mainly includes related controlled management activities and contents, e.g., permit-to-work management, process and equipment management, management of change and contractor management.

3.5.4 HSE knowledge, methods and tools

According to the training requirements set for an HSE management system, all employees are required to

Table 3 JSA before well shut-in (example)

Unit	Unit name	All hazards and risks and the corresponding measures to eliminate, mitigate and control these hazards and risks have been discussed, understood and confirmed by all personnel involved in this job, including emergency response measures, so the work can be started.
		Duty holder

Brief description of the task

Procedure	Hazard(s)	Risk control measure(s)	Emergency response measure(s)
1. Check work preparation and recorded data	1. Mechanical injury	1. Wear necessary personal protective equipment (PPE)	1. First-aid kit available on the site and send to the medical clinic in the work area in serious cases
2. Notify central control room (CCR) and close secondary nozzles and then primary nozzles on site	1. Mechanical injury	1. Wear necessary PPE and stand aside while operating valves	1. First-aid kit available on the site and send to the medical clinic in the work area in serious cases
3. Remote/on-site well shut-in	1. Suffocation injury: Entry into cellar valve chamber may cause suffocation due to gas leakage 2. Mechanical injury	1. Entry into the valve chamber shall comply with *the Regulations on Entry into Cellar Valve Chamber by Gas Production Crew* 2. Wear necessary PPE; stand aside while operating valves; do not stand near Christmas tree and pipelines; do not open the back door of the hydraulic control cabinet	1. Once a suffocation injury occurs, the monitoring personnel shall immediately wear positive pressure breathing apparatus and enter the valve chamber to move the injured out and give first-aid treatment. 2. Once a mechanical injury occurs, immediately give first-aid treatment, and then send to the medical clinic in the work area
4. Long shut-in: Close all nozzles, hydraulic wing valves, manual wing valves, hydraulic master valves, manual master valves and downhole safety valves, and decide the branch pressure according to real situations; report well shut-in to CCR and record valve switch status	1. Mechanical injury	1. Wear necessary PPE and stand aside while operating valves	1. First-aid kit available on the site and send to the medical clinic in the work area in serious cases

understand or master related national, industrial and corporate HSE requirements which mainly include basic HSE knowledge, tools and methods, e.g., JSA, job cycle analysis (JCA), safety observation and communication, and safety experience sharing. Besides, all employees should be able to apply HSE management methods and tools to routine HSE work using with their HSE knowledge.

3.6 Training requirements and training matrices development

Training requirements set up training goals and performance indicators. Four points of every post, training requirement, namely duration, frequency and trainers (Table 5), should be defined and specific HSE training matrices should be established based on job

responsibilities, training requirement, courses and trainees as shown in Table 6.

3.7 Training outline (if required)

Training outlines should be consistent, relatively complete, simple and concise (Table 2). Materials like video, audio and photographs can be used to illustrate HSE principles and practices in order to attract trainees to interactive participation.

3.8 Optimization of operating procedures

Training contents should be determined based on operating procedures. It is recommended to use tools and methods such as JCA and JSA to analyze the effectiveness of job-

Table 4 Job skills for oil production (partial)

S/N	(Job) Training content	Well group — Group leader	Juniors	Night shift	Data recorder	Pump station — Group leader	Pump	Boiler	Geology group — Group leader	Laboratory technician	Tester	Mechanic group — Group leader	Mechanic	Electrician	Electric welding	Logistics — Driver	Other	Remarks
2	*Job-related basic skills*																	
2.1	Installation, operation, adjustment and maintenance of pumping unit																	
2.1.1	Inspection of pumping unit	•	•	•	•				•		•	•	•	•	•			
2.1.2	Well start-up/shut-in	•	•	•	•				•		•	•	•	•				
2.1.3	Replacing belt	•	•	•	•							•	•	•				
2.1.4	Collision avoidance distance adjustment	•										•	•	•	•			
2.1.5	Balance adjustment											•	•	•	•			
2.1.6	Stroke adjustment											•	•		•			
2.1.7	Stroke frequency adjustment											•	•	•	•			
2.1.8	Walking beam adjustment											•	•		•			
2.1.9	Replacing crane pin											•	•		•			
2.1.10	Replacing wire rope											•	•		•			
2.1.11	Brake replacing (repair)											•	•		•			
2.1.12	Maintenance (filling lubricants and fastening screws)											•	•		•			
2.1.13	Oil well operation, monitoring and maintenance	•	•	•								•	•					
2.2	*Oil well inspection*																	
2.2.1	Oil well start-up and shut-in	•	•	•	•				•			•	•		•			
2.2.2	Gas measurement in oil well	•	•	•	•				•			•	•					
2.2.3	Casing pressure measurement	•	•	•	•				•			•	•					

Table 4 continued

S/N	Group/post (Job) Training content	Well group				Pump station			Geology group			Mechanic group				Logistics group		Remarks
		Group leader	Juniors	Night shift	Data recorder	Group leader	Pump	Boiler	Group leader	Laboratory technician	Tester	Group leader	Mechanic	Electrician	Electric welding	Driver	Other	
2.2.4	Oil well sampling	•	•	•	•				•									
2.2.5	Pressure buildup	•	•	•	•				•									
2.2.6	Washing well	•	•	•	•				•			•	•					
2.2.7	Disposal of coagulated pipes	•	•	•	•							•	•					
2.2.8	...	•		•														

related procedures which should be revised when necessary. To keep continuous improvement in training work, training matrices should be improved in accordance with the revision of operating procedures.

4 Application and improvement in HSE training matrices templates

4.1 Results of templates development

Using the technical methods above, 239 grassroots posts in 22 primary disciplines in petroleum and petrochemical industries (based on an investigation of 12 petroleum and petrochemical enterprises) were studied, and based on which 983 management units covering 4566 activities were proposed and defined, and accordingly, 239 HSE training matrices and 1726 training outlines were established.

4.2 Templates application

Application of these HSE training matrices templates was used in an oil field company and obtained the following results:

1. These HSE training matrices templates for grassroots posts can standardize job requirements and competency building, make HSE training for grassroots posts standard, systematic and scientific, enhance on-site risk management and control levels, improve job operating procedures and enlarge the application of HSE management tools and methods.
2. These HSE training matrices embody job-related basic HSE training requirements. Training is given by line managers. Training contents focus on safety and environmental risks on specific work and are concise, understandable, operable and interesting.
3. With appropriate incentive mechanism, all employees at all levels are active to pursue their initiatives and participate in development and revision of operating procedures and training matrices. This practice is supported and understood by all employees.
4. Dynamic management and sustainable improvement in the HSE training matrices make them adapt to changing management structure processes.

4.3 Future improvement

Analysis of the actual applications of these HSE training matrices templates indicates that future improvement can be carried out in the following aspects:

Table 5 Typical HSE training matrices for oil production juniors (partial)

S/N	Training description	Duration (hours)	Frequency	Method	Requirement	Trainer
1	*General HSE knowledge*					
1.1	HSE responsibilities, rights and obligations	0.50	Every 1 year	Lecturing or meeting	Mastery	Group leader/HSE officer
1.2	Crude oil HSE knowledge	0.25	Every 3 years	Lecturing	Mastery	Group leader/HSE officer
1.3	Gas HSE knowledge	0.25	Every 3 years	Lecturing	Mastery	Group leader/HSE officer
1.4	Emergency escape knowledge	0.50	Every 1 year	Lecturing + field	Mastery	Group leader/HSE officer
1.5	...	...	...	...	...	...
2	*Basic job skills*					
2.1	Operation, maintenance and adjustment of pumping unit					
2.1.1	Start-up and shutdown of pumping unit	0.50	3 years	Lecturing + field	Mastery	Group leader/HSE officer
2.1.2	Replacing the belt of a pumping unit	0.5	3 years	Lecturing + field	Mastery	Group leader/HSE officer
2.1.3	Pumping unit maintenance (fill lubricants and fasten screws)	0.50	3 years	Lecturing + field	Mastery	Group leader/HSE officer
2.1.4	...	...	...	...	...	...
3	*Controlled management process*					
3.1	Permit to work	0.50	3 years	Lecturing + field	Understanding	Squad leader or HSE officer
3.2	Lock out and tag out	0.50	3 years	Lecturing + field	Mastery	Squad leader or HSE officer
3.3	...	...	...	...	...	...
4	*HSE knowledge, methods and tools*					
4.1	JSA	0.25	3 years	Lecturing or meeting	Understanding	Squad leader or HSE officer
4.2	Site management	0.50	3 years	Lecturing or meeting	Mastery	Squad leader or HSE officer
4.3	...	...	...	...	...	...

Table 6 Training outlines

Outlines for basic job skills	Outlines for controlled management procedure	Outlines for HSE knowledge, methods and tools
1. Pre-job preparation	1. Basic definitions	1. Basic definitions and explanations
2. Procedures (hazard identification and risk control)	2. Management process and requirements	2. Why promote?
	3. Typical cases	3. How to implement effectively?
3. Emergency response and accident prevention (cases)	4. Cautions	4. Cautions and summary

1. The department in charge of training should fully play a leading role in strengthening the compliance with the enterprise's overall human resource management policies, unifying the management of output interfaces and effectively integrating various resources.

2. As different enterprises have different organizations, production techniques, equipment, facilities, allocations of posts and responsibilities, businesses and activities, they have different requirements for job-related HSE competency. The generality of the HSE

training matrices templates may not meet specific requirements. Therefore, as equipment, technologies and processes upgrade, the templates should be improved and revised accordingly.

3. Less professional technologies and human resources for developing exclusive training materials result in repeated development of general materials in enterprises. To increase the utilization of training resources, it is necessary to establish resource sharing and incentives and performance assessment policies.

5 Conclusions

1. HSE training matrices templates were developed for 239 grassroots posts in 22 primary disciplines of petroleum and petrochemical industries. The development of these HSE training matrices templates for grassroots posts complies with the training requirement of HSE management and is a re-innovation in HSE training for Chinese oil enterprises.

2. The application of these HSE training matrices templates for grassroots posts strengthens risk management and control, optimizes operating procedures, promotes the application of HSE systems and tools, highlights HSE responsibilities and improves training management mechanisms in grassroots units.

3. Future work should focus on dynamic and sustainable improvement in HSE training matrices templates, development of training resources sharing platforms and promotion and application technologies of training matrices, to meet the ever-developing enterprises, processes and technologies, equipment and facilities, laws, regulations, standards and specifications, and stringent job-related risks and control requirements.

References

Ahmed I, Kolachi NA. Employee payroll and training budget: Case study of a non-teaching healthcare organization. J Bus Econ Res. 2013;11(5):229. doi:10.19030/jber.v11i5.7838.

Bush VD, Ingram T. Adapting to diverse customers: a training matrices for international marketers. Ind Market Manag. 1996;25(5):373–83. doi:10.1016/0019-8501(96)00039-9.

Carr JE, Nicolson AC, Higbee TS. Evaluation of a brief multiple-stimulus preference assessment in a naturalistic context. J Appl Behav Anal. 2000;33:353–7. doi:10.1901/jaba.2000.33-353.

Falkenstein M, Gajewski PD, Michael F, Patrick DG. Changes of electrical brain activity after cognitive training in old adults and older industrial workers. In: Strobach T, Karbac J, editors. Cognitive training. Berlin: Springer International Publishing; 2016. p. 177–86. doi:10.1007/978-3-319-42662-4_17.

Finke M. GMP aspects in practice: topical items concerning GMP regulations-training demands on staff in GMP conforming areas—Establishing a training system. Die Pharmazeutische Industrie. 2004;66(6):765–8.

Gitomer J. The hard side of training and the soft side of learning. Enterp Salt Lake City. 2012;41(47):8.

Glushenkov DA. Applying systems of video conference communication based on a software endec in the training of state employees. Sci Techn Inf Process. 2009;36(2):90–1. doi:10.3103/S014768820902004X.

Graduate School of Chinese Academy of Social Sciences. Harvard model training management, vol. 1–3. Beijing: People's Daily Press; 2002. p. 1–27 (in Chinese).

Graff RB, Karsten AM. Evaluation of a self-instruction package for conducting stimulus preference assessments. J Appl Behav Anal. 2012;45(1):69–82. doi:10.1901/jaba.2012.45-69.

He H. Tailored training based on unique post needs: brief analysis on the application of HSE training matrices in the Tahe Oilfield. China Sci Technol Inf. 2013;17:124–5 (in Chinese).

Hogan DA, Greiner BA, O'Sullivan L. The effect of manual handling training on achieving training transfer, employee's behavior change and subsequent reduction of work-related musculoskeletal disorders: a systematic review. Ergonomics. 2014;57(1):93–107. doi:10.1080/00140139.2013.862307.

Hu YT. Training matrices management: an effective model for improving the effectiveness of training. J Beijing Pet Manag Train Inst. 2012;5:72 (in Chinese).

Jing WR. HSE-based workover job skill requirements and training design of an oil and gas field. Master Thesis Jilin University; 2012. p. 15 (in Chinese).

Kernberg OF. The twilight of the training analysis system. Psychoanal Rev. 2014;101(101):151–74. doi:10.1521/prev.2014.101.2.151.

Li J. Application of training matrices in HSE management of pipeline construction contractor. Health Saf Environ. 2010;10(7):10–2 (in Chinese).

MacDuffie JP. Human resource bundles and manufacturing performance: organizational logic and flexible production systems in the world auto-industry. Ind Lab Relat Rev. 2007;48(2):197–221. doi:10.2307/2524483.

Montreuil S. Ergonomics training for managers, employees and designers involved in the design and organization of work systems. Saf Sci. 1996;23(2):97–106. doi:10.1016/0925-7535(96)00035-5.

Nor NM. A requirements model for employees training management system: applying WAE-UML. Information Management and Engineering, 2009. ICIME'09. International Conference on. IEEE, 2009. p. 569–73. 10.1109/ICIME.2009.67.

Puhakainen P, Siponen M. Improving employees' compliance through information systems security training: an action research study. MIS Q. 2010;34(4):757–78.

Shadlovskiy EL, Kovaliov YT. Cognitive training in older employees: a comparison between office and industrial workers. J Psychophysiol. 2013;27:69. doi:10.1027/0269-8803/a000095.

Shafiabady N, Teshnehlab M, Shooredeli MA. Training matrices parameters by particle swarm optimization using a fuzzy neural network for identification. In: International conference on intelligent & advanced systems, 2007. p. 188–93. doi:10.1109/ICIAS.2007.4658372.

Shevchenko DI, Kudryavtsev AA. The "Oil and Gas Enterprise" integrated training system. Oil Gas Eur Mag. 2012;38(4):218–9.

Wang XX, Su GS, Zhang Y. Inspiration on HSE education and training of petrochemical enterprises. Saf Health Environ. 2008;8(1):14–5 (in Chinese).

Weldy CR, Rapp JT, Capocasa K. Training staff to implement brief stimulus preference assessments. J Appl Behav Anal. 2014;47(1):214–8. doi:10.1002/jaba.98.

Wen ZG, Dong PJ, Liu B. Matrices-type HSE training for oil

depot employees. Saf Health Environ. 2012;12(9):52–3 **(in Chinese)**.

Yan XY. Instructional system design in employee training in the industrial area: take research on the "Blue-Collar Talent" project in Xiasha Economic and Technological Development Zone of Hangzhou as the example. Appl Mech Mater. 2011;121–126: 912–7. doi:10.4028/www.scientific.net/AMM.121-126.912.

Yang DL, Liu SY, Yang B, et al. Research the training system of the oil field staff simulation based on virtual reality technology. Adv Mater Res. 2013;807–809:2863–7. doi:10.4028/www.scientific.net/AMR.807-809.2863.

Yao GY. HSE training practice and work direction in petrochemical enterprises. Safety, Health and Environment. 2011;11(2):11 **(in Chinese)**.

Yao GY, Liu ZH, Zhao LQ. Establishment of HSE training matrices in petrochemical enterprises. Saf Environ Eng. 2011;18(5):74–7 **(in Chinese)**.

Yao G, Wang J, Wu B. The establishment of HSE training matrices for HSE professional management staffs in refining-chemical enterprises. Technol Superv Pet Ind. 2012;1:24–9 **(in Chinese)**.

Zhang Q, Zhao G, Wang L, et al. Application of training matrices in the top level design of enterprise HSE Training. Environ Prot Oil Gas Fields. 2016;26(3):55–8 **(in Chinese)**.

Zia H, Ishaq HM, Zahir S, et al. To investigate the impact of training, employee empowerment and organizational climate on job performance. Res J Appl Sci Eng Technol. 2014;7(22):4832–7. doi:10.19026/rjaset.7.872.

4

Establishment of a multi-cycle generalized Weng model and its application in forecasts of global oil supply

Yi Jin[1] · Xu Tang[1] · Cui-Yang Feng[1] · Jian-Liang Wang[1] · Bao-Sheng Zhang[1]

Abstract Low oil prices under the influence of economic structure transformation and slow economic growth have hit the existing markets of traditional big oil suppliers and upgraded the conflict of oil production capacity and interest between OPEC producers and other big oil supplier countries such as the USA and Russia. Forecasting global oil production is significant for all countries for energy strategy planning, although many past forecasts have later been proved to be very seriously incorrect. In this paper, the original generalized Weng model is expanded to a multi-cycle generalized Weng model to better reflect the multi-cycle phenomena caused by political, economic and technological factors. This is used to forecast global oil production based on parameter selection from a large sample, depletion rate of remaining resources, constraints on oil reserves and cycle number determination. This research suggests that the world will reach its peak oil production in 2022, at about 4340×10^6 tonnes. China needs to plan for oil import diversity, a domestic oil production structure based on the supply pattern of large oil suppliers worldwide and the oil demand for China's own development.

Keywords Oil production · Multi-cycle · Generalized Weng model · Energy strategy

1 Introduction

The US shale revolution has rapidly increased its oil and gas supply. Meanwhile, OPEC chose to maintain production to protect its market. The oil production of OPEC in 2015 was 3160×10^4 bbl/d, which had increased by 2.7% compared to that in 2014 (OPEC 2016; EIA 2016). The slowdown of world economic growth and transformation of economic structures in many countries have intensified the production contest among OPEC and other large oil suppliers like Russia and the USA. Under the dual effects of supply increase and demand decline, oil price fell continuously and sharply, which had serious influences on the investment and production capacity construction of oil resources and new energy resources. Some shale gas suppliers have withdrawn from the market due to high production costs. What is more, the special requirements of oil exploitation concerning geological conditions and construction make this exit irreversible in the long term.

The previous supply-dominated oil market has gradually turned into a demand-oriented situation. Oil market imbalance, which is manifested as the rapid decline of oil price, affects the short- and long-term production decisions of oil suppliers. However, huge differences in some inherent historical factors in different areas, such as production costs, resource conditions and stakes, can lead to completely different final supply decisions (Apergis et al. 2016). The global oil production trend will finally affect the strategies of various stakeholders.

Using the method considering key production constraints to undertake quantitative research into global oil supply volumes and provide information for national energy strategy, has become the focus and the difficulty in the present study. Since the shale revolution, a large number of studies have focused on unconventional oil and

✉ Xu Tang
tangxu2001@163.com

[1] School of Business Administration, China University of Petroleum, Beijing 102249, China

Edited by Xiu-Qin Zhu

gas production. The oil market, which has experienced great changes, is still unstable. It is now necessary to analyze the long-term oil supply trends of the global market.

In terms of existing model theory, most models are established with lack of consideration of the actual characteristics of oil and gas production. Among them, the existing generalized Weng model applications addressing national oil and gas production still stay on the stage of original model use and simple piecewise curve fitting, rather than extending the whole model to multi-cycle mode to fully reflect production trends. Resource depletion constraints have not used through specific functions to avoid unrealistic production growth forecasts which may occur in results. In terms of model implications, much of the existing research focuses on single fields or single countries, which does not reflect the global supply pattern, so it is not relevant to allow oil-consuming nations to develop strategic plans. Studies are rarely focused on international production, at the same time there is a lack of analysis combining future supply and demand situations, geographical features and development appeals of oil suppliers and consumers. In view of the defects above, this paper implements improvements in the aspects of objective function selection, production calculation, internal and external constraints and the frequency of multi-cycle fitting to model future world oil supplies.

2 Methodology

2.1 Existing oil production forecast models

The current oil prediction methods can be mainly divided into three categories: curve-fitting models, which are based on historical production data. These include the Hubbert, Gaussian and Logistic models (Reynolds 2014; Saraiva et al. 2014; Brandt 2007); system simulation methods, which are based on causal relationship of factors, such as the system dynamics method (Tao and Li 2007; Tang et al. 2010; Hosseini and Shakouri 2016); and econometric models based on economic theory (Kaufmann 1991; Pindyk and Rubinfeld 1998). The most widely used method is the curve-fitting model (Gallagher 2011; Sorrell and Speirs 2010; Nashawi et al. 2010; Ebrahimi and Ghasabani 2015).

Hubbert was the first scholar to use forward curve fitting. Hubbert pointed out bell-shaped curve regularity of fossil energy development (Hubbert 1949). In 1956, Hubbert used a hand-drawn bell-shaped curve to forecast oil production in the 48 contiguous states of the USA. According to his prediction, the US oil production would peak in the early 1970s and then decrease (Hubbert 1956). This prediction was confirmed by the actual oil production.

Because of this successful prediction and the social concern about oil shortage, using bell-shaped curves to predict oil production has become especially popular. More and more scholars have begun to join in the forecasting of oil production. The forecasting method used by Hubbert has been adopted by more and more people and is named the "Hubbert model." Although many scholars used this method, Hubbert had not given the specific formula of the method and its derivation. Until 1982, for the first time, Hubbert published the full formula and derivation process of the Hubbert model (Hubbert 1982). Since then, Hubbert model has been widely used.

Although the Hubbert model is the most widely used method in curve fitting, the model is not perfect. For instance, the model has poor accuracy when it is applied in the regions which have multiple oil production peaks. Therefore, many scholars began to improve the model. Current improvements mainly include two categories: Firstly, additional production cycles were added into model to fit multiple historical production peaks and improve the prediction accuracy, which is called the multi-cycle model; second, the Generalized Hubbert model was established by extending the typical Hubbert model. Wang et al. (2011) pointed out that the multi-cycle model is the most widely used model.

After many modifications, the forecasting accuracy of Hubbert's model has been improved significantly. Even so, many inherent problems remain unresolved. The curve shape of the Hubbert model is completely symmetrical. However, the reality is that in many oilfields, the production grows fast at the beginning and then declines slowly after reaching the peak. It is mainly because many measures are always taken to prevent rapid decline of oil production, such as improving recovery efficiency. It means that the production curve shapes of many oilfields are not completely symmetrical. Brandt (2007) analyzed 67 oil producing countries which have passed the production peak and found that most of these production curves follow a positive skewness distribution.

The generalized Weng model is the most widely used oil production prediction method in China. The curve shape of generalized Weng model is positive skewness. Wang et al. (2011) established a multi-cycle generalized Weng model on the basis of the generalized Weng model and compared it with the multi-cycle Hubbert model. This shows that the forecasting accuracy of multi-cycle generalized Weng model is better than that of the multi-cycle Hubbert model.

However, both the multi-cycle generalized Weng model and the multi-cycle Hubbert model lack a quantitative basis for choosing the number of cycles. Generally, the fitting effect of models would be better if the production cycles are increased. But meanwhile, excessive production cycles may cause overfitting. Overfitting could reduce the forecasting function of the model. Therefore, how to determine the optimal number of production cycle is very important. Wang and Feng (2016) proposed a quantitative method to quantify production cycle numbers, namely the F test. But this method has not been applied to multi-cycle generalized Weng models. In addition, many scholars have pointed out that the depletion rate of residual resources would also have significant impact on prediction. Therefore, the depletion rate of residual resource was suggested to be added into the model as a constraint parameter (Wang et al. 2013; Wang and Feng 2016).

Based on the multi-cycle generalized Weng model proposed by Wang et al. (2011), this study will establish a new multi-cycle generalized Weng model by adding the F test and residual resource depletion rate and apply this new model to forecast future global oil production.

2.2 Traditional Weng model and its characteristics

Among many oil production prediction methods, the Weng model takes into account the life-cycle characteristics of non-renewable resources and that "for many life limited systems, such as non-renewable resources, their whole life process can be imaged as a Poisson distribution probability function" (Weng 1984). This method can improve the measuring accuracy due to its full reflection of the known oil and gas resources in a short time. On this basis, Chen (1996) further derived a generalized Weng model which can be used to predict oil field production, final recoverable reserves and peak production based on a gamma distribution. The prediction model is as follows:

$$Q = at^b e^{(-t/c)} \tag{1}$$

$$a = \frac{N_R}{C^{b+1}\Gamma(b+1)} \tag{2}$$

Take logarithm for both sides:

$$\log\frac{Q}{t^b} = \log a - \frac{1}{2.303c}t \tag{3}$$

Let:

$$A = \log a, \quad B = 1/2.303c \tag{4}$$

Then:

$$\log\frac{Q}{t^b} = A - Bt \tag{5}$$

where Q represents the production; t represents relative development time of the oil field; a, b, c are unknown parameters; N_R represents recoverable reserves of the oil-field. The simplified Eq. (5) can be solved by using a linear differential method. In particular, first, different values of $\log(Q/t^b)$ can be obtained by plugging into different b values. Second, the correlation coefficient between

$\log(Q/t^b)$ and t can be obtained, select the b value which maximizes the correlation coefficient to fit the straight line represented by Eq. (5), and then the two values of A and B can be obtained. Finally, the two values a and c can be obtained, and Eq. (1) is identified.

2.3 Establishment of a multi-cycle generalized Weng model

The oil production at the regional level is affected by many factors such as politics, economy and technology. The historical yield curve of many regions showed multi-cycle phenomena. A single generalized Weng model cannot accurately describe this characteristic and causes large deviation in production estimation. This paper expands the single generalized Weng model to a multi-cycle generalized Weng model, which is established through stages as follows:

$$q(t) = \frac{URR}{c^{b+1}\Gamma(b+1)}t^b e^{-(t/c)} \tag{6}$$

where $q(t)$ represents production; b and c are unknown parameters.

In terms of goodness-of-fit tests, most scholars like Chen and Hu (1996) adopt a decision coefficient as a measure gauge. If the decision coefficient is close to 1, the fitting effect is better. But the determination coefficient represents the interpretation of the independent variable on the dependent variable; if the production fluctuation is large, even if the determination coefficient value is high, the gap between predicted values and real values may not be minimized. Root-mean-square error (RMSE) directly measures the deviation between predicted values and real values. The prediction goal is to minimize the gap between predicted values and real values. So in this paper, RMSE is used instead of R^2 to evaluate the predictive ability of model; RMSE is expressed as follows:

$$RMSE = \sqrt{\frac{\sum_{i=1}^{n}(q_{act} - q_{for})^2}{n}} \tag{7}$$

where n represents the number of the empirical data, q_{act} represents actual historical production, q_{for} represents forecast production, the target of the model is minimizing RMSE.

The analysis of constraints is as follows: external URR (ultimate recoverable resources) are used to constrain production. The F test is used to determine the number of production cycles.

In many traditional oil production forecasting models, URR is usually regarded as an internal variable, together with oil production, becoming the production variable of the prediction model. The disadvantage of this approach is that URR cannot constrain production. It is likely to overestimate or underestimate future oil production. This paper uses URR as an external variable to constrain production. The value of URR can be obtained by summing cumulative production and reserves. This constraint is expressed in the following equation:

$$\text{URR}_{\text{en}} = \text{URR}_{\text{ex}} \qquad (8)$$

The left side is internal URR, and the right side is external URR.

F test is established as follows.

First, the variance of sample sequence can be obtained from Eq. (7).

$$S^2 = \frac{\sum_{i=1}^{n}(q_{\text{act}} - q_{\text{for}})^2}{n - m - 1} = \frac{\text{RMSE}^2 \times n}{n - m - 1} \qquad (9)$$

where *m* represents the number of unknown parameters in Eq. (9). $n - m - 1$ represents the degrees of freedom. Then in terms of the prediction results in two groups (one group is established before an additional cycle is added, and another group is established after an additional cycle is added); then the *F* statistic is established as follows:

$$F_{\text{value}} = \frac{S_1^2}{S_2^2} = \frac{\frac{\text{RMSE}_1^2 \times n}{n - m_1 - 1}}{\frac{\text{RMSE}_2^2 \times n}{n - m_2 - 1}} = \frac{\text{RMSE}_1^2}{\text{RMSE}_2^2} \frac{n - m_2 - 1}{n - m_1 - 1} \qquad (10)$$

where RMSE_1 and RMSE_2 represent the mean square root before and after an additional production cycle is added, respectively. In general, $\text{RMSE}_1 > \text{RMSE}_2$; m_1 and m_2 represent the number of free variables in the model before and after an additional production cycle is added, respectively. In general, $m_1 < m_2$; *n* represents the number of empirical data.

A production cycle can be added only when the following conditions are met:

$$F_{\text{value}} > F_{\alpha}(n - m_1 - 1, \ n - m_2 - 1) \qquad (11)$$

where α represents significance level, whose value is 0.01 in this paper.

The significance of the *F* test is that a new production cycle is allowed only when it can significantly improve the goodness of fit.

In reality, under the influence of economy, technology and other factors, the remaining resource depletion rate cannot grow without limit. Further, extremely high depletion rates mean destructive exploitation of underground resources, which is unfavorable for long-term development. Therefore, in actual production, the residual resource depletion rate has a maximum ceiling. The residual resource depletion rate is expressed as follows:

$$d(t) = \frac{q(t)}{\text{URR} - Q(t)} \qquad (12)$$

where $d(t)$ represents the residual resource depletion rate, $q(t)$ represents annual production, and $Q(t)$ represents cumulative production.

Above all, the multi-cycle generalized Weng model can be expressed as follows:

$$\text{Min RMSE} = \sqrt{\frac{\sum_{i=1}^{n}(q_{\text{act}} - q_{\text{for}})^2}{n}}$$

$$\text{st.} \begin{cases} q(t) = \dfrac{\text{URR}}{c^{b+1}\Gamma(b+1)}t^b e^{-(t/c)} \\[2mm] Q(t) = \sum_{i=1}^{k} q(t)_i \\[2mm] \text{URR}_{\text{en}} = \text{URR}_{\text{ex}} \\[2mm] F_{\text{value}} > F_{\alpha}(n - m_1 - 1, n - m_2 - 1) \\[2mm] d(t) = \dfrac{q(t)}{\text{URR}_{\text{en}} - Q(t)} \le d_{\max} \\[2mm] b > 0, \ c > 0 \end{cases} \qquad (13)$$

where $Q(t)$ is the annual forecast production, whose value is the summation of the forecast production of all cycles. $d_{\max}$ represents the maximum residual resource depletion rate which is extracted by combining the existing research literature with the investigation into the current oil production situation. This model is solved by using Excel VBA programming.

3 Application of a multi-cycle generalized Weng model in forecasts of global oil supply

3.1 Current situation of global oil supply

The world's main oil sources are OPEC and some other traditional large oil suppliers, such as Russia and the USA. Venezuela, Saudi Arabia, Iran and Iraq have abundant oil reserves, with more than 2×10^{10} tonnes for each country. The production gap between OPEC and non-OPEC's total oil production is less than 5%. The reserve-production ratios of Venezuela, Libya, Iran or Iraq are more than 100. The overall OPEC reserve-production ratio is 91, which proves that OPEC has strong oil supply potential under the current oil production situation.

At the same time, the shale oil revolution has significantly boosted the traditional oil market in recent years, and changes have taken place in oil market patterns. However, the rapid price fall not only made the oil market cool down, but also curtailed the unconventional oil and gas revolution which had just arisen. The global oil market has entered a stable phase recently after huge short-term fluctuations.

Table 1 Value of parameters in model	Index	URR_{en}	RMSE	Number of years	Number of production cycles
	Value	$398,500 \times 10^6$ tonnes	114	50	5

Root-mean-square error (RMSE) measures the deviation between predicted values and real values

Oil price decline due to global oil being excessively supplied has resulted in the cessation of drilling in a large number of oilfields. Oil stocks continued to decline, which reduced the global oil surplus. But the oversupply situation still exists and the status of many traditional oil-rich countries is being challenged. Along with the conventional and unconventional oil production tending to be stable, what pattern will global oil supply evolve into? This has profound influence on the main oil suppliers and consumers who have just experienced sharp fluctuations of oil price. This paper forecasts the global oil supply using the multi-cycle generalized Weng model.

3.2 Data

In this paper, the oil data from 1965 to 2014 have been chosen for analysis; the data on annual oil production, proven oil reserves and relative exploitation time (the base year is 1965) are obtained from the BP Statistical Review of World Energy 2015.

3.3 Results

The value of some key parameters in the multi-cycle generalized Weng model which is applied to forecast global oil supply is listed as follows (Table 1).

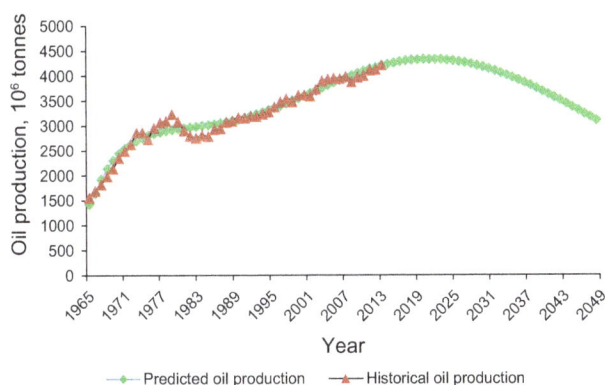

Fig. 1 Prediction of global oil production

The analysis and prediction on global oil production are carried out based on the multi-cycle generalized Weng model established above, as shown in Fig. 1.

Large fluctuations in oil production happened in the 1970s and 1980s, mainly caused by turmoil in the Middle East. This multi-cycle model can well reflect the multi-modal phenomena of oil supply. Before 2022, oil supply will slowly rise on the basis of status quo; after the peak, it will continue to drop.

So far, Saudi Arabia has occupied the main position in oil supply for a long time. Meanwhile, the oil supply of Persian Gulf is still the focus of the world. In the future, global oil supply and social situations will be more closely linked in this region due to resource depletion and the global competition for energy. Instability in this region will rapidly affect global development through energy chains.

The oil production of some countries is going to change significantly. On the one hand, China should make corresponding preparations in advance and expand diverse oil import channels; On the other hand, strategic oil cooperation with Africa is still a key support for China's economic and social development in the short term. In addition to cooperation with Nigeria, cooperation with Sudan, Congo and other countries in the field of energy must be strengthened to achieve win–win situations and energy security.

Table 2 lists the various peak time and the productions in four time points of the world.

The world's peak time is estimated to happen in 2022 from this research, and it is close to the results of Tang et al. (2009) and Shell (2011).

4 Conclusion

To overcome the shortcomings in existing oil production forecast models, this paper establishes a multi-cycle generalized Weng model and predicts global oil production based on data from the BP Statistical Review of World Energy, 2015. This model includes parameter selection from a large sample, depletion rate of remaining resources, constraint of oil reserves and cycle number determination,

Table 2 Peak time and peak yield, 10^6 tonnes	Peak time	Peak yield	Yield in 2020	Yield in 2030	Yield in 2040	Yield in 2050
	2022	4340	4330	4215	3760	3110

not only to better fit curves but to strengthen the forecast capacity of model.

In the process of model establishment and application, it is found that the number of model cycles can dramatically affect the prediction outcome. So, an appropriate number of cycles determined by the fitting error can effectively avoid excessive fitting, promoting model prediction reliability. The residual resource depletion rate can effectively avoid unrealistic production changes in many models; recoverable reserves will have a significant impact on future oil supply. At present, the global oil supply exceeds demand, however, its peak is going to be reached following China's "13th Five-Year plan," and then oil supply is predicted to decline. Therefore, during this period, China should accelerate the conventional and unconventional oil exploration and imports to ensure the future oil demand can be satisfied.

Acknowledgements The authors appreciate the financial support from the National Natural Science Foundation of China (Grant Nos. 71303258, 71373285, and 71503264), National Social Science Funds of China (13&ZD159), MOE (Ministry of Education in China) Project of Humanities and Social Sciences (13YJC630148, 15YJC630121), and Science Foundation of China University of Petroleum, Beijing (ZX20150130).

References

Apergis N, Ewing BT, Payne JE. A time series analysis of oil production, rig count and crude oil price: evidence from six U.S. oil producing regions. Energy. 2016;97:339–49. doi:10.1016/j.energy.2015.12.028.

Brandt AR. Testing Hubbert. Energy Policy. 2007;35:3074–88. doi:10.1016/j.enpol.2006.11.004.

Chen YQ, Hu JG. Review and derivation of Weng model. China Offshore Oil Gas (Geol). 1996;10(5):41–8 (**in Chinese**).

Chen YQ. Derivation and application of Weng's prediction model. Nat Gas Ind. 1996;16(2):22–6 (**in Chinese**).

Ebrahimi M, Ghasabani NC. Forecasting OPEC crude oil production using a variant Multicyclic Hubbert Model. J Pet Sci Eng. 2015. doi:10.1016/j.petrol.2015.04.010.

Gallagher B. Peak oil analyzed with a logistic function and idealized Hubbert curve. Energy Policy. 2011;39(2):790–802. doi:10.1016/j.enpol.2010.10.053.

Hosseini SH, Shakouri HG. A study on the future of unconventional oil development under different oil price scenarios: a system dynamics approach. Energy Policy. 2016;91:64–74. doi:10.1016/j.enpol.2015.12.027.

Hubbert MK. Energy from fossil fuels. Science. 1949;109(2823): 103–9.

Hubbert MK. Nuclear energy and the fossil fuels. Report presented before the Spring Meeting of the Southern District. Plaza Hotel. American Petroleum Institute, San Antonio, Texas, 7–9 Mar 1956.

Hubbert MK. Techniques of prediction as applied to the production of oil and gas, oil and gas supply modelling. In: Proceedings of a symposium at the Department of Commerce, Washington DC, 18–20 Jun 1982.

Kaufmann RK. Oil production in the lower 48 states: reconciling curve fitting and econometric models. Resour Energy. 1991;13(1):111–27. doi:10.1016/0165-0572(91)90022-U.

Nashawi IS, Malallah A, Al-Bisharah M. Forecasting world crude oil production using multicyclic Hubbert model. Energy Fuels. 2010;24(3):1788–800. doi:10.1021/ef901240p.

OPEC Monthly Oil Market Report 10 February 2016, 2016. http://www.opec.org/opec_web/static_files_project/media/downloads/publications/MOMR%20February%202016.pdf.

Pindyk RS, Rubinfeld DL. Econometric models and economic forecasts. Boston: McGraw-Hill; 1998.

Reynolds DB. World oil production trend: comparing Hubbert multicycle curves. Ecol Econ. 2014;98:62–71. doi:10.1016/j.ecolecon.2013.12.016.

Saraiva TA, Szklo A, Lucena AFP. Forecasting Brazil's crude oil production using a multi-Hubbert model variant. Fuel. 2014;115:24–31. doi:10.1016/j.fuel.2013.07.006.

Shell. Signals and signposts: Shell energy scenarios to 2050. The Hague: Royal Dutch Shell PLC; 2011.

Sorrell S, Speirs J. Hubbert's legacy: a review of curve-fitting methods to estimate ultimately recoverable resources. Nat Resour Res. 2010;19(3):209–30. doi:10.1007/s11053-010-9123-z.

Tang X, Feng LY, Zhao L. Prediction and analysis of world oil supply pattern based on a Generalized Weng Model. Resour Sci. 2009;2:238–42. doi:10.3321/j.issn:1007-7588.2009.02.009 (**in Chinese**).

Tang X, Zhang BS, Höök M, et al. Forecast of oil reserves and production in Daqing oilfield of China. Energy. 2010;35:3097–102. doi:10.1016/j.energy.2010.03.043.

Tao ZP, Li MY. System dynamics model of Hubbert Peak for China's oil. Energy Policy. 2007;35:2281–6. doi:10.1016/j.enpol.2006.07.009.

U.S. Energy Information Administration (EIA). Monthly energy review: March 2016. 2016. https://www.eia.gov/totalenergy/data/monthly/pdf/sec11_5.pdf.

Wang JL, Feng LY, Zhao L, et al. A comparison of two typical multicyclic models used to forecast the world's conventional oil production. Energy Policy. 2011;39:7616–21. doi:10.1016/j.enpol.2011.07.043.

Wang JL, Feng LY. Curve-fitting models for fossil fuel production forecasting: key influence factors. J Nat Gas Sci Eng. 2016;32:138–49. doi:10.1016/j.jngse.2016.04.013.

Wang J, Feng L, Davidsson S, et al. Chinese coal supply and future production outlooks. Energy. 2013;60(7):204–14. doi:10.1016/j.energy.2013.07.031.

Weng WB. Prediction theory basis. Beijing: Petroleum Industry Press; 1984 (**in Chinese**).

Features and genesis of Paleogene high-quality reservoirs in lacustrine mixed siliciclastic–carbonate sediments, central Bohai Sea, China

Zheng-Xiang Lü[1,2] · Shun-Li Zhang[1] · Chao Yin[1] · Hai-Long Meng[1] · Xiu-Zhang Song[1] · Jian Zhang[1]

Abstract The characteristics and formation mechanisms of the mixed siliciclastic–carbonate reservoirs of the Paleogene Shahejie Formation in the central Bohai Sea were examined based on polarized light microscopy and scanning electron microscopy observations, X-ray diffractometry, carbon and oxygen stable isotope geochemistry, and integrated fluid inclusion analysis. High-quality reservoirs are mainly distributed in Type I and Type II mixed siliciclastic–carbonate sediments, and the dominant pore types include residual primary intergranular pores and intrafossil pores, feldspar dissolution pores mainly developed in Type II sediments. Type I mixed sediments are characterized by precipitation of early pore-lining dolomite, relatively weak mechanical compaction during deep burial, and the occurrence of abundant oil inclusions in high-quality reservoirs. Microfacies played a critical role in the formation of the mixed reservoirs, and high-quality reservoirs are commonly found in high-energy environments, such as fan delta underwater distributary channels, mouth bars, and submarine uplift beach bars. Abundant intrafossil pores were formed by bioclastic decay, and secondary pores due to feldspar dissolution further enhance reservoir porosity. Mechanical compaction was inhibited by the precipitation of pore-lining dolomite formed during
early stage, and oil emplacement has further led to the preservation of good reservoir quality.

Keywords High-quality reservoirs · Mixed sediments · Paleogene Bohai Sea

1 Introduction

In addition to carbonate and clastic reservoir rock types, magmatic, metamorphic, shale and mixed siliciclastic–carbonate sedimentary reservoirs can also be considered as important targets for oil and gas exploration and development (Ge et al. 2011; Tong et al. 2012; Xiao et al. 2015; Palermo et al. 2008). The concept of "mixed sediments" was firstly proposed by Mount (1984) and is commonly referred to as sediments that are composed of mixtures of siliciclastic and carbonate material (including allochemical particles) (Lubeseder et al. 2009; Brandano et al. 2010; Xu et al. 2014). Many Chinese and foreign scholars have made in-depth studies of the formation mechanisms of this type of sediment and suggested that it can be developed in both marine and lacustrine environments. Influenced by sea (lake)-level fluctuations, structural changes, storm, current and tidal actions, mixed siliciclastic–carbonate sediments are widely distributed in transitional marine-terrestrial, continental shelf, and slope environments (García-Hidalgo et al. 2007; Zonneveld et al. 2012). Under certain conditions, mixed siliciclastic–carbonate sediments may be rich in oil and gas. For example, hydrocarbon accumulations have been discovered in the high-quality mixed siliciclastic–carbonate reservoirs in China, such as the Bohai Bay Basin, the Qaidam Basin and the Sichuan Basin (Feng et al. 2011a, b, 2013; Zhang et al. 2006; Liu et al. 2011; García-Hidalgo et al. 2007). Although carbonate and clastic

✉ Shun-Li Zhang
  1205799554@qq.com

[1] College of Energy Resources, Chengdu University of Technology, Chengdu 610059, Sichuan, China

[2] State Key Laboratory of Oil-Gas Reservoirs Geology and Exploitation, Chengdu University of Technology, Chengdu 610059, Sichuan, China

Edited by Jie Hao

reservoirs have been the subject of intensive study by a large number of researchers, mixed siliciclastic–carbonate reservoirs have received less attention. Previously, studies of mixed siliciclastic–carbonate reservoirs have mainly focused on petrography, structure, classification, the establishment of depositional models (Caracciolo et al. 2012; Sha 2001; Zand-Moghadam et al. 2013; Zonneveld et al. 2012; Ma and Liu 2003), and the reconstruction of the sedimentary environment on the basis of sequence stratigraphy, sea level change and paleoclimate (Anan 2014; Campbell 2005; Moissette et al. 2010). However, the microscopic features and the formation mechanisms of high-quality reservoirs have not been well investigated, which has restricted the exploration and development of mixed siliciclastic–carbonate reservoirs.

The Bohai Bay Basin is an important petroliferous basin in North China. In the Paleogene, steep slope zones were well developed and are represented by a series of high steep fault noses (Lu 2005; Guan et al. 2012). The tectonically induced physiographic changes controlled the distribution and areal extension of mixed siliciclastic–carbonate sediments. For example, typical mixed sediments composed of lacustrine carbonate and siliciclastic material are widely distributed in the Shijiutuo Uplift in the central basin, and a large number of high-quality reservoirs are developed in them (Liu et al. 2011; Song et al. 2013). Statistics suggest that reservoir quality is one of the key controls on prospectivity during petroleum exploration and production. The study of the characteristics and formation mechanisms of mixed siliciclastic–carbonate reservoirs is therefore of significant importance for guiding the oil and gas exploration and production in the Bohai Sea. The purpose of this paper is to compare different types of mixed siliciclastic–carbonate reservoirs, to describe the main features of high-quality reservoirs, and to determine the formation mechanisms of high-quality reservoirs by integrating geological and geochemical data.

2 Geological setting

The Bohai Bay Basin is a Cenozoic rift basin superimposed on the Paleozoic basement of the North China platform (Lu 2005). The study area is located in the Shijiutuo Uplift in the central Bohai Bay Basin and bounded by two large hydrocarbon generation sags—Bozhong Sag and Qin'nan Sag (Fig. 1). The hydrocarbon accumulation condition is excellent with high-quality source rocks and a series of Paleogene high steep fault nose traps developed (Guan et al. 2012). The mixed sedimentary reservoirs in the study area are mainly developed in the Paleogene Shahejie Formation (E_2s). The Shahejie Formation is 300–400 m thick with burial depths >3000 m and conformably overlies the Kongdian Formation (E_2k) and underlies the Dongying Formation (E_2d). Because economically significant hydrocarbon accumulations have been found in the mixed reservoirs, the mixed siliciclastic–carbonate reservoirs have been the focus of study in recent years (Wang et al. 2015).

3 Samples and experimental methodology

In this study, 240 mixed siliciclastic–carbonate sediment samples from 12 wells in the Shijiutuo Uplift in the central Bohai Bay Basin, such as Well HD2 and Well HD5, were selected for porosity and permeability measurements. The locations of sample wells are shown in Fig. 1. The microscopic features, such as petrology, pore space types, and diagenesis, were obtained from 122 thin sections with different physical properties. Multi-purpose thin sections were prepared with blue-dyed epoxy impregnation and double-sided polishing. The mineral composition was identified by polarized light microscopy, and X-ray diffraction (XRD) analyses were carried out on twenty-two bulk samples and <2 μm size fractions using a Rigaku DMAX-3C diffractometer. The chemical composition of grain-coating and dissolved minerals was determined quantitatively by electron microprobe analysis (EMPA) using a Shimadzu EPMA-1720 and a JEOL JXA-8100 electron microprobes (operating conditions: 15 kV accelerating voltage, 10 mA current, 1 μm beam diameter). Fifteen double-thickness polished thin sections were selected for microthermometric measurements. Homogenization temperatures were measured using a Linkam THMS-600 heating/cooling stage. Only primary fluid inclusions with both aqueous and hydrocarbon phases were selected from authigenic minerals to determine their minimum precipitation temperatures (Liu et al. 2005; Lü et al. 2015; Guo et al. 2012; Tian et al. 2016). In-situ carbon and oxygen isotope analysis was performed using an Nd:YAG laser microprobe. Laser probe microsampling of C and O from carbonate cements for isotopic analysis was achieved by focusing a laser beam with a wavelength of 1064 nm and a diameter of 20 μm onto a sample situated in a vacuum chamber to ablate a small area on the sample and liberate CO_2 gas. After purification, the CO_2 gas was led directly into a Finnigan MAT 252 mass spectrometer for isotopic analysis. After obtaining the isotopic values, the dolomite formation temperature (T) was calculated using the empirical formula proposed by Hu et al. (2012):

$$T = 16.5 - 4.3(\delta C - \delta W) + 0.14(\delta C - \delta W)(\delta C - \delta W)$$

where δC is the $\delta^{18}O$ of a dolomite precipitate, and δW is the $\delta^{18}O$ of parent water.

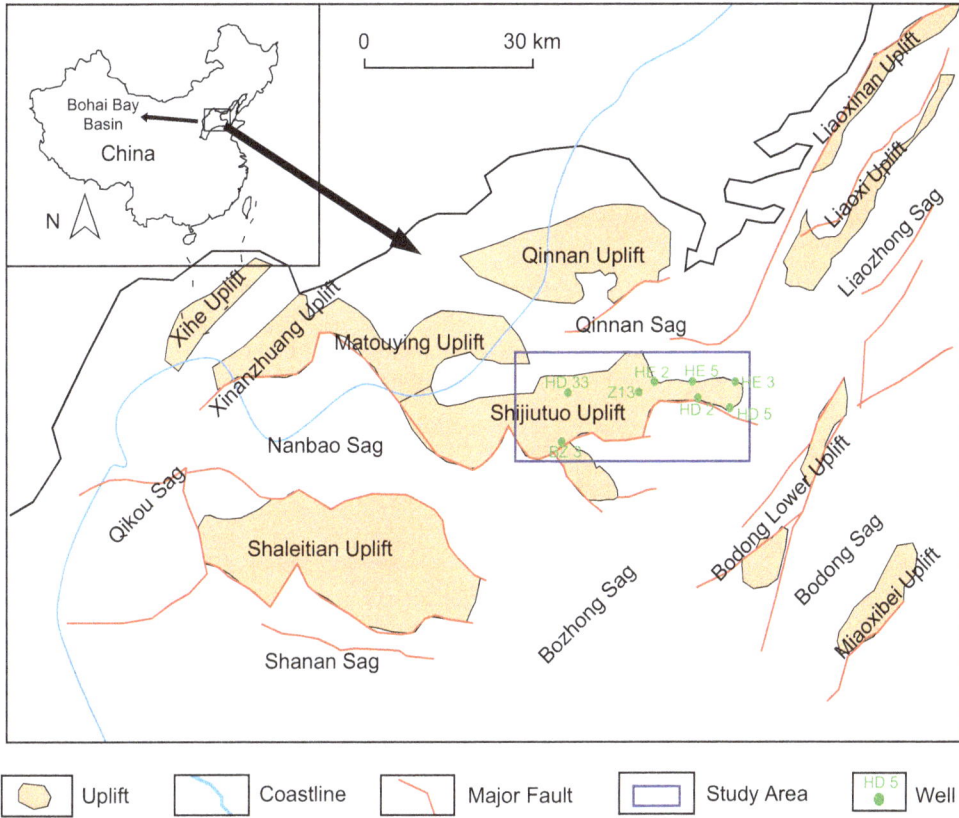

Fig. 1 Location map and tectonic elements of the central Bohai Sea

The timing of feldspar dissolution was mainly determined based on the fluid inclusion temperatures of the dissolution products (authigenic quartz). Sixty-two samples were observed under a DM4500P fluorescence microscope in order to identify possible petroleum inclusions. Fifteen inclusions were also examined using a Renishaw inVia laser Raman microprobe with a wavelength of 514.5 nm to document the existence of hydrocarbons.

The mixed siliciclastic–carbonate sediments were deposited in a fan delta environment (Guan et al. 2012; Zhang et al. 2015; Ni et al. 2013). In order to illustrate the relationship between petrophysical properties and sedimentary microfacies, the sedimentary facies were identified by analyzing rock textures and well log data for Well HD2 and Well HD5, in which core porosity and permeability were measured.

4 Results

4.1 Rock types

The E_2s mixed sediments are composed of siliciclastic and lacustrine carbonate rocks. For the siliciclastic grains, carbonate grains, matrix, and micrite that constituted the

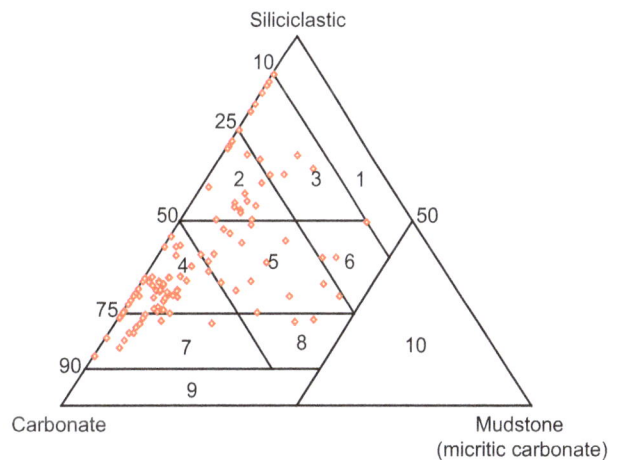

Fig. 2 Rock types of E_2s mixed siliciclastic–carbonate sediments. 1: sand (gravel) rock, 2: carbonate siliciclastic mixed sedimentary rocks, 3: carbonate-bearing siliciclastic mixed sedimentary rocks, 4: siliciclastic carbonate mixed sedimentary rocks, 5: carbonate/siliciclastic mixed sedimentary rocks, 6: carbonate-bearing argillaceous siliciclastic mixed sedimentary rocks, 7: siliciclastic-bearing carbonate mixed sedimentary rocks, 8: siliciclastic-bearing micrite carbonate mixed sedimentary rocks, 9: carbonate, 10: mudstone (micritic carbonate)

mixed sediments, the content of the former two was, respectively, not less than 10%, while the latter two accounted for less than 50%. The identification results of

122 thin sections show that (Fig. 2) E_2s mixed siliciclastic–carbonate sediments were divided into three classes. Class I was mainly composed of siliciclastic carbonate mixed sedimentary rocks and siliciclastic-bearing carbonate mixed sedimentary rocks. It represented up to 55% with carbonate particles content of more than 50% (4, 7 area in Fig. 2). Carbonate grains were mainly bioclasts, accounting for 65% (103 sampling points), followed by oolites and arenes; Class II was mainly composed of carbonate siliciclastic mixed sedimentary rocks and carbonate-bearing siliciclastic mixed sedimentary rocks, accounting for 30%, with siliciclastic particles content of more than 50% (2, 3 area in Fig. 2); Class III was uniformly with less than 50% of siliciclastic grains and of carbonate grains (5, 6 and 8 area in Fig. 2). It was in the lowest content, only accounting for 16%. The interstitial material was mainly dolomite, followed by calcite and small amounts of argillaceous matrix, which was well-sorted and sub-rounded to rounded.

4.2 Diagenetic features

4.2.1 Compaction

From the contact relationship of grains in E_2s mixed siliciclastic–carbonate sediments, it showed that the compaction was not strong, mainly composed of point-line contact (Fig. 3a, b).

4.2.2 Precipitation of authigenic minerals

There were numerous types of authigenic minerals formed in E_2s mixed siliciclastic–carbonate sediments. As with the different proportions of siliciclastics and carbonate, it led to the differences of authigenic mineral content in the mixed siliciclastic–carbonate sediments. In the mixed siliciclastic–carbonate sediments with a high proportion of carbonate, authigenic dolomite, calcite and other carbonate minerals were in high proportions and authigenic clay was in small proportions. However, in the mixed siliciclastic–carbonate sediments with a high proportion of siliciclastic rocks, the authigenic minerals were dominated by kaolinite, illite, and quartz, and the authigenic carbonate minerals were in minor amounts.

Among authigenic carbonate minerals, dolomite made up the largest share, followed by calcite; in addition, there were minor amounts of ankerite and ferroan calcite. The occurrence states of dolomites were a pore liner (Fig. 3a), pore fillings (Fig. 3a, c) and replacement particles (Fig. 3d, e). Calcite mainly occurred as local replacement particles. Authigenic clay minerals included kaolinite (Fig. 3f) and a small amount of illite (Fig. 3f). Authigenic quartz was

distributed in the pores in the form of small crystals (Fig. 3f), and pyrite can be occasionally seen.

4.2.3 Dissolution

Dissolution was well developed in the E_2s mixed siliciclastic–carbonate sediments, and it effectively improved the quality of reservoirs with high proportion of siliciclastic rocks. The dissolved minerals were mainly feldspar, especially albite and K-feldspar (Fig. 3d, e). A small amount of carbonate minerals, such as dolomite and ankerite, were dissolved but this made little contribution to pores.

4.3 Reservoir space features

The reservoir space of E_2s mixed siliciclastic–carbonate sediments was dominated by residual primary intergranular pores and dissolved pores, with minor intercrystalline porosity. Primary pores mainly included residual primary intergranular pores and intrafossil pores (Fig. 3a). Dissolved pores mainly included intergranular dissolved pores in feldspars and rock fragments (Fig. 3d) and intercrystalline pores mainly included intercrystalline pores in kaolinite (Fig. 3f).

4.4 Petrophysical features

The porosity of E_2s mixed siliciclastic–carbonate sediments ranged between 0.45% and 36%. In the 240 samples, 76% samples had a porosity of over 15% (Fig. 4). Permeability mainly ranged between 0.014 and 11259 mD. Most samples had a permeability of over 10 mD, accounting for 53% of the total samples (Fig. 5).

4.5 Features of sedimentary microfacies

The E_2s in the study area was deposited in a continental offshore lacustrine and near-source fan delta depositional environment (Guan et al. 2012; Zhang et al. 2015; Ni et al. 2013). The mixed siliciclastic–carbonate sediments were mainly developed in delta front sandbars and shallow lacustrine underwater uplift beach bars, followed by delta front underwater distributary channels. Front sandbars were divided into mouth bar and distal bar microfacies with reverse grain size grading and funnel-shaped gamma-ray (GR) curves, but the former showed lower GR curves. Underwater uplift beach bars were characterized by fine grain size, good sorting, low content of matrix and micrite and a box-shaped GR curve. Underwater distributary channels abruptly contacted with underlying strata, with coarse grain size at the bottom and minor gravel (Fig. 6).

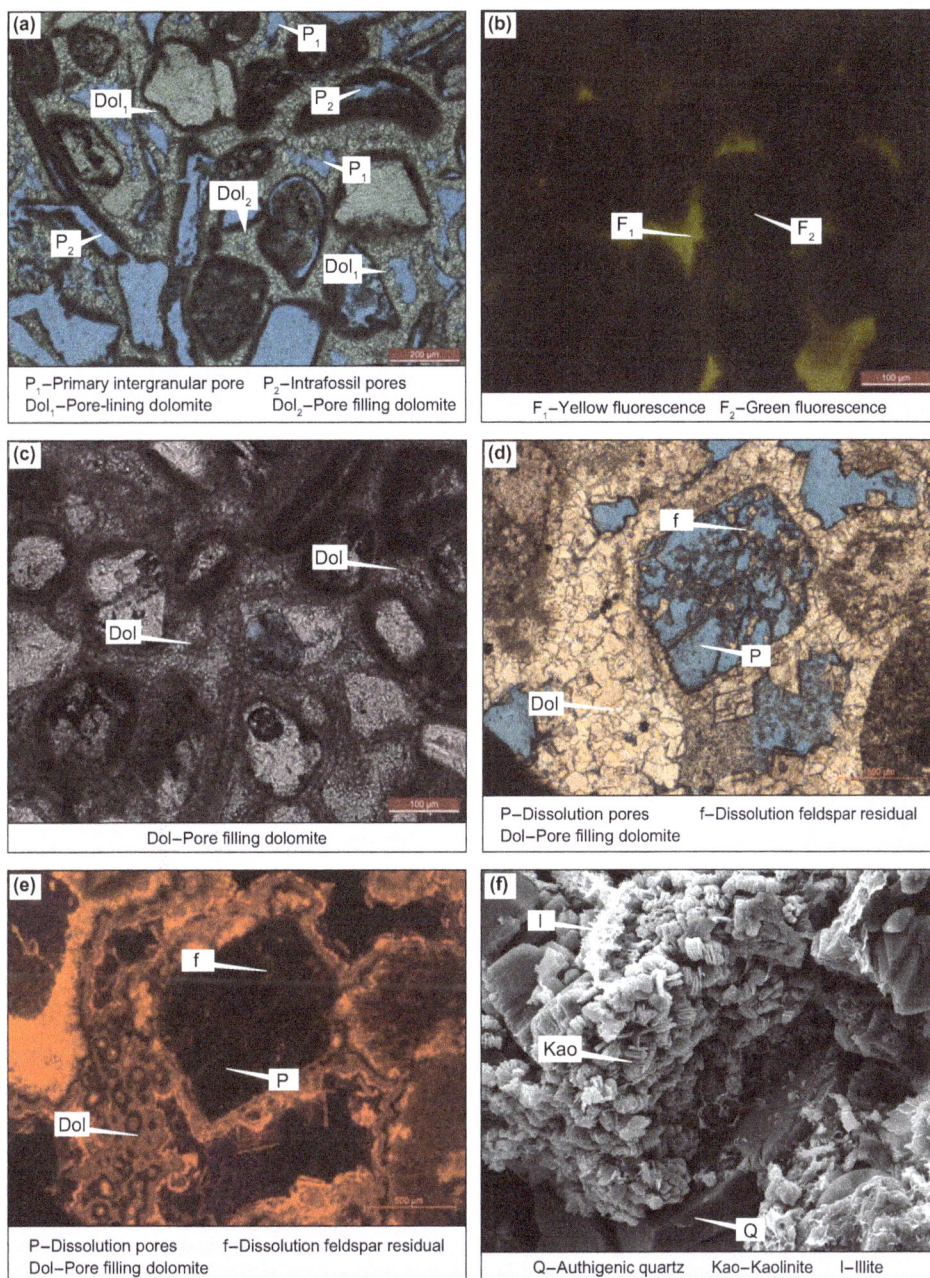

Fig. 3 Photomicrographs of **a** residual intergranular primary pore and intrafossil pores, pore-lining dolomite and filling dolomite, point-line contact, Well HD2, 3762.6 m, polarized light. **b** Two phases of hydrocarbon charging in intergranular dissolved pores and residual intergranular primary pore, Well HD2, 3774.33 m, fluorescence microscope. **c** Pore filling dolomite, pore is poorly developed, Well HD2, 3774.33 m, polarized light. **d** Multiphased authigenic dolomite, dolomite replaces feldspar, feldspar (EPMA: Na_2O: 0.2%, K_2O: 16.5%, Al_2O_3: 18.3%, SiO_2: 64.6%) dissolution, Well HD5, 3382.1 m, polarized light. **e** multiphased authigenic dolomite, dolomite replaces feldspar, feldspar dissolution, Well HD5, 3382.1 m, Cathodoluminescence. **f** Kaolinite, authigenic quartz, illite, Well HD5, 3486.5 m, Scanning electron microscope

According to the sedimentary microfacies and the statistics of components in the 156 samples, the content of siliciclastic particles decreased from 83% to 18% from delta front mouth bar—distal bar—shallow lake underwater uplift beach bar facies.

4.6 Features of high-quality reservoirs

The reservoirs with porosity of >15% and permeability of >10 mD were generally referred to as high-quality reservoirs in this paper.

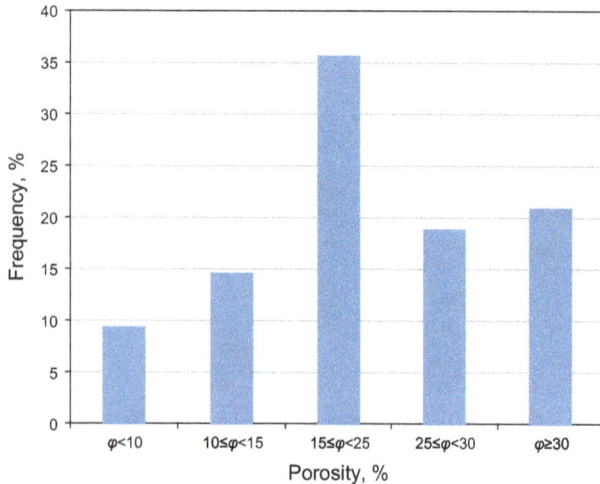

Fig. 4 Porosity distribution histogram of E_2s mixed siliciclastic–carbonate sediments

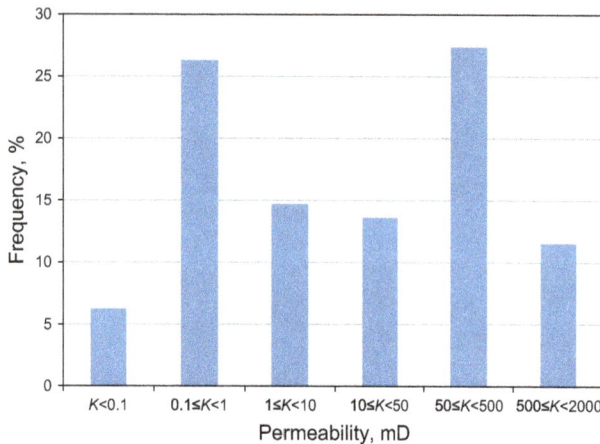

Fig. 5 Permeability distribution histogram of E_2s mixed siliciclastic–carbonate sediments

The sedimentary microfacies and the corresponding 106 groups of petrophysical data of the coring interval in Well HD2 and HD5, as well as the petrophysical data of the 50 sidewall cores of the other wells showed that the mixed siliciclastic–carbonate sediments developed in mouth bar, distributary river channel, and underwater beach bar microfacies had good physical properties, but high-quality reservoirs were basically not developed in the other microfacies (Fig. 7). Seventy-six percentage of the high-quality reservoirs were developed in Class I mixed siliciclastic–carbonate sediments, and their porosity and bioclastic content had good positive correlation (Fig. 8). Seventeen percentage of the high-quality reservoirs occurred in Class II mixed siliciclastic–carbonate sediments, and only 7% high-quality reservoirs occurred in Class III mixed siliciclastic–carbonate sediments. For diagenetic features, the vast majority of high-quality reservoirs were composed of pore-lining dolomite (Fig. 3a),

with minor authigenic calcite. The mixed siliciclastic–carbonate sediments with pore-filling dolomite (Fig. 3c) and calcite were poor in physical properties. Dissolution was common in Class II mixed siliciclastic–carbonate sediments. Through comparing the reservoir space of high-quality reservoirs and poor-quality reservoirs, it can be seen that Class I mixed siliciclastic–carbonate sediments were dominated by primary porosity, such as intrafossil pores, followed by residual intergranular primary pores (Fig. 3a), while Class II mixed siliciclastic–carbonate sediments were dominated by residual intergranular primary and dissolved porosity.

5 Discussion

5.1 Genesis of primary pore development

Primary pores were pervasive in high-quality mixed siliciclastic–carbonate sedimentary reservoirs, especially in the reservoirs of Class I mixed siliciclastic–carbonate sediments. According to statistics of the microscopic pore type and plane porosity of the 87 cast thin sections of Class I mixed siliciclastic–carbonate sediments and 20 cast thin sections of Class II mixed siliciclastic–carbonate sediments, the primary plane porosity of Class I accounted for 90% of the total, while the primary plane porosity of Class II accounted for 42% of the total. The sedimentary microfacies of different types of mixed siliciclastic–carbonate sediments indicate that rocks were formed due to the mixed deposition of the siliciclastic grains and matrix in fan delta facies, and the carbonate particles and micrite deposited in the lacustrine facies. Carbonate particles mainly occurred in lacustrine high-energy underwater beach bars far away from terrigenous provenance, so that Class I mixed sediments with low micrite content and high primary intergranular porosity were well developed, whereas terrigenous clastics were common in the mouth bars near terrigenous provenance, so that Class II mixed sediments with low matrix content were developed.

Sixty-four percentage of carbonate particles were bioclasts. The bioclastic content and reservoir porosity showed a positive correlation in the study area (Fig. 8), since a large amount of intrafossil pores were formed due to biological decay. Most intrafossil pores were well preserved during burial process, so intrafossil pores are well developed in rocks (Fig. 3a). Thus, the bioclasts in the high-quality reservoirs in Class I mixed sediments contributed largely to the primary porosity.

Pore-lining dolomites were common in high-quality reservoirs, and they represented the formation features of vadose zone–phreatic zone as indicated by blade- and overhang-shaped distribution features. The analytical

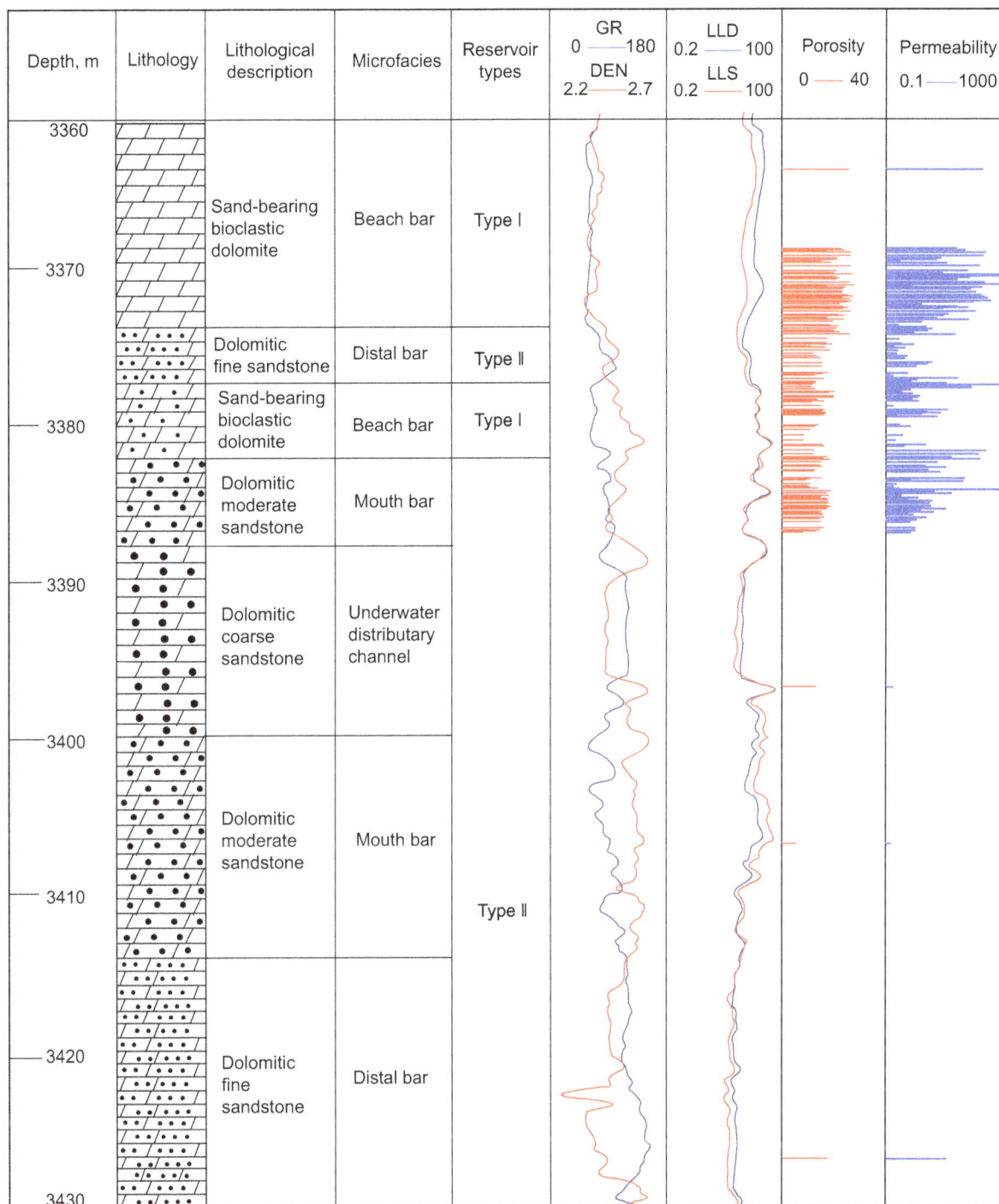

Fig. 6 Sedimentary microfacies of HD5 mixed siliciclastic–carbonate sedimentary interval (3360–3430 m)

results of isotopic temperatures (Table 1) showed that pore-lining dolomite was formed at a temperature of 29–83°C; together with the geothermal gradient of this region (Liu et al. 2012), it is inferred that the pore-lining dolomite in stage 1 (the earliest stage) was formed at a paleoburial depth of less than 150 m and the liner dolomite in stage 3 (the latest stage) was formed at a paleoburial

depth of less than 1700 m and the pore-lining dolomites were formed early. It is indicated by the microscopic observation of early pore-lining dolomite development in reservoirs that the compaction was not strong. Under the burial condition of 4000 m, the point-line contact in grains was ubiquitously seen and the primary pores were well developed (Fig. 3a). By comparing the mixed siliciclastic–

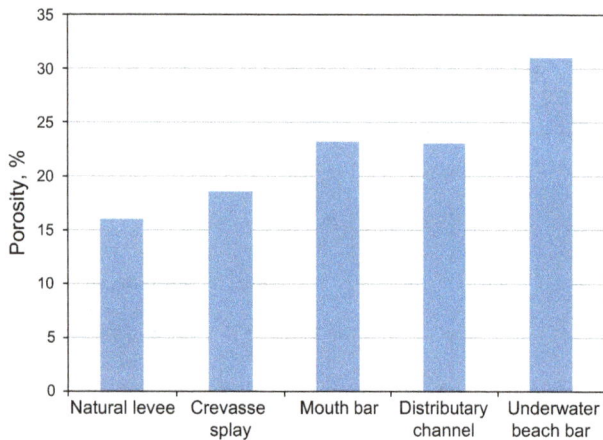

Fig. 7 Physical properties of different sedimentary microfacies of E$_2$s mixed siliciclastic–carbonate sediments

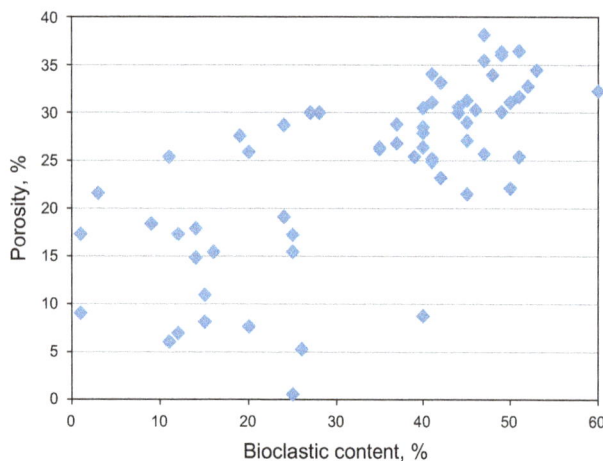

Fig. 8 Relation between the bioclastic content and porosity of E$_2$s mixed siliciclastic–carbonate sediments

carbonate sediments with and without pore-lining in early stage, it can be seen that the rocks without development of pore-lining in early stage mostly represented line contact, with low primary porosity (Fig. 3c). Therefore, the formation of the pore-lining dolomites in early stage effectively weakened the destruction of compaction on pores and was favorable to the preservation of intergranular primary pores.

5.2 Genesis of dissolved pore development

The dissolved pores were well developed in E$_2$s Class II mixed siliciclastic–carbonate sediments. The statistics of the pore types and content of 20 cast thin sections showed that the plane porosity of dissolved pores accounted for 58%. The crystal optical features of dissolved minerals showed that the dissolved minerals were mainly feldspar. Furthermore, the microprobe component analysis results of erosion remnants confirmed that the dissolved minerals were mainly albite and K-feldspar (Table 2). In addition, the authigenic clay minerals were kaolinite and illite, indicating that dissolution took place under a K-rich condition, resulting in the further transformation from kaolinite to illite (Zhang et al. 2007). The inclusion temperature of the authigenic quartz in mixed sedimentary reservoirs ranged between 122–143°C, and the authigenic quartz was formed due to the dissolution of feldspar. Thus, it is inferred that the dissolution of feldspar took place from late middle diagenetic stage to early epidiagenetic stage. The temperature ranges coincided with the temperature ranges of organic matter maturity stage, indicating that the abundant acidic fluids were discharged during organic matter evolution which had created conditions for the formation of dissolved pores in feldspar (Meng et al. 2010; Cao et al. 2014).

5.3 Early phase and multiphased hydrocarbon charging on pores

The microscopic fluorescence features of E$_2$s mixed sediments reflected multistage hydrocarbon charging features. For example, residual primary pores and intragranular dissolved pores had two types of completely different fluorescence, indicating at least two stages of hydrocarbon charging. The early stage residual intergranular primary pore showed yellow fluorescence, and the late stage showed green fluorescence, which was mainly from the dissolved pores in oolite (Fig. 3b). Hydrocarbon components were detected in the inclusions in temperature range of 73–87°C and 119–129°C with laser Raman (Table 3), indicating at least two stages of hydrocarbon charging.

Table 1 C and O isotope distribution of the pore-lining dolomite in E$_2$s mixed siliciclastic–carbonate sediments

Well	Well depth, m	Sample attribute	$\delta^{13}C$ PDB, ‰	$\delta^{18}O$ PDB, ‰	Formation temperature, °C	Formation buried depth, m
HD2	3382.1	Pore-lining dolomite in Stage 1	4.7	−0.76	29.4	126.70
HD2	3762.6	Pore-lining dolomite in Stage 2	5.41	−3.99	47.3	636.58
HD5	3375.06	Pore-lining dolomite in Stage 2	1.88	−4.01	47.4	639.99
HD5	3380.25	Pore-lining dolomite in Stage 2	2.02	−4.97	53.3	807.78
HD5	3375.65	Pore-lining dolomite in Stage 3	−0.42	−9.35	83.3	1666.86

Table 2 Probe composition distribution of the dissolved feldspar remnants in E_2s mixed siliciclastic–carbonate sediments

Well	Well depth, m	Na_2O	K_2O	Cr_2O_3	Al_2O_3	CaO	MnO	MgO	SiO_2	FeO	NiO	TiO_2	Mineral
HD5	3340.8	0.6	14.0	0.0	19.1	0.4	0.8	0.8	62.0	2.0	0.1	0.2	K-feldspar
HD5	3367.5	0.3	16.1	0.0	18.7	0.0	0.1	0.1	64.4	0.3	0.0	0.0	K-feldspar
HE3	3321.4	0.8	15.7	0.1	17.9	0.0	0.0	0.0	65.4	0.1	0.0	0.0	K-feldspar
HE3	3321.4	3.5	12.0	0.0	18.3	0.1	0.0	0.0	65.9	0.2	0.0	0.0	K-feldspar
HE3	3320.8	0.4	16.9	0.0	18.5	0.0	0.0	0.0	64.1	0.1	0.0	0.0	K-feldspar
HD2	3324.4	0.9	15.8	0.0	19.4	0.0	0.0	0.0	63.9	0.0	0.0	0.0	K-feldspar
HD2	3324.4	0.8	15.9	0.0	19.2	0.0	0.0	0.0	64.0	0.0	0.1	0.0	K-feldspar
HD2	3326	0.4	16.3	0.0	18.0	0.0	0.0	0.0	65.2	0.1	0.0	0.0	K-feldspar
Z13	3762.6	11.6	0.0	0.0	19.2	0.0	0.0	0.0	69.2	0.0	0.0	0.0	Albite
Z13	3762.6	11.8	0.0	0.1	19.2	0.0	0.0	0.0	68.9	0.0	0.0	0.0	Albite
Z13	3762.6	11.4	0.1	0.0	19.1	0.0	0.0	0.0	69.4	0.0	0.0	0.0	Albite
Z13	3762.6	11.6	0.1	0.0	19.2	0.0	0.0	0.0	69.1	0.0	0.0	0.0	Albite
BZ3	3779.2	11.7	0.1	0.0	19.3	0.2	0.0	0.0	68.6	0.1	0.0	0.0	Albite
BZ3	3779.2	11.5	0.0	0.0	19.1	0.2	0.0	0.0	69.2	0.0	0.0	0.0	Albite
BZ3	3779.2	11.6	0.0	0.0	19.4	0.1	0.0	0.0	68.9	0.0	0.0	0.0	Albite

Table 3 Gas phase components of the inclusions in E_2s mixed siliciclastic–carbonate sedimentary reservoirs

Well	Well depth, m	Gas phase, %						Host minerals	Homogenization temperature, °C
		CO_2	H_2S	CH_4	N_2	H_2	Total		
HD5	3382.1	0	0	35.7	0	64.3	100.0	Pore-lining dolomite of Stage 3	119
HD5	3370.05	0	16.1	20.8	63.1	0	100.0	Filling dolomite within oolite	73
HD5	3382.1	0	0	9.5	90.5	0	100.0	Pore-lining dolomite within intergranular pores	87
HD5	3383.1	78.1	0	21.9	0	0	100.0	Filling dolomite within intergranular pores	129
HD2	3454.98	51.2	0	7.3	41.5	0	100.0	Quartz enlarging	122

Combined with the paleogeothermal gradient in the study area, it is inferred that the reservoirs were buried at less than 1500 m when there was hydrocarbon charging at the earliest time. Generally, the early hydrocarbon charging can inhibit cementation and also reduced further compaction. Thus, the pores in reservoirs were effectively preserved (Meng et al. 2010; Cao et al. 2014).

6 Conclusions

The E_2s high-quality mixed siliciclastic–carbonate sedimentary reservoirs in the central Bohai Sea were deposited in a fan delta-lacustrine environment. The rocks were formed due to the mixed deposition of the siliciclastic material in fan deltas and carbonate particles deposited in lacustrine environments. The mixed sediment content of the carbonates gradually increased from a near provenance region to lacustrine underwater high-energy beach bars. The E_2s high-quality mixed siliciclastic–carbonate

sedimentary reservoir rocks are mainly developed in Class I, followed by Class II. The development of the high-quality reservoirs of Class I siliciclastic–carbonate sediments was mainly controlled by a high-energy depositional environment, high bioclastic content and pore-lining dolomite and hydrocarbon charging in the early stage. Primary pores were developed in the underwater uplift beach bars with strong hydrodynamic conditions and low micrite content. Intrafossil pores were common due to soft biological decay, forming the main reservoir space of the high-quality reservoir rocks of Class I. The development of early stage pore-lining dolomite effectively weakened the destruction of mechanical compaction on pores. The hydrocarbon charging in the early stage effectively preserved reservoir pores. The development of the high-quality reservoirs of Class II mixed siliciclastic–carbonate sediments was mainly controlled by high-energy depositional environments, feldspar dissolution, pore-lining dolomite and hydrocarbon charging in the early stage. The intergranular primary pores were formed in a high-energy environment, such as fan delta front mouth bars and underwater distributary channels. Feldspar dissolution further improved reservoir properties. The hydrocarbon charging in the early stage and the formation of pore-lining dolomites effectively reduced the destruction of mechanical compaction on pores. Therefore, the E_2s mixed siliciclastic–carbonate sediments in the central Bohai Sea had good geological conditions for high-quality reservoir accumulation, and it is prospective for exploration and development.

Acknowledgements This work was financially supported by the National Science & Technology Specific Project (Grant No. 2011ZX05023-006).

References

Anan TI. Facies analysis and sequence stratigraphy of the Cenomanian–Turonian mixed siliciclastic–carbonate sediments in west Sinai, Egypt. Sediment Geol. 2014;307:34–6. doi:10.1016/j.sedgeo.2014.04.006.

Brandano M, Tomassetti L, Bosellini F, et al. Depositional model and paleodepth reconstruction of a coral-rich, mixed siliciclastic–carbonate system: the Burdigalian of Capo Testa (northern Sardinia, Italy). Facies. 2010;56(3):433–44. doi:10.1007/s10347-009-0209-1.

Campbell AE. Shelf-geometry response to changes in relative sea level on a mixed carbonate–siliciclastic shelf in the Guyana Basin. Sediment Geol. 2005;175(1–4):259–75. doi:10.1016/j.sedgeo.2004.09.003.

Cao YC, Yuan GH, Li XY, et al. Characteristics and origin of abnormally high porosity zones in buried Paleogene clastic reservoirs in the Shengtuo area high porosity zones in buried Paleogene clastic reservoirs in the Shengtuo area, Dongying Sag, East China. Pet Sci. 2014;11(3):346–62. doi:10.1007/s12182-014-0349-y.

Caracciolo L, Gramigna P, Critelli S, et al. Petrostratigraphic analysis of a Late Miocene mixed siliciclastic–carbonate depositional system (Calabria, Southern Italy): implications for mediterranean paleogeography. Sediment Geol. 2012;284–285:117–32. doi:10.1016/j.sedgeo.2012.12.002.

Feng JL, Cao J, Hu K, et al. Dissolution and its impacts on reservoir formation in moderately to deeply buried strata of mixed siliciclastic–carbonate sediments, northwestern Qaidam Basin, northwest China. Mar Pet Geol. 2013;39(1):124–37. doi:10.1016/j.marpetgeo.2012.09.002.

Feng JL, Cao J, Hu K, et al. Formation mechanism of middle-deep mixed rock reservoirs in the Qaidam basin. Acta Pet Sin. 2011a;27(8):2461–72 (**in Chinese**).

Feng JL, Hu K, Cao J, et al. A review on mixed rocks of terrigenous clastics and carbonates and their petroleum-gas geological significance. Geol J China Univ. 2011b;17(2):297–307 (**in Chinese**).

García-Hidalgo J, Gil J, Segura M, et al. Internal anatomy of a mixed siliciclastic–carbonate platform: the Late Cenomanian-Mid Turonian at the southern margin of the Spanish Central System. Sedimentology. 2007;54(6):1245–71. doi:10.1111/j.1365-3091.2007.00880.x.

Ge ZD, Wang XZ, Zhu M, et al. Reservoir characteristics of Archean magmatic rocks in the Dongying Sag. Lithol Reserv. 2011;23(4):48–52 (**in Chinese**).

Guan DY, Wei G, Wang YC, et al. Controlling factors of middle-to-deep reservoir in Bozhong depression, Bohai Sea: an example from Shahejie formation in the steep slope belt of eastern Shijiutuo uplift. Nat Gas Explor Dev. 2012;35(2):5–8 (**in Chinese**).

Guo XW, Liu KY, He S, et al. Petroleum generation and charge history of the northern Dongying Depression, Bohai Bay Basin, China: insight from integrated fluid inclusion analysis and basin modelling. Mar Pet Geol. 2012;32(1):21–35. doi:10.1016/j.marpetgeo.2011.12.007.

Hu ZW, Huang SJ, Li ZM, et al. Preliminary application of the dolomite-calcite oxygen isotope thermometer in studying the origin of dolomite in Feixianguan Formation, Northeast Sichuan, China. J Chengdu Univ Technol (Science & Technology Edition). 2012;39(1):1–9 (**in Chinese**).

Liu DL, Tao SZ, Zhang BM. Application and questions about ascertaining oil-gas pools age with inclusions. Nat Gas Geosci. 2005;16(1):16–9 (**in Chinese**).

Liu Z, Zhu WQ, Sun Q, et al. Characteristics of geotemperature-geopressure systems in petroliferous basins of China. Acta Pet Sin. 2012;27(2):1–17 (**in Chinese**).

Liu ZG, Zhou XH, Li JP, et al. Reservoir characteristics and controlling factors of the Paleogene Sha-2 member in the 36-3 structure, Eastern Shijiutuo uplift, Bohai Sea. Oil Gas Geol. 2011;32(54):832–8 (**in Chinese**).

Lubeseder S, Redfern J, Boutib L. Mixed siliciclastic-carbonate shelf sedimentation-Lower Devonian sequences of the SW Anti-Atlas, Morocco. Sediment Geol. 2009;215(1–4):13–32. doi:10.1016/j.sedgeo.2008.12.005.

Lu XL. Cenozoic faulting and its influence on the hydrocarbon-bearing systems hydrocarbon distribution in the Bohai Bay Basin. Pet Geol Recover Effic. 2005;12(3):31–5 (**in Chinese**).

Lü ZX, Ye SJ, Yang X, et al. Quantification and timing of porosity evolution in tight sand gas reservoirs: an example from the Middle Jurassic Shaximiao Formation, western Sichuan. China Pet Sci. 2015;12(2):207–17. doi:10.1007/s12182-015-0021-1.

Ma YP, Liu L. Sedimentary and diagenetic characteristics of paleogene lacustrine mixed siliciclastic–carbonate sediments in the beach district, Dagang. Acta Sedimentol Sin. 2003;21(4):607–13 (**in Chinese**).

Meng YL, Liang HW, Meng FJ, et al. Distribution and genesis of the anomalously high porosity zones in the middle-shallow horizons of the northern Songliao Basin. Pet Sci. 2010;7(3):302–10. doi:10.1007/s12182-010-0072-2.

Moissette P, Cornée J, Mannaï-Tayech B, et al. The western edge of the Mediterranean Pelagian Platform: a Messinian mixed siliciclastic–carbonate ramp in northern Tunisia. Palaeogeogr Palaeoclimatol Palaeoecol. 2010;285(1–2):85–103. doi:10.1016/j.palaeo.2009.10.028.

Mount JF. Mixing of siliciclastic and carbonate sediments in shallow shelf environments. Geology. 1984;12(7):432–5. doi:10.1130/0091-7613(1984)12<432:MOSACS>2.0.CO;2.

Ni JE, Sun LC, Gu L, et al. Depositional patterns of the 2nd member of the Shahejie Formation in Q oilfield of the Shijiutuo Uplift, Bohai Sea. Oil Gas Geol. 2013;34(4):491–8 (**in Chinese**).

Palermo D, Aigner T, Geluk M, et al. Reservoir potential of a lacustrine mixed carbonate/siliciclastic gas reservoir: the lower Triassic Rogenstein in the Netherlands. J Pet Geol. 2008;31(1):61–96. doi:10.1111/j.1747-5457.2008.00407.x.

Sha QA. Discussion on mixed deposits and mixed siliciclastic-carbonate rock. J Palaeogeogr. 2001;3(3):63–6 (**in Chinese**).

Song ZQ, Chen YF, Du XF, et al. Study on sedimentary characteristics and reservoir of second member of Shahejie Formation, a structural area, Bohai sea. Offshore Oil. 2013;33(4):13–8 (**in Chinese**).

Tian Y, Ying CC, Yan ZW, et al. The coupling of dynamics and permeability in the hydrocarbon accumulation period controls the oil-bearing potential of low permeability reservoirs: a case study of the low permeability turbidite reservoirs in the middle part of the third member of Shahejie. Pet Sci. 2016;13(2):204–24. doi:10.1007/s12182-016-0099-0.

Tong KJ, Zhao CM, Lü ZB, et al. Reservoir evaluation and fracture characterization of the metamorphic buried hill reservoir in Bohai Bay Basin. Pet Explor Dev Online. 2012;39(1):62–9. doi:10.1016/S1876-3804(12)60015-9.

Wang YB, Xue YA, Wang GY, et al. Shallow layer hydrocarbon accumulation characteristics and their exploration significances in Shijiutuo uplift, Bohai sea. China Offshore Oil Gas. 2015;27(2):8–16 (in Chinese).

Xiao XM, Wei Q, Gai HF, et al. Main controlling factors and enrichment area evaluation of shale gas of the Lower Paleozoic marine strata in south China. Pet Sci. 2015;12(4):573–86. doi:10.1007/s12182-015-0057-2.

Xu W, Cheng KY, Cao ZL, et al. Original mechanism of mixed sediments in the saline lacustrine basin. Acta Pet Sin. 2014;30(6):1804–16 (in Chinese).

Zand-Moghadam H, Moussavi-Harami R, Mahboubi A, et al. Comparison of tidalites in siliciclastic, carbonate, and mixed siliciclastic-carbonate systems: examples from Cambrian and Devonian deposits of East-Central Iran. ISRN Geol. 2013;2013:1–22. doi:10.1155/2013/534761.

Zhang JL, Jia Y, Du GL. Diagenesis and its effect on reservoir quality of Silurian sandstones, Tabei area, Tarim Basin, China. Pet Sci. 2007;4(3):1–13. doi:10.1007/s12182-007-0001-1.

Zhang NS, Reng XJ, Wei JX, et al. Rock types of mixed-sediment reservoirs and oil-gas distribution in Nanyishan of the Qaidam Basin. Acta Pet Sin. 2006;27(1):42–6 (in Chinese).

Zhang YK, Hu XQ, Niu T, et al. Controlling of paleogeomorpology to Paleogene sedimentary systems of the Shijiutuo uplift in the Bohai Basin. J Jilin Univ Earth Sci Ed. 2015;45(6):1589–96 (in Chinese).

Zonneveld JP, Gingras MK, Beatty TW et al (2012) Mixed siliciclastic/carbonate systems. In: Developments in sedimentology (Eds), vol 64, pp 807–3. doi:10.1016/B978-0-444-53813-0.00026-5.

Understanding aqueous foam with novel CO_2-soluble surfactants for controlling CO_2 vertical sweep in sandstone reservoirs

Guangwei Ren[1,2] · Quoc P. Nguyen[1]

Abstract The ability of a novel nonionic CO_2-soluble surfactant to propagate foam in porous media was compared with that of a conventional anionic surfactant (aqueous soluble only) through core floods with Berea sandstone cores. Both simultaneous and alternating injections have been tested. The novel foam outperforms the conventional one with respect to faster foam propagation and higher desaturation rate. Furthermore, the novel injection strategy, CO_2 continuous injection with dissolved CO_2-soluble surfactant, has been tested in the laboratory. Strong foam presented without delay. It is the first time the measured surfactant properties have been used to model foam transport on a field scale to extend our findings with the presence of gravity segregation. Different injection strategies have been tested under both constant rate and pressure constraints. It was showed that novel foam outperforms the conventional one in every scenario with much higher sweep efficiency and injectivity as well as more even pressure redistribution. Also, for this novel foam, it is not necessary that constant pressure injection is better, which has been concluded in previous literature for conventional foam. Furthermore, the novel injection strategy, CO_2 continuous injection with dissolved CO_2-soluble surfactant, gave the best performance, which could lower the injection and water treatment cost.

✉ Guangwei Ren
guangweiren@utexas.edu

[1] Petroleum and Geosystems Engineering Department, University of Texas at Austin, Austin, TX, USA

[2] Present Address: Total E&P R&T USA, Houston, TX, USA

Edited by Yan-Hua Sun

Keywords Foam · CO_2-soluble surfactant · Sweep efficiency · Gravity segregation · Optimal injection strategy

1 Introduction

Gases have been used as driving fluids in improved oil recovery processes since 1900 (Lake 1989), in which CO_2 flooding has attracted a lot of attention because of its proven miscible-like displacement (Stalkup 1983), high availability, and environmental concerns. However, this process frequently experiences viscous fingering, gravity override, and gas channeling because of reservoir heterogeneity as well as low density and viscosity of CO_2, which results in a decreased oil recovery (Rossen and Renkema 2007). Fortunately, the use of foam can reduce gas mobility and effects of heterogeneity and therefore increase sweep efficiency (Rossen 1995). This was first proposed in 1958 by Bond and Holbrook (1958). Carbon dioxide (CO_2) foams in porous media with aqueous soluble surfactants have been widely studied in connection with their application in enhanced oil recovery (EOR) (Lee and Heller 1988; Du et al. 2007). These experimental and theoretical studies have contributed to the success of several field foam applications (Patzek 1996), especially for carbonate reservoirs (Hoefner et al. 1995; Stevens 1995). Unfortunately, field experiences have shown that conventional foams with only aqueous soluble surfactants have some important limitations. For example, the injected surfactant slugs do not improve the CO_2-oil contact. Gravity override and macroscopic heterogeneity also challenge the success of surfactant placement into theft zones where the presence of foam is desired.

Gravity segregation leads to poor sweep efficiency and has received great attention because of its importance in

EOR processes involving gas injection. An analytical model developed by Stone (1982) for gravity segregation of water and gas provided a conceptual framework for understanding gravity segregation without foam. He assumed that gas was incompressible and there were negligible gradients of capillary pressure. After a steady state was established, there were three zones in the reservoir (Fig. 1): a gas zone at the top with gas, a water zone at the bottom with only water, and a mixed zone with both gas and water flowing. Jenkins (1984) extended this study and provided a solution to determine the saturation profile and shapes of three zones. Rossen and van Duijn (2004) showed that the theoretical justifications presented by Stone and Jenkins for their models were incorrect, but they can be derived rigorously with only some assumptions (Rossen et al. 2006; Rossen and Shen 2007; Jamshidnezhad et al. 2008a).

Stone's model may be applied to foam processes that obey the "fixed limiting capillary pressure" (Rossen et al. 1994, 1995b). Therefore, Shi and Rossen (1996) proposed that for foam in a cylindrical reservoir:

$$\left(\frac{R_g}{R_e}\right)^2 = VGR = \frac{1}{N_g}\frac{1}{R_L} = \left(\frac{\nabla P^f(R_g)}{\Delta\rho g}\right)\left(\frac{2R_gHk_x}{R_e^2k_z}\right) \quad (1)$$

where R_e is the reservoir radius; R_g is the radial position at which gas and water flows are completely segregated; N_g is the gravity number, the ratio of the vertical driving force for segregation to the horizontal pressure gradient; VGR represents the viscous-to-gravity ratio; R_L is a modified reservoir aspect ratio; N_g and R_L are evaluated at the radial position R_g at which gas and water flows are completely segregated; $\Delta\rho$ is the density difference between gas and liquid; $\nabla P^f(R_g)$ is the pressure gradient in the foam bank near the injection well in the absence of gravity segregation and is a simple function of water flow rate and water saturation (Friedmann et al. 1991); H is the reservoir height; k_z and k_x are the vertical and horizontal absolute permeabilities; and g is the gravity acceleration constant. This result implied that for a given reservoir with density differences between phases, the only way to increase the distance that gas and water travel together before complete segregation was to increase the horizontal pressure gradient

(Rossen and Shen 2007). Because of nonuniform mobility in the foam bank for surfactant solution alternating gas (SAG) and the differences between processes, the criteria for gravity override with co-injection (Shi and Rossen 1996) cannot be simply applied to SAG (Shi et al. 1998).

In addition to water flooding, pure gas flooding, water alternating gas (WAG) (Ma and Youngren 1994), and simultaneous WAG (SWAG) (Sanchez 1999), heretofore, some additional injection strategies in the presence of foam can be classified as:

(1)　*Co-injection: simultaneous injection of surfactant solution with gas*

Most of the laboratory experiments were conducted in this manner (Svorstol et al. 1996; Mohd Shafian et al. 2015), even though it was tried in only few fields (Blaker et al. 1999) since it may lead to fractures due to high back pressure. Chen et al. (2012) and Elhag et al. (2014) demonstrated that apparent viscosities of foams measured with a capillary viscometer were more than 8 cP at variable temperatures and foam qualities with a switchable ethoxylated cationic CO_2-philic surfactant. They found that the delivery media of CO_2-soluble surfactant imposed less impact. Later, further tests on a 1.2 Darcy glass bead pack and a 49-mD dolomite core gave apparent viscosities of foams as high as 390 and 100 cP, respectively (Chen et al. 2015). Xing et al. (2012) and McLendon et al. (2014) measured the pressure drop across a Berea sandstone core as the CO_2/surfactant solution was injected with selected branched ethoxylated CO_2-soluble surfactants, which gave a weak foam with a mobility reduction factor around five. Through simulation with an analytical model, Rossen et al. (2006) drew a series of conclusions concerning the optimal injection strategy for co-injection with conventional water-soluble surfactants, regarding longer gravity segregation length. Recently, Zeng et al. (2016) demonstrated a spreading effect caused by different partition coefficients of CO_2-soluble surfactants based on published data (Ren et al. 2013) through 1D simulation during co-injection. Surfactants were injected with brine even though they are CO_2-soluble.

(2)　*SAG or foam-assisted WAG (FAWAG)*

For SAG, surfactant is added to a water cycle and the actual diverting foam is generated in a subsequent gas cycle. Some experiments on laboratory scale have been conducted (Lawson and Reisberg 1980; Xu and Rossen 2003). It was shown that a higher injection rate will promote stronger foam generation (Mohd Shafian et al. 2015). Rossen et al. (1995a) pointed out SAG foam processes can combine high gas injectivity with low mobility at the front of the foam bank, which offers an escape from the dilemma posed by early modeling, i.e., improved vertical sweep of

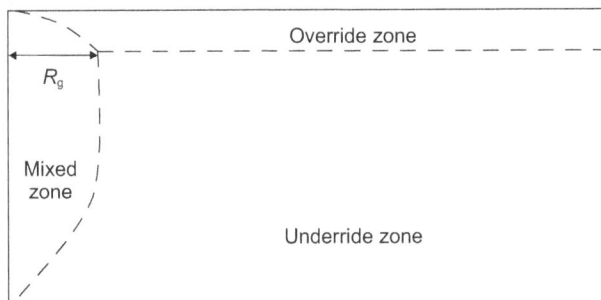

Fig. 1 Three zones during gas injection

gas with foam requires an increase in injection pressure (Shi et al. 1998). Shan and Rossen (2004) proposed an optimal injection strategy for overcoming gravity override with foam in a homogeneous reservoir. Kloet et al. (2009) extended the study of Rossen and Renkema (2007) and developed design criteria for the optimal foam strength and slug size for a given permeability contrast between layers. SAG injection also minimizes contact between water and gas in surface facilities and piping (Heller 1994). Relative to co-injection, SAG could overcome gravity override better and improve injectivity (Patzek 1996; Shi and Rossen 1996; Blaker et al. 1999). Sagir et al. (2014c, d) conducted pure CO_2 injection after surfactant solution flooding to mimic FAWAG with a new CO_2-soluble surfactant using Berea cores, during which a mobility reduction factor of 3.1 was achieved. Similar results were demonstrated by Xing et al. (2012) and McLendon et al. (2014) relative to pure CO_2 injection on carbonate and sandstone cores. Recently, 80 cP of foam apparent viscosity was achieved by Chen et al. (2015) in a 49-mD dolomite core with a switchable ethoxylated cationic CO_2-philic surfactant.

(3) WAG with dissolved surfactant (WAGS)

Relative to the conventional SAG, WAGS delivers the surfactant in the CO_2 phase, which may increase the CO_2–oil contact. Le et al. (2008) conducted both SAG and WAGS at the same conditions, which resulted in similar ultimate oil recoveries and pressure drops.

(4) Novel CO_2 injection: CO_2 continuous injection with dissolved CO_2-soluble surfactant

CO_2 continuous injection with dissolved CO_2-soluble surfactant is another novel concept, which is the extreme case of WAGS and may not require brine injection. Foam is created in situ as CO_2 and dissolved surfactant propagate through the formation mixing with reservoir brine to maximize the benefit of CO_2-miscible displacement and improve the injectivity. Similar to WAGS, this method improves in situ foam generation, drastically lowers the injection costs, and reduces the loss of surfactant onto the rock surface due to adsorption. Xing et al. (2012) and McLendon et al. (2014) showed that the mobility reduction factor was around two, which was slightly higher than pure CO_2 injection on pre-saturated cores. In Le et al. (2008), experimental and field scale simulation results show this is the most promising injection scenario which gave highest oil recovery and injectivity. Field trials indicated a 30% sweep efficiency improvement (Sanders et al. 2012).

Supercritical carbon dioxide (scCO$_2$) with its easily attainable critical temperature and pressure (31.1 °C and 7.38 MPa) can be viewed as an ideal chemical processing solvent because it is nontoxic, inexpensive, volatile,

nonflammable, readily available in large quantities, and environmentally benign (Eastoe et al. 2001). However, because of its very low dielectric constant with weak inter-molecular forces and low polarizability per volume and correspondingly weak van der Waals forces (O'Shea et al. 1991), CO_2 is a poor solvent for high molecular weight, hydrophilic molecules, and polar compounds. In the past decades, several approaches have been explored to enhance the solubility of polar substances in scCO$_2$ (Hoefling et al. 1993; McHugh and Krukonis 1994). Eastoe et al. (2003) reported stability and aggregation structures of various economically viable surfactants for CO_2. Results confirmed the affinity of methyl-branched tails for CO_2 but still contributing limited solubility. A review article from Eastoe et al. (2006) told the story of small-molecule CO_2-active surfactants, from fluorinated compounds to oxygenated amphiphiles. Xing et al. (2012) screened solubility of several commercially available nonionic surfactants in CO_2, and the most stable foams were obtained with branched alkylphenol ethoxylates which exhibited 0.01wt%–0.1wt% solubility in CO_2. A lot of effort has put into obtaining low toxicity and low price CO_2-soluble surfactants, of which non-fluorinated AOT (sodium bis(2-ethylhexyl)-sulfosuccinate) (Eastoe et al. 2001; Liu and Erkey 2001) and nonionic surfactants (Liu et al. 2001; Xing et al. 2012) were of most interest. Dhanuka et al. (2006) noted that DOW Tergitol TMN 6 was an effective foaming agent characterized by stable, white, and opaque foams formed at 25 °C and 345 bar. Fan et al. (2005) established that oligo vinyl acetate (OVAc) is extremely CO_2-philic and suitable for incorporation into CO_2-soluble ionic surfactants. Tan and Cooper (2005) used polyethylene oxide (PEO) as the hydrophile during their design of tri-block OVAc-b-PEO-b-OVAc surfactants capable of stabilizing CO_2 foam. Sanders et al. (2010) reported the design and synthesis of a new class of twin-tailed surfactants based on glycerin and designed for the scCO$_2$–water interface, whose performance was better than a linear secondary alcohol CO_2-soluble surfactant. Adkins et al. (2010a, b) and Chen et al. (2010) have proven that a branched hydrocarbon nonionic surfactant can effectively reduce the contact of CO_2 and water phases and raise the surface pressure and the surfactant efficiency (the concentration to produce 20 mN/m interfacial tension reductions). Chen et al. (2012) developed a switchable ethoxylated cationic CO_2-philic surfactant which was able to stabilize CO_2/water foams up to 182 g/L at 120 °C, 3400 psia. Those hybrid surfactants combined the high cloud points of ionic surfactants with high solubility in CO_2 of nonionic surfactants. The adsorption characteristics of this surfactant were described by Cui et al. (2014), and the interfacial tension (IFT) between CO_2/water was around 5 mN/m (Elhag et al. 2014). Sagir et al. (2014a, b) synthesized several CO_2-philic surfactants using maleic anhydride with either 4-tert-butylbenzyl alcohol or dipropylene tertiary butyl alcohol. The IFT between CO_2/brine could reach 1.93 to 4.2 mN/m. The surfactant used here was a new branched nonionic hydrocarbon surfactant with suitable combination of PPO (poly-(propylene oxide)) and PEO (poly(ethylene oxide)).

The notion of applying a CO_2-soluble surfactant during an EOR process to generate C/W mobility control foams was suggested by Bernard and Holm (1967). Soong et al. (2009) probed two strategies for using CO_2-soluble compounds to decrease the mobility of $scCO_2$, "direct thickening" of CO_2 which is accomplished by a macroemulsion formed by an associated thickener in $scCO_2$, and in situ foam generation. Several laboratory experiments with distinct CO_2-soluble surfactants have been conducted with variable injection strategies, which will be reviewed below. Either liquid or CO_2 phase could be used to delivery those novel surfactants. A field trial was carried out in west Texas using surfactant injection in the CO_2 phase to create a CO_2-in-water emulsion or foam to improve vertical conformance and create in-depth mobility control (Sanders et al. 2012). Results indicated a 30% CO_2 trapping improvement in situ.

In a previous publication (Le et al. 2008), a novel foam concept was proposed and a surfactant concentration of 0.1wt% in CO_2 at ambient temperature and 1800 psi was roughly determined. Oil recoveries with variable injection strategies were presented briefly. In our earlier work (Ren et al. 2014), solubility and partition coefficients of a series of nonionic CO_2-soluble surfactants have been tested at varying pressures, temperatures, and salinity in our laboratory. Preliminary probes have revealed superiorities of CO_2-soluble surfactant foam over conventional aqueous soluble surfactant foam through laboratory core floodings of Silurian dolomite carbonate and field scale simulations (Ren et al. 2013). However, the conclusions drawn previously deserve to be further examined with broader rock types and injection strategies. Moreover, some conclusions from prior literature based on conventional surfactants, such as optimal injection strategy, could be updated or modified in the presence of CO_2-soluble surfactants. Through laboratory experiments and field scale simulations, in the current paper, we will peruse the following goals: demonstrate the remarkable advantages of CO_2-soluble surfactant on the laboratory scale with co-injection, alternating injection, and novel pure CO_2 injection with dissolved surfactant; with field scale simulation, exhibit the considerable superiorities of CO_2-soluble surfactant over conventional aqueous surfactant through SAG and co-injection with variable perforation interval or slug size; investigate the unique characteristics of the novel CO_2 foam, including surfactant delivery media, optimal injection strategy, and some additional considerations; and then, examine whether previous conclusions in the literature for conventional surfactants were still valid for this novel foam with our practical postulations.

2 Experimental section

2.1 Materials

A 2-ethylhexanol (2-EH) alkoxylate nonionic hydrocarbon surfactant, which has been used in a previous study (Ren et al. 2013) and named S, and a commercially available anionic surfactant (CD 1045) which is not soluble in CO_2, were used in this study. The properties of S at variable pressures, temperatures, and salinity, such as solubility in CO_2, partition coefficient between brine and CO_2, and aqueous stability have been studied in earlier work (Ren et al. 2014). The adsorptions of the used surfactants are neglected due to negative surface charges of sandstone samples at neutral pH (Lawson 1978; Mannhardt et al. 1993) and without the presence of clay in used outcrop. Except for novel CO_2 continuous injection with dissolved CO_2-soluble surfactant, in all other core flood experiments, the surfactant solution containing 0.2wt% surfactant and 3wt% NaCl (analytical grade quality) was used to stabilize supercritical CO_2 foam generated in 1-ft-long Berea sandstone cores. The same NaCl concentration was used to saturate the core with brine before injection of the surfactant solution and CO_2. The purity of the liquid CO_2 was 99.5%. The rock permeability to brine was around 300 mD.

2.2 Experimental apparatus and procedures

A schematic of the core flood setup is shown in Fig. 2. It is comprised of three main modules: a fluid injection system, core holder and pressure transducers, and a back pressure and effluent collection system.

Fluid injection system A TELEDYNE ISCO Model 500D syringe was used to directly inject brine or surfactant solution into the cores. CO_2 was displaced into the core by deionized (DI) water through a high pressure accumulator that had a piston to separate water from CO_2.

Core holder and pressure transducers A Phoenix Hassler-type core holder with capacity for 2-inch-diameter core was mounted vertically, and fluids were injected from the top to the bottom. Hydraulic oil was used as an overburden fluid, which compressed and sealed the 0.25-inch-thick rubber sleeve to assure the axial flow of the injection fluids, and to prevent leakage. There were five pressure taps along the side of the core holder in the vertical direction, which connected two absolute pressure transducers (Channel 1 and 5) and three differential transducers (Channel 2, 3 and 4). The differential transducers detected

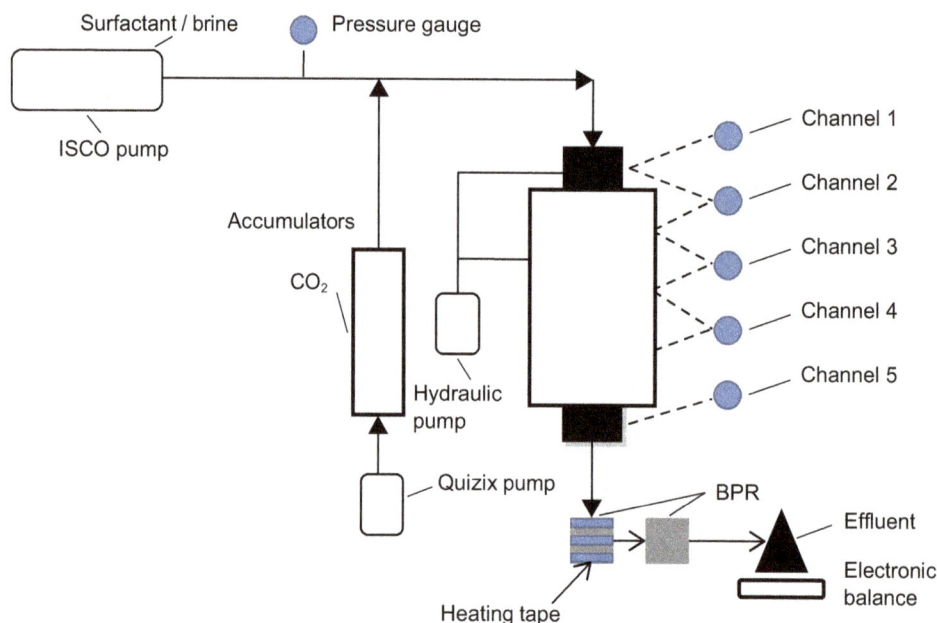

Fig. 2 Schematic of an experimental setup for core flooding

the pressure drops over sections along the core from the top, whose lengths were 2, 4, and 4 inches and denoted as Sect. 1, 2, and 3, respectively.

Back pressure regulator (BPR) and effluent collector Two BPRs were used in series to maintain a constant back pressure of 1500 psig during core flooding. The first BPR placed immediately at the outlet of the core holder was set at 1500 psig, and the second BPR set at 1100 psig.

Core preparation The core was cleaned and dried in a convection oven at 110 °C for 48 h. It was then wrapped in three layers of aluminum foil and a thin Teflon heat shrink tube to prevent CO_2 diffusion and penetration. The wrapped core was placed in the core holder and evacuated for 10 h before the core was saturated with brine (3wt% NaCl) for porosity measurement. The permeability of the brine-saturated core was determined from Darcy's law.

Foam flooding All core floods were conducted at 35 °C and 1500 psi back pressure. Three injection strategies were examined without using the pre-generator. These were simultaneous injection of CO_2 and surfactant solution, alternating injection, and CO_2 continuous injection with dissolved CO_2-soluble surfactant. Except for the last one, surfactants were always injected with brine even though the novel surfactant is CO_2-soluble. It will partition into CO_2 instantaneously when two phases contact. The impact of delivery media on the novel foam performance is out of the scope of this study and will be discussed in a separate publication. To obtain a fixed injection foam quality of 75% for co-injection, the injection rates of the surfactant solution (containing 0.2 wt% surfactant) and CO_2 were fixed at 0.1 cc/min and 0.3 cc/min, respectively. Through

adjusting injection time individually, slug sizes of the surfactant solution and gas in alternating injection were kept at 0.1 PV and 0.2 PV, respectively. For the third novel strategy, 0.6 cc/min was employed for CO_2 injection. The surfactant needed in CO_2 in the container was determined by the known container volume, CO_2 density (0.494 g/cc under experimental conditions), and fluid injection rates, to maintain the mass injection rate the same as in other scenarios. After calculation, 0.1wt% in CO_2 was used to maintain the same amount of surfactant per minute to be injected in different strategies. Pressure drops over the three sections of the core were recorded. Water saturation was determined based on the difference in cumulative mass between the injected and the produced waters.

3 Simulation description

3.1 Reservoir model

A 15° sector of a cylindrical homogenous reservoir, 100 ft thick and 440 ft in radius, was used for all simulations in this work. Porosity is 20%. The vertical and horizontal permeabilities are 400 and 200 mD, respectively. The reservoir model was numerically constructed using 100 grid blocks in the radial direction and 20 grid blocks in the vertical direction. A vertical injector is placed at the center of the reservoir and fully completed over 100 ft along, while a parallel fully penetrating producer is placed in the outer boundary grids whose permeability is set to 10,000 Darcy to simulate an open boundary reservoir and prevent artificial gas back flow (Namdar Zanganeh and Rossen 2013). The radial grid size increases from 3 ft for the first 30 grids from the injector to 5 ft for the remaining grid blocks. Leeftink et al. (2013) found that a fine grid resolution near the injection well is important to prevent underestimating the effects of dry-out increasing injectivity in a SAG process in finite-difference simulations. The schematic is shown in Fig. 3. The reservoir is isothermal at 35 °C, and the initial reservoir pressure is 1500 psi. For the sake of simplification, only the water phase is present in the reservoir initially (Kloet et al. 2009) since foam is only beneficial to sweep efficiency. All simulations were conducted with the Computer Modeling Group's STARS simulator.

The heuristic foam model built in CMG/STARS has been introduced in the literature (Zeng et al. 2016) and widely used in foam process simulation (Farajzadeh et al. 2015). Model parameter values fitting through different algorithms have been discussed by several researchers (Ma et al. 2014; Rossen and Boeije 2015). In our current work, the surfactant concentration and dry-out effects are considered and some typical values are chosen, as shown in Table 1. Here, we chose 100 for *fmmob*, which is less than these employed by earlier researchers, such as 1000 (Rossen et al. 2006), 3000 (Rossen and Shen 2007; Jamshidnezhad et al. 2008a), or 5000 (Cheng et al. 2000; Rossen and Renkema 2007; Kloet et al. 2009), because we believe too strong foam used previously masked some details in the foam process, which is crucial for the novel foam. Meanwhile, it is also much less than the result of coreflood matching (Ma et al. 2013) because the foam in 3D is much weaker than in 1D, which has been demonstrated by Li et al. (2006). In simulations, we made *fmsurf* equal to the injected surfactant concentration (3.34×10^{-5}, molar fraction) (Hanssen et al. 1994; Rossen and Renkema 2007). A linear dependence between foam strength and surfactant concentration was chosen (*epsurf* = 1) (Rossen and Renkema 2007). Here, we set the

Table 1 Foam model parameters used in field scale foam process simulation

Parameter	*fmmob*	*fmsurf*	*epsurf*	*fmdry*	*epdry*
Value	100	3.34×10^{-5}	1	0.15	1

epdry regulates the slope of *krg* curve near *fmdry*

epsurf regulates foam strength for surfactant concentration below *fmsurf*

fmdry critical water saturation at which foam experiences significant coalescence

fmmob reference mobility reduction factor

fmsurf surfactant concentration for full-strength foam

fmdry as 0.15, which is more than irreducible water saturation. The reasons that we do not attempt to derive those parameters from laboratory scale history matching are threefold. The first, heretofore, the empirical model used in STARS is based on pseudo-steady-state assumption that the local equilibrium is achieved instantaneously without accounting for transient behavior of foam (Fisher et al. 1990; Rossen and Renkema 2007; Jamshidnezhad et al. 2008a). Therefore, it is suspected that the model may be more suitable for field scale simulation but not for laboratory scale. The second, there is no consensus on how to scale up foam behavior and corresponding parameters from laboratory to field. Hence, typical values are chosen. The third, it is of most importance to examine the relevant foam behavior discrepancy between different injection strategies and surfactants, rather than absolute performances, as long as the same parameter values are used.

3.2 Injection scheme

Injection schemes were composed of two modules, in which constant rate mode is examined followed by

(a)

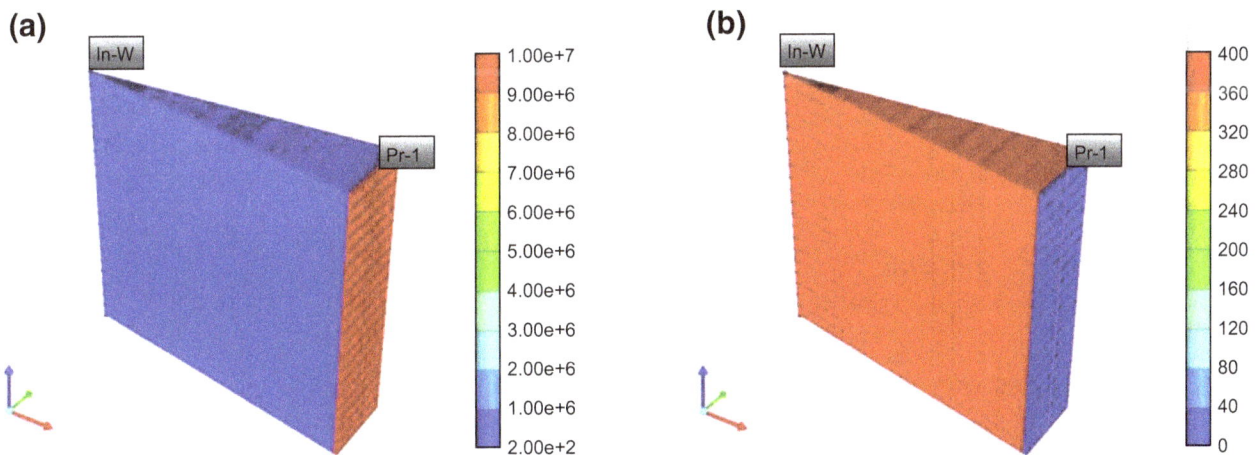

(b)

Fig. 3 Cylindrical reservoir model employed in field scale simulations. **a** Permeability I and J directions. **b** Permeability K direction

constant pressure mode. In each mode, SAG, co-injection, and CO_2 continuous injection with dissolved CO_2-soluble surfactant are presented, as summarized in Table 2. Relative to a previous publication (Le et al. 2008), it is the first time that the measured surfactant partition coefficients of CO_2-soluble surfactant between two phases (Ren et al. 2014) have been used in field scale simulations.

The first strategy is the alternating injection of the surfactant solution and gas (surfactant alternating gas, SAG), in which the novel surfactant is injected with brine even though it is CO_2-soluble. The liquid/gas slug size ratio is kept at 1:1 in volume. Two different slug sizes are tested, 36.5 and 182.5 days, respectively.

Then, co-injection is examined, either two phases in the same intervals (simultaneous injection through all perforation, SIAP, or simultaneous injection through partial perforation, SIPP), or water into the upper part while gas into the lower part (separate injection no barrier, SINB, or separate injection with barrier, SIWB). Stone (2004a, b) proposed injection of water in an interval above the gas to increase reservoir sweep, which is called "modified SWAG" by Algharaib et al. (2007). The main goal of separate injection is to reduce the effect of gravity segregation commonly encountered in gas–liquid flow in reservoirs with high vertical communication (Rossen et al. 2006; Liu et al. 2011).

A schematic of the four strategies is shown in Fig. 4. For constant pressure mode, only the best case selected from the constant rate mode, SINB, is displayed.

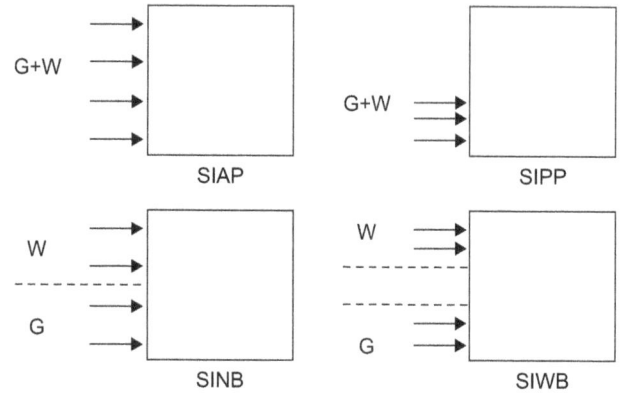

Fig. 4 Four different injection strategies for simultaneous injection of the surfactant solution and CO_2 (*G* gas, *W* water)

At last, the novel strategy, CO_2 continuous injection with dissolved CO_2-soluble surfactant, was conducted for both modes with variable perforations. This is a unique one in which the surfactant concentration in CO_2 after splitting between phases during injection should be lower than its maximum solubility (Ren et al. 2014).

Table 2 summarizes the design parameters for the injection strategies described above. CO_2 and water injection rates are chosen so as to achieve approximately 75% foam quality under reservoir conditions. Doubled injection time is employed to inject the same amount of fluids. The selection of injection pressure at the constant pressure injection mode is discussed in details in the corresponding section.

Table 2 Simulation scenarios for variable injection modes and strategies

Injection strategies			Water/CO_2 cycle cycle ratio	Water cycle, day	Water rate, bbl/d	CO_2 rate, scf/d	Water injection pressure, psi	CO_2 injection pressure, psi	Injection surfactant concentration, molar fraction	Total injected, PV/ Years
Constant rate mode	Alternating injection	SAG	1:1	36.5 and 182.5	45	90,000	–	–	3.34×10^{-5}	5.76/16
	Co-injection	SIAP	–	–	45	90,000	–	–	3.34×10^{-5}	5.76/8
		SIPP	–	–	45	90,000	–	–	3.34×10^{-5}	5.76/8
		SINB	–	–	45	90,000	–	–	3.34×10^{-5}	5.76/8
		SIWB	–	–	45	90,000	–	–	3.34×10^{-5}	5.76/8
	Novel CO_2 injection	All perforation	–	–	–	115,714.3	–	–	9.54×10^{-5}	5.76/8
		Partial perforation	–	–	–	115,714.3	–	–	9.54×10^{-5}	5.76/8
Constant pressure mode	Alternating injection	SAG	1:1	36.5 and 182.5	–	–	1547	1581	3.34×10^{-5}	5.76/16
	Co-injection	SINB	–	–	–	–	1598	1608	3.34×10^{-5}	5.76/8
	Novel CO_2 injection	All perforation	–	–	–	–	–	1585	9.54×10^{-5}	5.76/8

We employ the CO_2 storage, R_g (gravity segregation length), and CO_2 utilization ratio as evaluation criteria. The CO_2 storage is defined as CO_2 staying in the reservoir at the end of injection under surface conditions, which directly reflects the sweep efficiency. CO_2 utilization ratio is defined as the ratio of CO_2 storage over cumulative CO_2 injection, which would be more useful to reflect the economic concern.

4 Results and discussion

4.1 Experimental results

4.1.1 Co-injection

The sectional pressure drops of two types of foams are shown in Fig. 5. For the conventional foam, a strong foam started to propagate into Sect. 2 at 8 IPV (total injected pore volume) and then Sect. 3 as late as 15 IPV. On the contrary, the novel foam displayed higher foam strength and earlier pressure response in each section, at 5 IPV and 12 IPV, respectively. However, this is more attributed to the superior and essential ability of the novel surfactant to stabilize the bubble film than improved foam propagation (Ren et al. 2013) or surfactant spreading effect (Zeng et al. 2016) since foam will not directly affect the liquid propagation and the surfactant already spreads to the whole core with current injection quality. In prior publications, it was found that intermediate partition ability of the novel surfactant could significantly improve foam propagation owing to higher mobility of the gas phase than the aqueous phase. However, too high a partition coefficient may adversely impact foam propagation due to the local surfactant concentration being lower than the critical value, which is the so-called spreading effect. Nevertheless, this

effect may play a less important role compared with surfactant stabilization capacity on bubbles since more than 1 PV liquid has been injected when a significant pressure drop is observed. Adkins et al. (2010b) reported that the CO_2-philic nonionic surfactant used here could lower the gas/water interfacial tension to 5.6 mN/m (2000 psia @ 24 °C, 0.01wt%). This remarkably outruns highly commercialized surfactant CD-1045 that has an interfacial tension of 9.5 mN/m under similar conditions (Grigg 2004). Specifically, with approximate calculation, the apparent viscosity exhibited by the tested novel surfactant was either comparable (Chen et al. 2015) or at least two magnitudes higher than other CO_2-soluble surfactants (Sanders et al. 2010; Xing et al. 2012; McLendon et al. 2014). Outstanding stability of a bubble film contributes the earlier pressure drop response and stronger foam, which is also revealed by water saturation curves, shown in Fig. 6. For the novel foam, preceding the strong foam presence, the weak foam begin to displace water immediately after gas breakthrough and, then, gradually toward the residual water saturation (0.2). On the contrary, for the conventional foam, a lack of high pressure drop causes the displacement curve to level off earlier and the residual water saturation is much higher (0.46). Relative to carbonate, sandstone tends to give more uniform pore size distribution and smaller pores, which contributes to higher water saturation at gas breakthrough and residual values.

4.1.2 Alternating injection

Figure 7 demonstrates the sectional pressure drop performances for conventional and novel foams. For both types of foams, we observed that the pressure drops built up during liquid injection and declined in the following gas injection, then they tended to fluctuate from cycle to cycle (Lawson and Reisberg 1980). As we observed above for

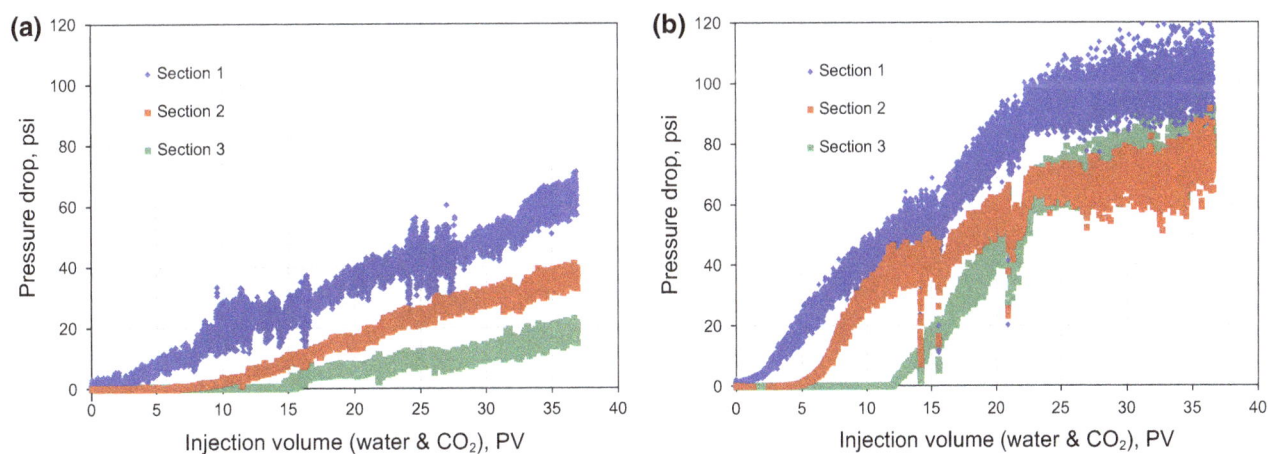

Fig. 5 Pressure drops across the core during simultaneous injection. **a** Conventional foam. **b** Novel foam

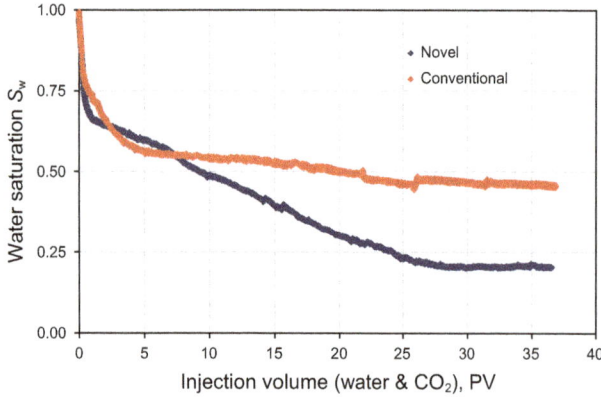

Fig. 6 Average water saturation in the core during simultaneous injection

Fig. 8 Average water saturation in the core during alternating injection

co-injection, here the contrast became more significant both in pressure drop magnitude and in strong foam propagation. For the novel foam, it only took 1 IPV into Sect. 2 and then 5 IPV to reach Sect. 3. On the contrary, double the amount of fluids were required for the conventional foam to obtain some response. It is well known that the sandstone holds a negative surface charge under the normal formation pH (6–7.5). In turn, the anionic (CD 1045) and nonionic (novel surfactant) surfactants should be close to the adsorption level without the presence of a large amount of clay. Therefore, the novel CO_2-soluble surfactant really improves the strong foam propagation without any concern about surfactant adsorption. Meanwhile, the magnitude of the pressure drop for the novel foam is 5 to 10 times higher than that of the conventional foam, which also proves the superior ability of the CO_2-soluble surfactant to stabilize the bubble film over only aqueous soluble surfactant. Furthermore, the apparent viscosity of the novel foam achieved here is either comparable to (Chen et al. 2015) or almost two magnitudes higher than those in

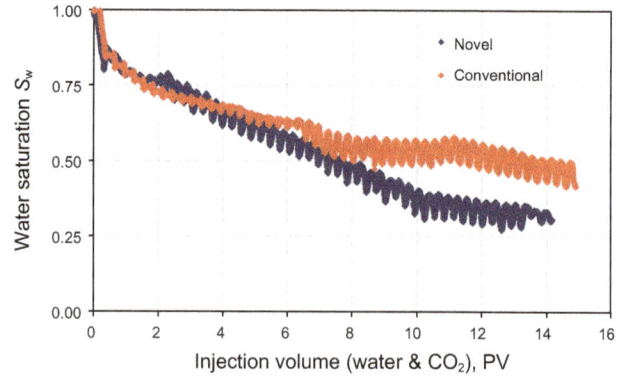

the literature with other CO_2-soluble surfactants (Xing et al. 2012; McLendon et al. 2014; Sagir et al. (2014c, d).

Correspondingly, the residual water saturation after foam propagation has been lowered from 0.42 (conventional) to 0.31 (novel), as shown in Fig. 8, and the displacement efficiency has been improved almost 20%. At the same time, we also notice that relative to simultaneous injection (Fig. 6), the alternating injection does promote the foam generation and injectivity (Li and Rossen 2005) for both types of foams indicated by the quick strong foam propagation and lower pressure drops.

4.1.3 CO_2 continuous injection with dissolved CO_2-soluble surfactant

The pressure drops and water saturation in the displacement process are shown in Fig. 9. As mentioned in the prior section, the same amount of surfactant per time was injected in the co-injection and the current novel strategy, which eliminated possible bias during comparison. Propagation of the strong foam was accordingly observed in

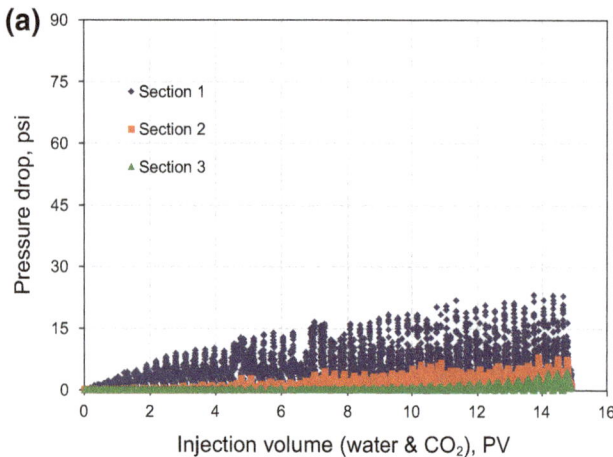

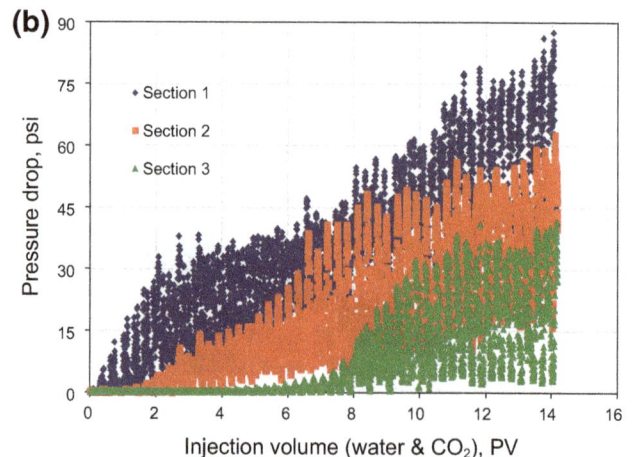

Fig. 7 Pressure drops across the core during alternating injection. **a** Conventional foam. **b** Novel foam

every section. As early as 0.5 IPV, the strong foam propagated into Sect. 2 and then toward into Sect. 3 after 1.1 IPV. Much quicker foam propagation is attributed to the ability of the surfactant to dissolve in CO_2 without interference from water injection as well as inlet gas trapping. Foam collapsed when water saturation reached the critical value regionally. However, a huge amount of gas trapped in the core indicated by the residual pressure drops as high as 4 psi across the whole core is beneficial enough to the gas mobility control. Correspondingly, after gas breakthrough and weak foam propagation at 0.5 IPV, the strong foam developed in following sections drops the water saturation to 0.25 as early as 2.4 IPV and then levels off. This novel injection strategy really displays the superior surfactant transportation and foam propagation ability of this CO_2-soluble surfactant. In addition, the foam strength here in apparent viscosity was at least one magnitude higher than published data of other soluble surfactants (Xing et al. 2012; McLendon et al. 2014), which confirmed the superior capacity to stabilize the bubbles by the currently employed novel surfactant.

4.2 Simulation

4.2.1 Constant rate injection mode

4.2.1.1 Alternating injection   From the gas production rate file (Fig. 10), early CO_2 breakthrough due to gravity segregation can be clearly observed at almost the same time for both types of foams. However, thereafter, the gas production rates differ significantly. For the conventional foam (zero partition coefficient), the gas rate abruptly increases and almost levels off quickly; while it decreases for the novel CO_2-soluble surfactant foam as the partition coefficient becomes nonzero. The distributions of gas saturation (Fig. 11a, b) in the reservoir at the end of injection

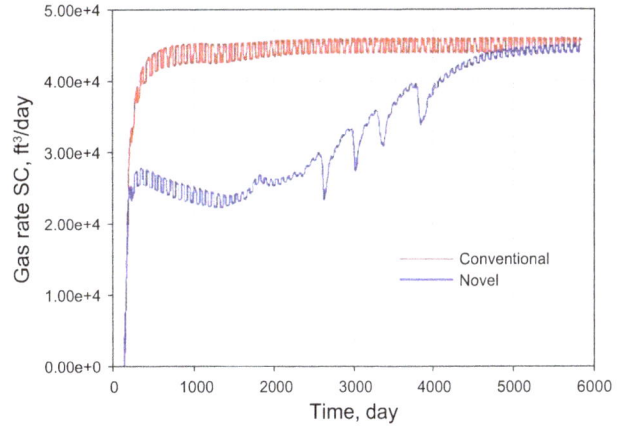

Fig. 10 Gas production rate of small slug size (36.5 days)

show that once gas reached the edge of the surfactant front at the top of the reservoir, it rapidly segregated upwards and reaches the production well in a thin override zone (Rossen and Renkema 2007). On the contrary, the novel foam increases the override zone dramatically even though the two types of foams gave the same gravity segregation length (R_g), which were read from profiles approximately as 130 ft. The tremendously different performances of the two foams essentially come from surfactant properties and can be seen on the concentration distribution profiles (Fig. 12a, b). There is a surfactant vacuum zone in the top layer for the conventional foam, while it exists only near the wellbore for the novel foam. The key reason of fluid segregation leading to the foam process losing efficiency is not only gas override but also surfactant slumping with water. Rossen and Renkema (2007) obtained the similar observation that surfactant has slumped toward the bottom of the reservoir and there was no surfactant ahead of the foam front at the top of the reservoir. The novel surfactant would be chased by the following gas slug and is delivered to the top layer with CO_2 override. Then, the gas mobility

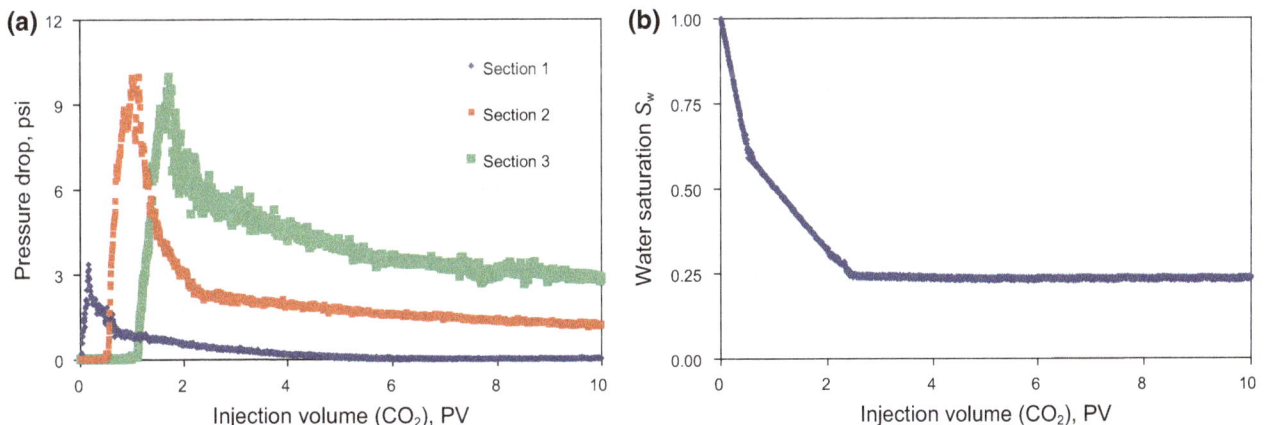

Fig. 9 Pressure drops across the core (**a**) and average water saturation in the core (**b**) during CO_2 continuous injection with dissolved CO_2-soluble surfactant

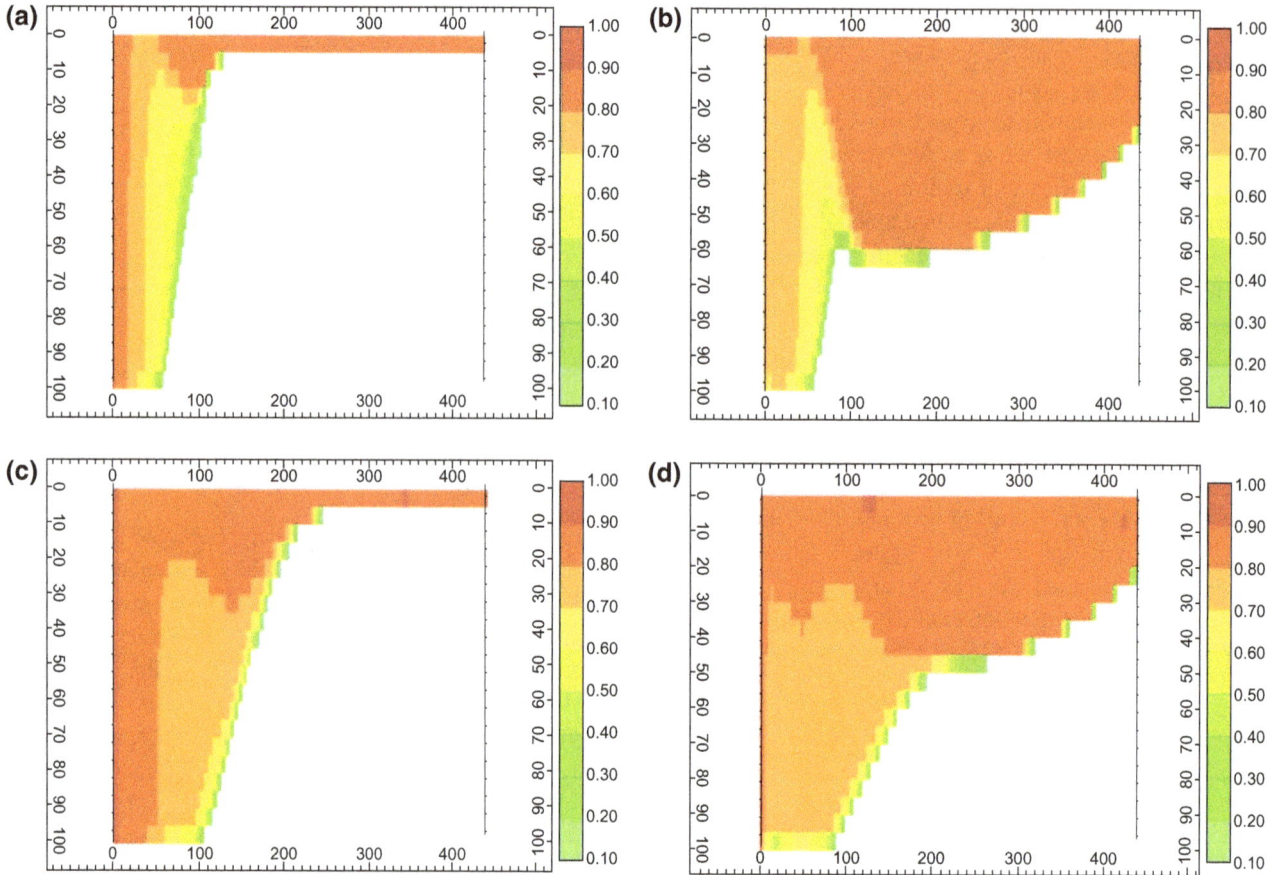

Fig. 11 Gas saturation of small slug size (36.5 days) for **a** CD1045, **b** novel surfactant and of larger slug size (182.5 days) for **c** CD1045, **d** novel surfactant during alternating injection

in the override zone has been reduced greatly and effective diversion occurred. Furthermore, this particular surfactant partitioning improves not only sweep efficiency, but also well injectivity with significantly reduced surfactant concentration near the wellbore. Here, one may have the suspicion that whether the superiority of the novel foam comes from the foam model effect. It is true that we employ the surfactant concentration in the water phase as the scale for gas mobility reduction in the simulation. Theoretically, we should supervise the concentration in the whole cell (*global*) because surfactant will act at the interface regardless of its partition in the gas phase. In reality, a comparison between corresponding plots (Figs. 12 vs. 13) tells us they display exactly the same trend except for the magnitude, which is attributed to the mass conservation and constant partition coefficient of injected fluids. Hence, in the following parts, we will only employ and illustrate the surfactant concentration in the aqueous phase.

With a larger injection cycle (182.5 days), for the conventional foam (Fig. 11a, c), an increase in the slug size significantly improves the vertical sweep efficiency by extending the distance R_g that the injected gas–water mixture flows before complete segregation but at the cost

of much lower injectivity (Ren et al. 2013). It was observed that the average bottom-hole pressure (BHP) of both foams increased with larger slug size (Ren et al. 2013). Indeed, the well bottom-hole pressure decreases for the novel foam over the conventional one (Fig. 14). It has been established that R_g increases with foam strength (i.e., reduced total relative fluid mobility, λ_{rt}) for the conventional foam as described by Eq. (2) (Rossen et al. 2006). Therefore, with higher injection pressure, the larger slug size yields stronger foam because of larger contact between the injected CO_2 and surfactant slugs. It is much clearer to observe the variation of R_g from Fig. 12a, c with the shrinkage of the low surfactant concentration zone in the top layers for the conventional foam. However, a comparison between Fig. 11b, d reveals that the novel foam is actually weaker with larger slug size as indicated by the overall reduction in the gas saturation. Different from the performance of CD1045, a larger slug size expands the low surfactant concentration area near the wellbore for the novel foam (Fig. 12b, d), which tends to lower the injection pressure and may decrease the sweep efficiency. Reservoir pressure distribution, as shown in Fig. 15, can be a more direct way to correlate surfactant transport with foam

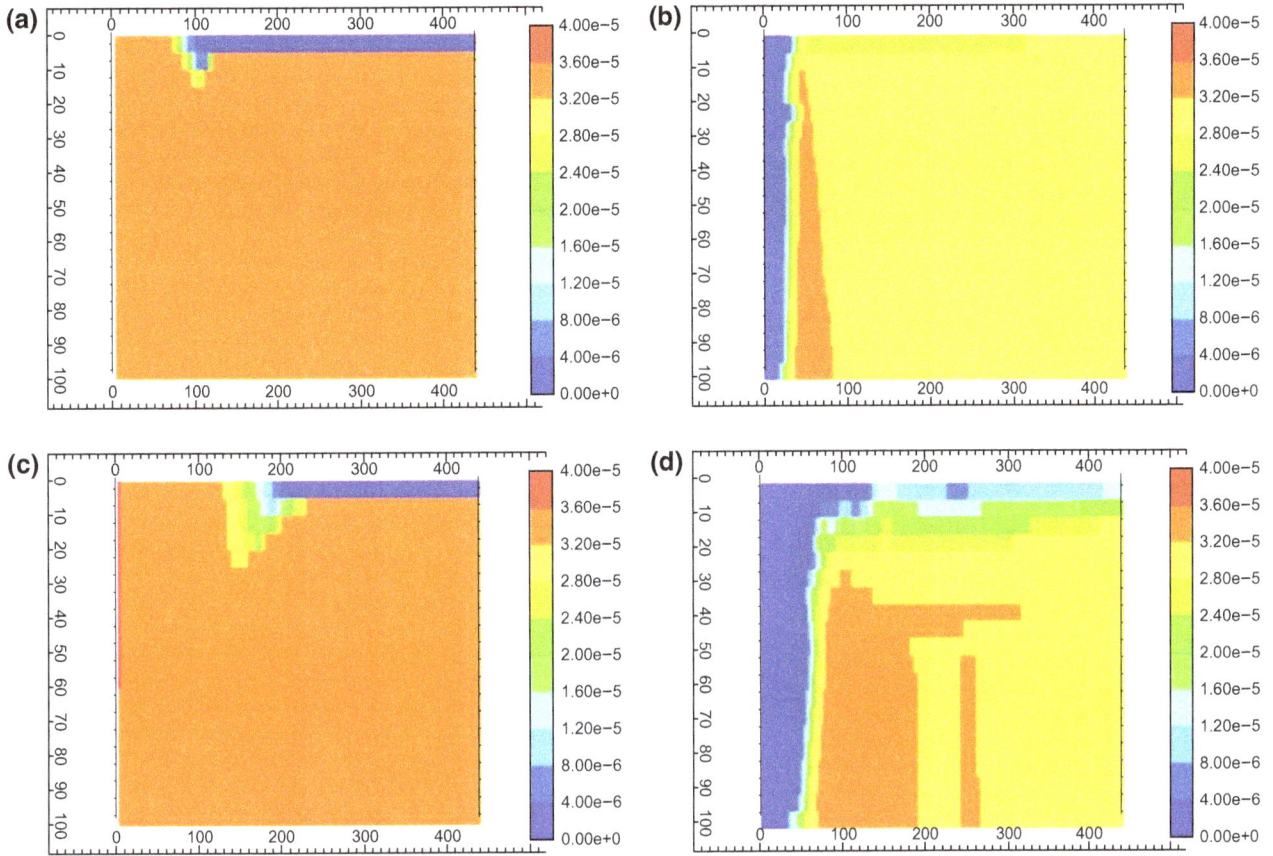

Fig. 12 Surfactant concentration in the water phase of small slug size (36.5 days) for **a** CD1045, **b** novel surfactant and of larger slug size (182.5 days) for **c** CD1045, **d** novel surfactant during alternating injection

propagation and fluid redistribution. For the conventional surfactant, the high pressure gradient is concentrated only within the near-wellbore region and expands somewhat from the wellbore as the fluid cycle increases (Shan and Rossen 2004). However, it spreads much further into the reservoir for CO_2-soluble surfactant indicating by a more even pressure gradient distribution. The variations of CO_2 storage and R_g with slug size are summarized in Table 3 as well as Fig. 16. With slug size increasing, the gravity segregation lengths do enlarge for both types of foams. However, the sweep efficiency varies differently, 11% reduction for the novel foam and 111% improvement for the conventional one. Therefore, we confirm the previous conclusion for the conventional foam the larger slug size was beneficial to the foam process for cylindrical homogeneous reservoirs (Shan and Rossen 2004; Rossen and Shen 2007; Rossen and Renkema 2007). For the novel foam, insensitivity to injected slug size gives less restriction for operation. With CO_2-soluble surfactant, the contradiction between gravity segregation length and CO_2 storage tells us again that R_g is only one criterion for fighting gravity segregation and not the sufficient condition.

$$R_g = \sqrt{\frac{Q}{\pi k_z (\rho_w - \rho_g) g \lambda_{rt}^m}} \qquad (2)$$

4.2.1.2 Co-injection

(1) *Water and gas injection through the same intervals (SIAP and SIPP)*

Figure 17 shows gas production rates for two different simultaneous injection strategies with distinct perforation locations. For the full completion (SIAP), both the novel and conventional foams give unsatisfactory performances even though the former is fluctuating to approach the steady state after gas breakthrough in a short time. With partial completion in the lower interval (SIPP), the improvement in the conventional foam is almost unnoticeable, while a distinct reduction in gas production is showed by CO_2-soluble surfactant foam. More clear comparisons can be seen in gas saturation distributions (Fig. 18). For SIAP, the injected gas concentrates highly in the top layers. The huge difference between gas and water mobility contributes to the poor performance for both foams. Eventually, from this point of view, the novel surfactant makes the situation worse because the gas

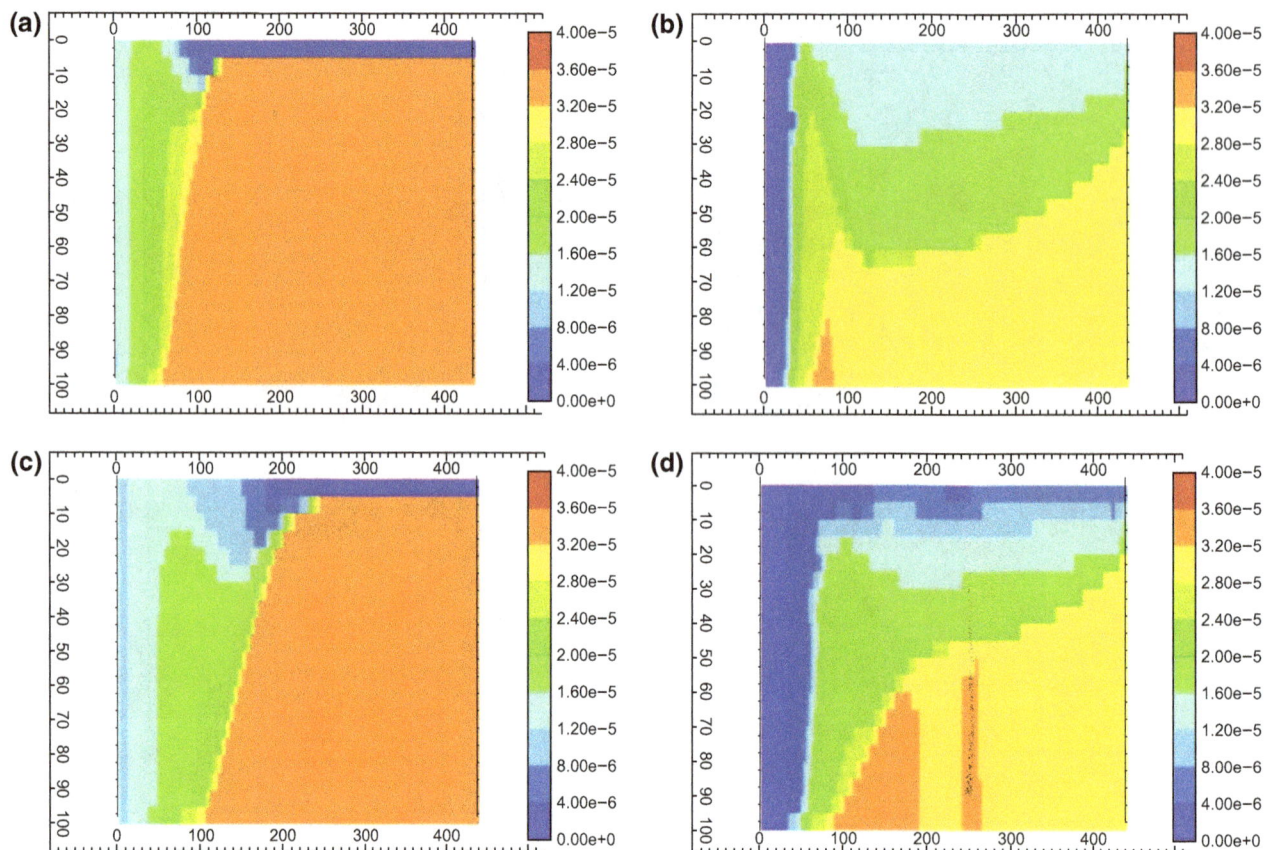

Fig. 13 Global surfactant concentration of small slug size (36.5 days) for **a** CD1045, **b** novel surfactant and of larger slug size (182.5 days) for **c** CD1045, **d** novel surfactant, during alternating injection

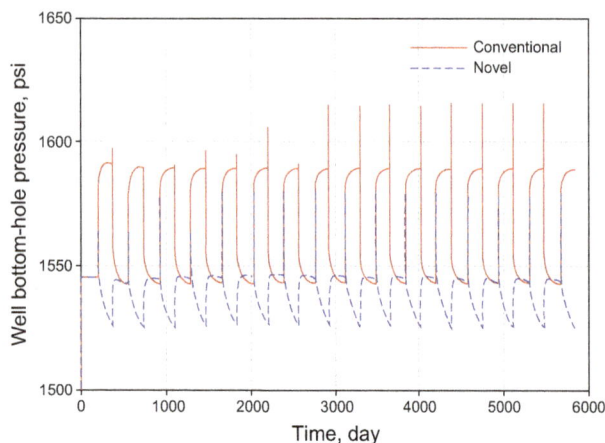

Fig. 14 Gas well BHP of large slug size (182.5 days)

extraction effect CO_2-soluble surfactant decreases the surfactant concentration on the gas escape path further. Even though the chased surfactant is transported to the top layer far from the injector, now it fails to reduce the gas mobility because too much gas flow results in a very low surfactant concentration, as shown in Fig. 19a, b, which fails the effective gas diversion. On the other hand, for

SIPP, injection from the lower part does improve gas storage for the conventional foam near the wellbore (Fig. 18c). A conventional aqueous soluble surfactant will slump with water and a lack of ability to migrate to the top layer will not heal the surfactant scarce zone (Fig. 19c). Oppositely, the novel foam highly expands the override zone vertically (Fig. 19d) since the increased water and gas contact will retain more CO_2-soluble surfactant in the upper zone and the gas escape path does not exist anymore (Fig. 19d). Meanwhile, we confirm that the partial perforation does significantly increase the bottom-hole pressure relative to the full completion (Rossen et al. 2006; Rossen and Shen 2007; Jamshidnezhad et al. 2008b) even though the novel foam always gives a lower value, as shown in Fig. 20. In addition, compared with alternating injection, simultaneous injection does lower the injectivity for the conventional foam (Rossen et al. 1995) while this problem has been greatly mitigated with the novel foam (SIAP) even though the sweep efficiency is relatively poor. The pressure distribution in the reservoir, Fig. 21, provides a direct evidence for injection strategy screening. High pressure gradient compresses near the wellbore for both completion schemes although the novel foam is still

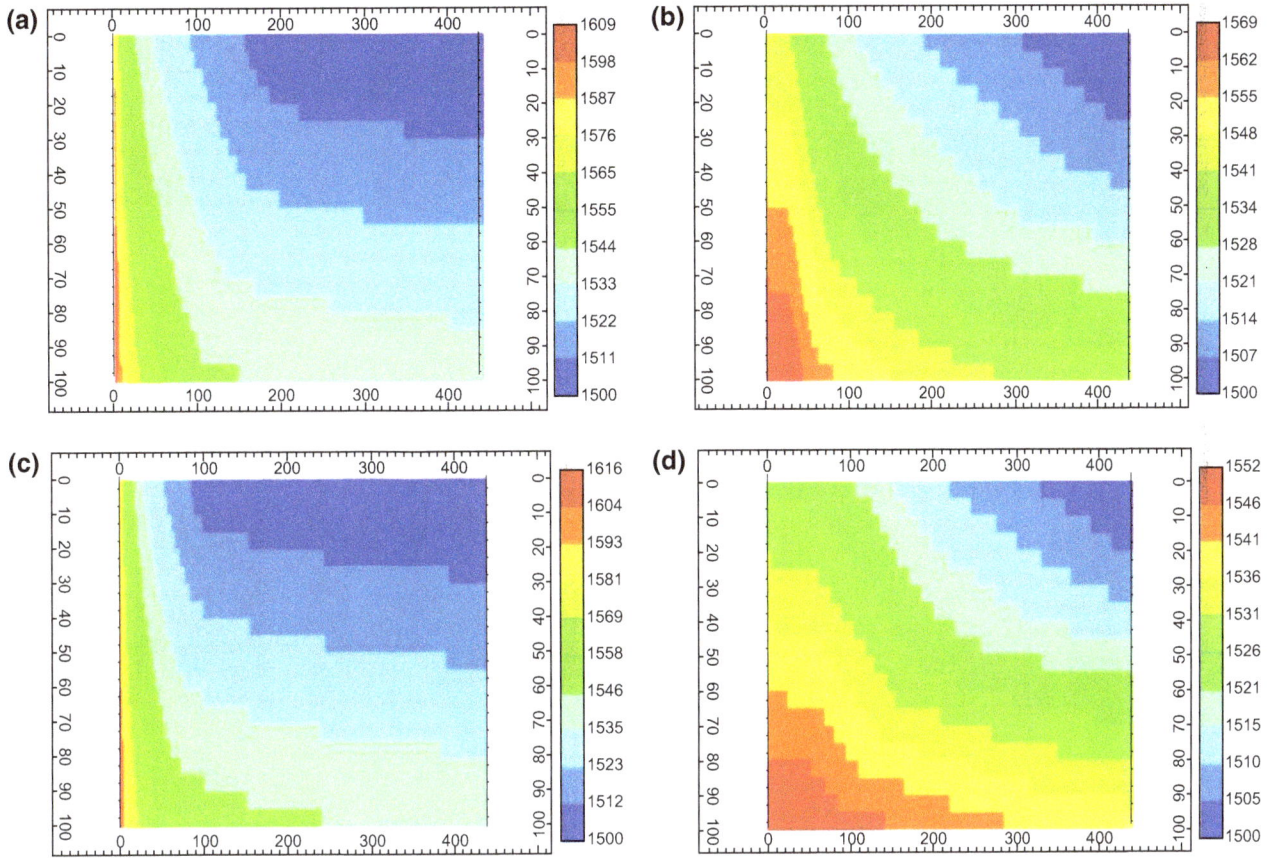

Fig. 15 Pressure distribution of small slug size (36.5 days) for **a** CD1045, **b** novel surfactant and of larger slug size (182.5 days) for **c** CD1045, **d** novel surfactant during alternating injection

Table 3 CO_2 storage and gravity segregation length

	CO_2 storage, 10^7 scf		R_g, ft	
	36.5-day slug size	182.5-day slug size	36.5-day slug size	182.5-day slug size
Novel	7.36	6.54	130	200
Conventional	1.16	2.45	130	165

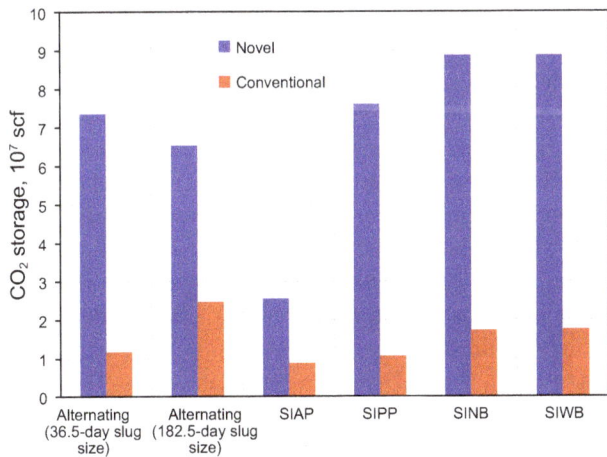

Fig. 16 Summary of CO_2 storages for alternating and simultaneous injection with constant rate injection mode

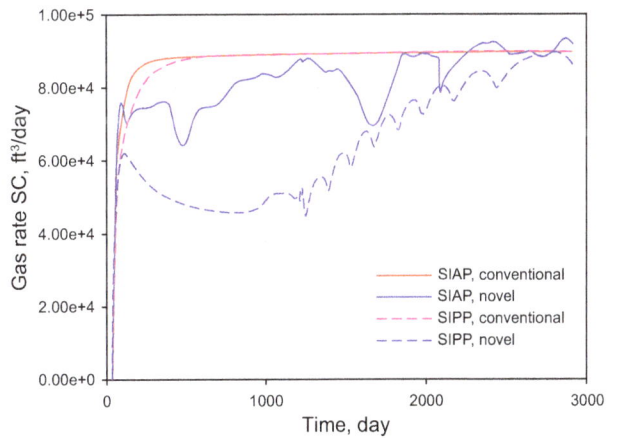

Fig. 17 Gas production rate during SIAP and SIPP

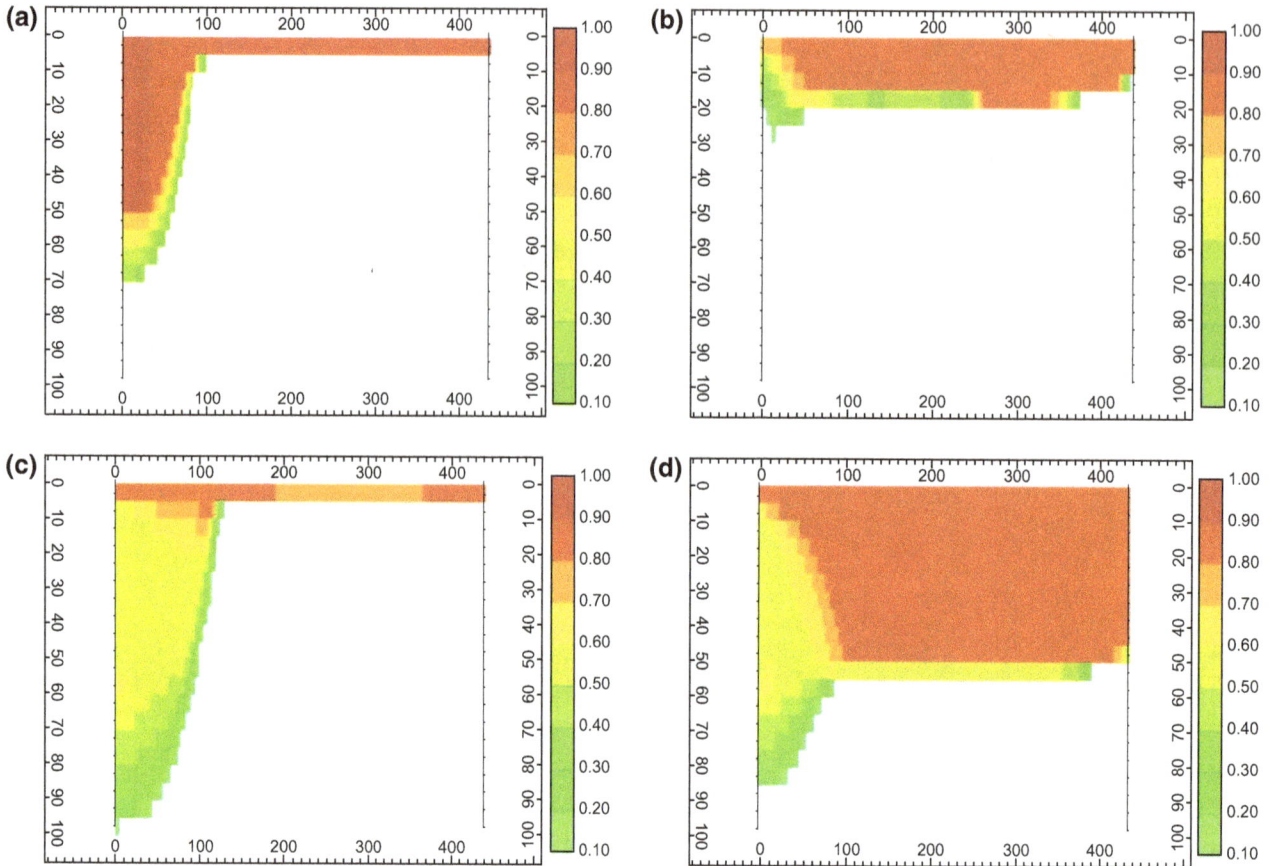

Fig. 18 Gas saturation during SIAP of **a** convectional foam and **b** novel foam and during SIPP of **c** conventional foam and **d** novel foam

superior over the conventional one with respect to sweep efficiency and injectivity. This is in accordance with the previous conclusion that relative to SAG with constant injection rate, a foam process with continuous foam injection performs even worse, because most of the well-to-well pressure drop was dissipated in the near-well region (Shi et al. 1998; Shan and Rossen 2004; Rossen and Shen 2007). The results in this case tell us gas override takes precedence in importance over water slumping in fighting gravity segregation (Shan and Rossen 2004).

Table 4 summarizes CO_2 storage (Fig. 16) and gravity segregation length for SIAP and SIPP. It is obvious that the novel foam gives much higher sweep efficiency than the conventional one. Here, we do confirm the close R_g with alternating injection for SIPP, but not for SIAP, which is expected as the cycle size decreases (Shan and Rossen 2004). In addition, we do not reach the conclusion that R_g for constant rate injection is not sensitive to the simultaneous injection of gas and water into either a partially (SIPP) or a fully completed well (SIAP) (Rossen et al. 2006; Rossen and Shen 2007; Jamshidnezhad et al. 2008b).

(2)   *Injection of water into the top part and gas into the bottom part (SINB and SIWB)*

Figure 22 shows the gas production rates for another two different simultaneous injection strategies with water injection through the top and gas into the bottom (SINB and SIWB). Except for injectivity (Fig. 23), it is hard to tell the difference between partial and full completions as regards of gas production rate, gas saturation, and surfactant distribution for both foams. Therefore, only plots of SINB are shown for those parameters. For the conventional foam, again, injectivity reduction is observed as a typical characteristic of the partial completion. On the other hand, the novel foam really reduces the difference caused by perforation locations, but not to the same extent as in the above two strategies (SIAP and SIPP, Fig. 20). The gas–water mixed zones, as shown in Fig. 24, do not expand vertically for the conventional foam, while opposite is observed for the CO_2-soluble surfactant foam. The injection of water above gas increases the travel distance for both gas and water in the vertical countercurrent flow (Rossen et al. 2006) which is in turn resisted by foam

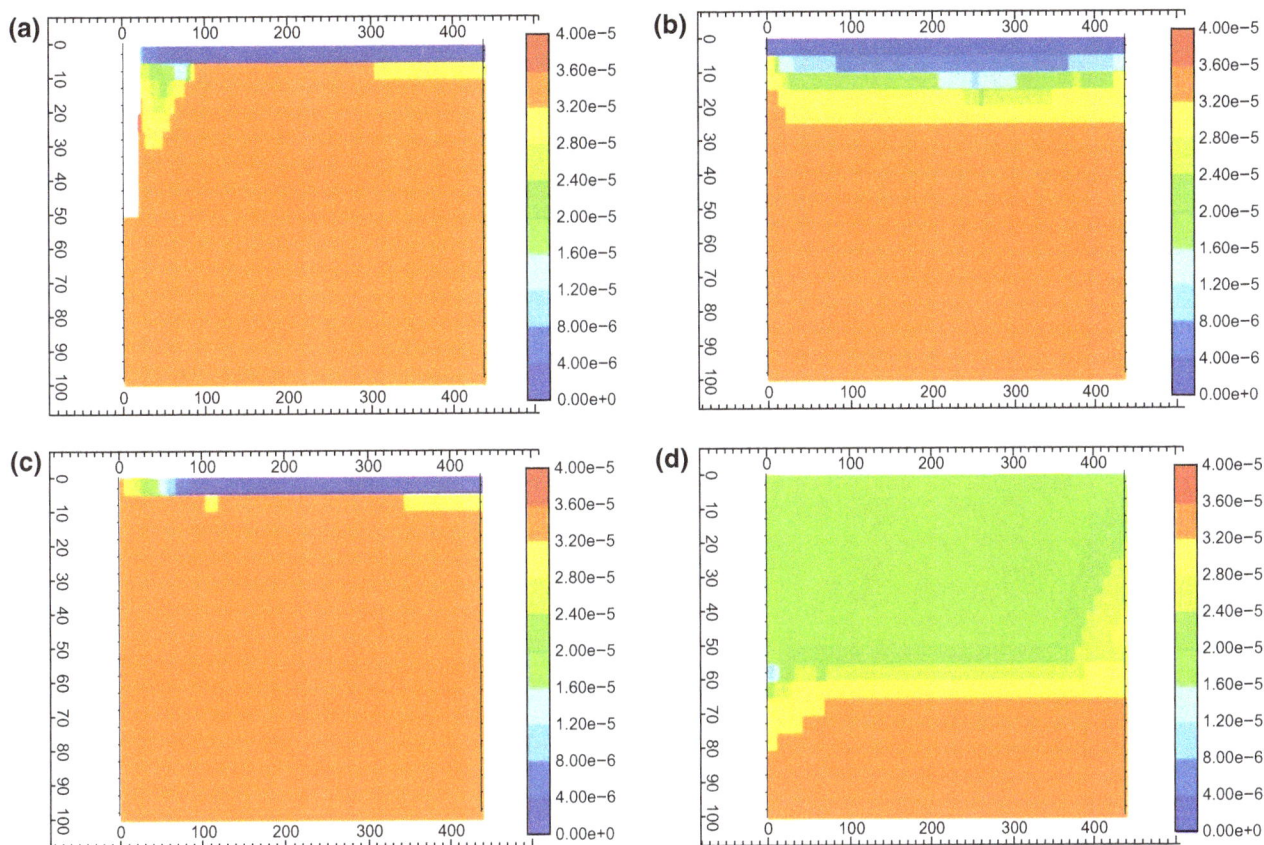

Fig. 19 Surfactant concentration during SIAP of **a** convectional foam and **b** novel foam and during SIPP of **c** conventional foam, **d** novel foam

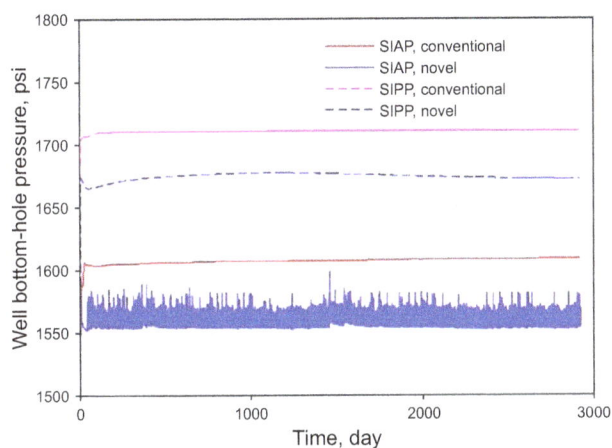

Fig. 20 Well bottom-hole pressures during SIAP and SIPP

formation. The advantage is further enhanced with the CO_2-soluble surfactant as this injection strategy allows more surfactant to be carried with CO_2 into the override zone. As a result, better foam propagation in the upper part of the reservoir can be achieved. This is illustrated by the surfactant concentration distribution (Fig. 25) and pressure distribution in the reservoir (Fig. 26). Now, relative to the conventional foam, the novel surfactant extends the high pressure gradient much further into the reservoir with more even pressure distribution.

Table 4 also summarizes CO_2 storage (Fig. 16) and gravity segregation lengths for those two separate injections. It is obvious that the novel foam still gives much higher sweep efficiency than the conventional one. We do achieve the same R_g for SINB and SIWB as well as the higher injection pressure for the latter (Rossen et al. 2006; Rossen and Shen 2007; Jamshidnezhad et al. 2008b). However, the sweep efficiency improvement was not remarkable, particularly for the novel foam. Relative to water and gas injection through the same intervals (SIAP and SIPP), the distance to the point of complete segregation R_g increases by a factor of about 1.5 and higher injectivity (Figs. 20, 23) has been achieved for separate injection (SINB and SIWB). This result agrees with the theoretical prediction of R_g as a function of water fractional flow reported in the literature (Rossen et al. 2006; Rossen and Shen 2007). From above analysis we can find, for novel foam, gravity segregation length is a less precise representative parameter of sweep efficiency.

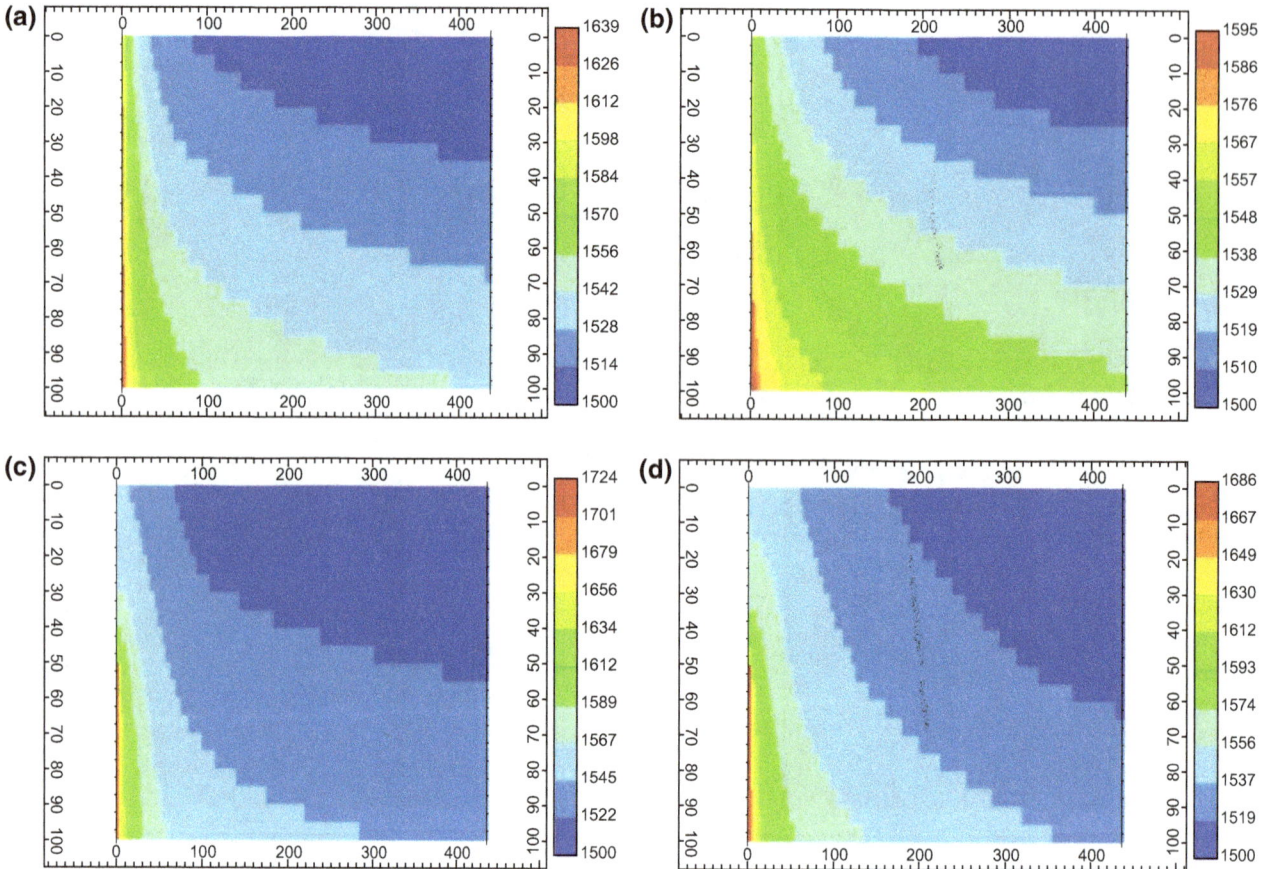

Fig. 21 Pressure distribution during SIAP of **a** convectional foam and **b** novel foam and during SIPP of **c** conventional foam, **d** novel foam

Table 4 Comparison of CO_2 storage and gravity segregation length among alternating and simultaneous injection

	CO_2 storage, 10^7 scf		R_g, ft	
	Novel	Conventional	Novel	Conventional
Alternating injection (36.5-ay slug size)	7.36	1.16	130	130
SIAP	2.54	0.874	54	30
SIPP	7.61	1.06	115	115
SINB	8.9	1.73	215	190
SIWB	8.9	1.74	215	190

4.2.2 Constant pressure injection mode

The performance difference between novel and conventional foams is illustrated below. Also, we will address the validation of the previous conclusion that this mode was more efficient than the constant rate mode and whether this is still valid for the novel foam (Shan and Rossen 2004)

4.2.2.1 Alternating injection
Let us look at alternating injection with a 36.5-day slug size first. In earlier studies (Shan and Rossen 2004; Rossen et al. 2006; Rossen and Shen 2007; Rossen and Renkema 2007; Kloet et al. 2009), conservation of injection fluids including water, gas, and surfactant was not maintained between two injection

modes, which causes the quandary that whether the superiority of constant pressure mode comes from more injected surfactant over constant rate mode. Therefore, we try to pursue the conservation through trial and error on injection pressure. However, owing to different injectivity between novel and conventional foams, we decide to use the values close to those for the conventional one for the first manipulation. 1581 psi for the gas well and 1547 psi for the water well were determined through strictly equalizing amounts of injection phases for the conventional foam between two modes. Then, the same pressures were used for the novel foam. It is straightforward that much more fluid will be injected for the novel foam owing to higher injectivity, as shown in Fig. 27a. The CO_2 storages are

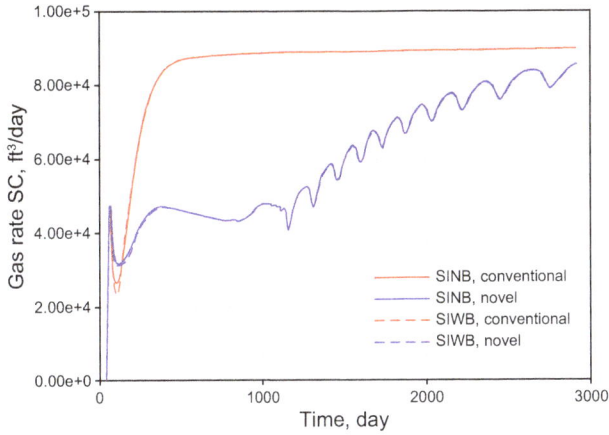

Fig. 22 Gas production rate during SINB and SIWB

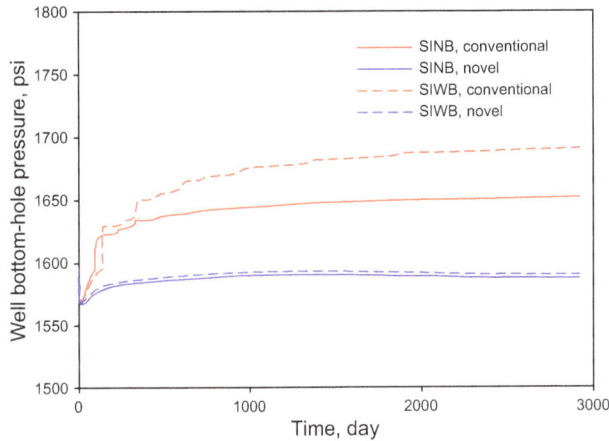

Fig. 23 Well bottom-hole pressure during SINB and SIWB

listed in Table 5, which are improved 3.6% and 50.7% for conventional and novel foams, respectively, relative to the constant rate mode. For the conventional foam, the constant pressure mode can increase the injection rate when the foam near the wellbore becomes weak owing to water

saturation approaching the critical value (Rossen et al. 1995); thus with relatively high mobility ahead of and behind the displacement front, a pressure-constrained SAG process can force the entire reservoir pressure drop into the region of low mobility at the displacement front, i.e., more even pressure drop distribution is expected instead of most of them dissipating in a short distance. However, our results show that the improvement is very limited, with respect to gas saturation (Fig. 28a), surfactant concentration (Fig. 29a), and pressure distribution (Fig. 30a). All of those are extremely similar to the results before with the constant rate mode (Figs. 11a, 12a). Hence, we deduce the injection mode may not be the crucial parameter as long as injection mass conservation is honored. On the other hand, from the point view of sweep efficiency, alternating injection with constant pressure mode tends to amplify the superiority of the novel foam over the conventional one, characterized by a vertically expanded gas saturation profile (Fig. 28b), uniform surfactant distribution (Fig. 29b), and much deeper extended high pressure gradient (Fig. 30b), even though more gas has been produced (Fig. 31a). In addition, it seems that the constant pressure mode does able the enhancement of the sweep efficiency tremendously for the novel foam (Figs. 11b, 28b), which deserves further discussion below. Meanwhile, unequal amount of gas injection requires us to employ another parameter, CO_2 utilization ratios, for the economic consideration, which are also listed in Table 5. Therefore, there is no question that for alternating injection, constant pressure injection mode is beneficial to conventional foam with respect to both sweep efficiency and CO_2 utilization ratio even though the improvement is trivial. For the novel foam, operators need to balance the extra profits from 50% sweep efficiency improvement against the ascending injection cost from 29% deducted gas utilization even though it is still more than four times higher than that of conventional foam.

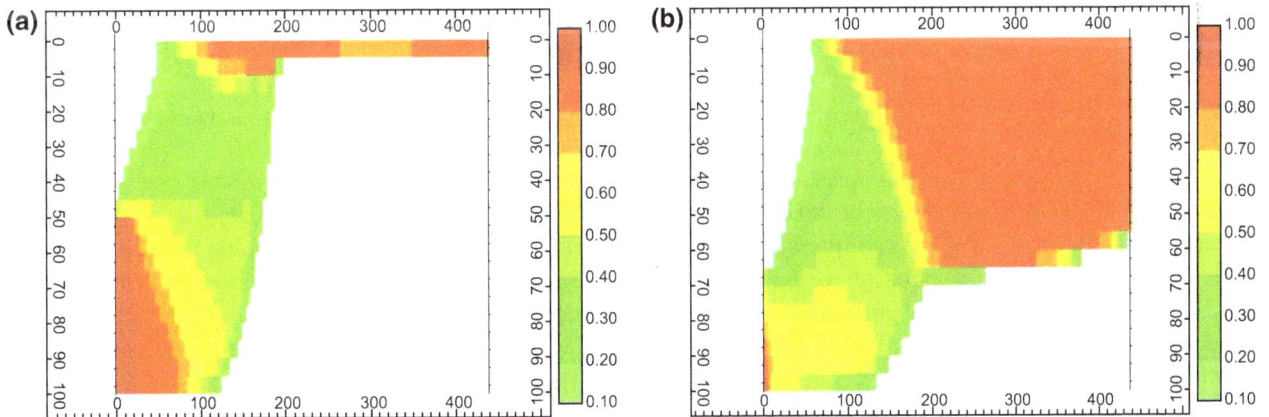

Fig. 24 Gas saturation during SINB of **a** conventional foam and **b** novel foam

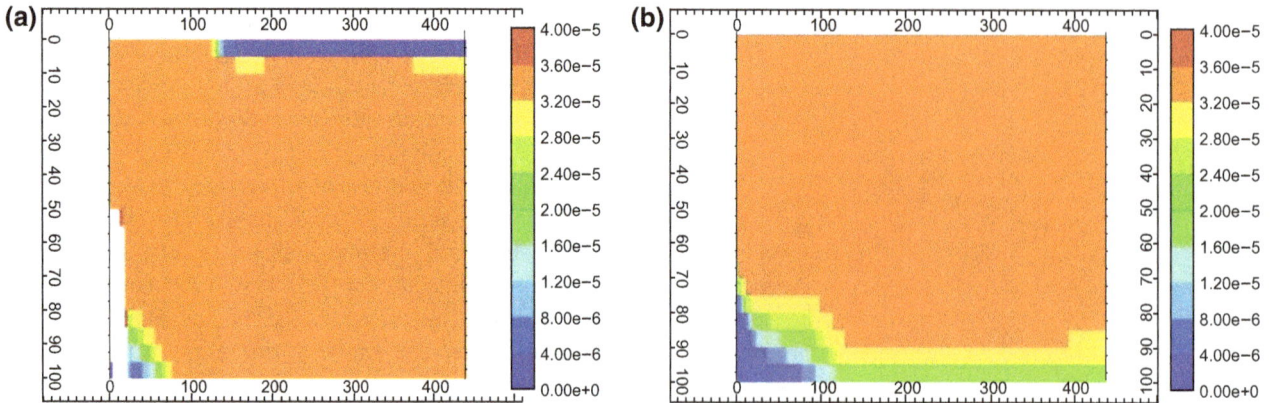

Fig. 25 Surfactant concentration during SINB of **a** conventional foam and **b** novel foam

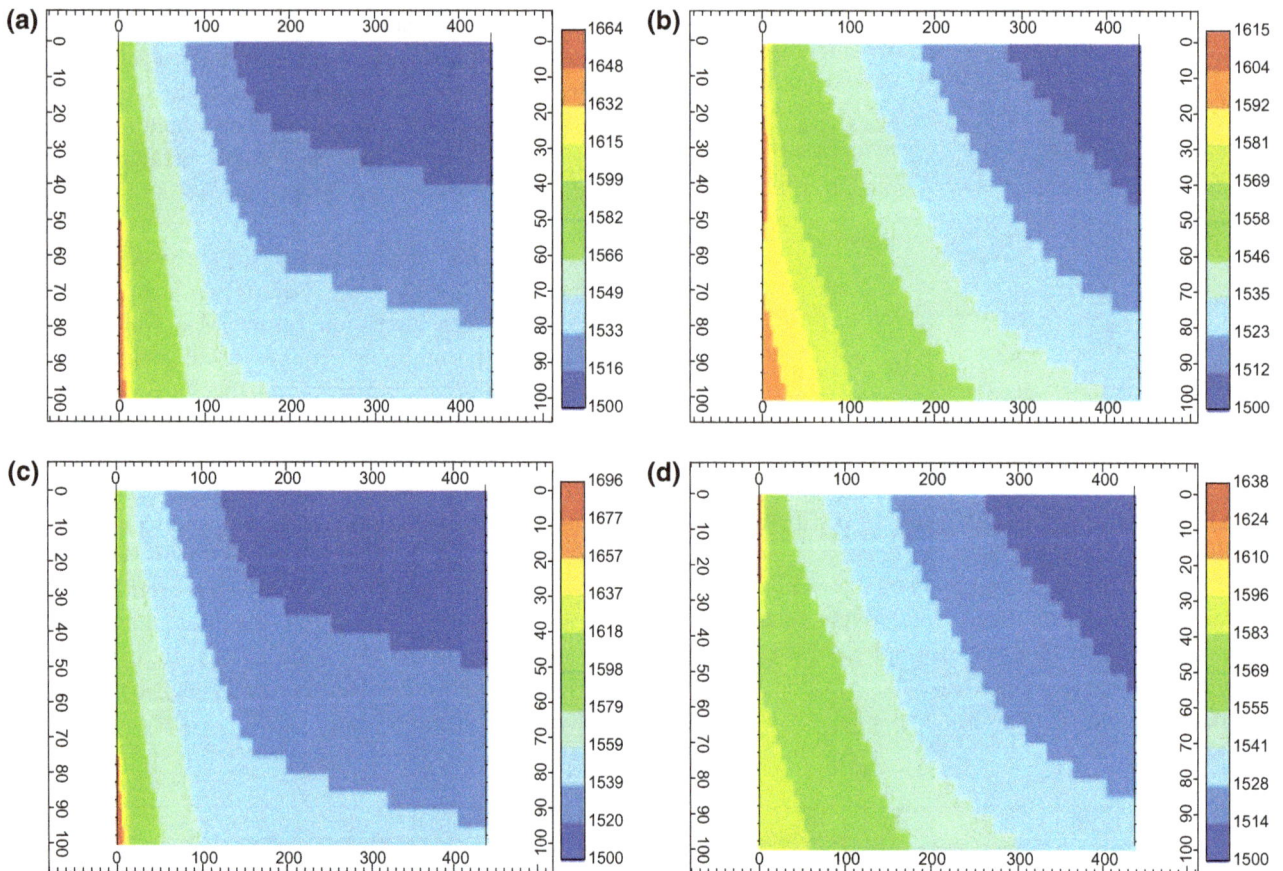

Fig. 26 Pressure distribution during SINB of **a** convectional foam and **b** novel foam and during SIWB of **c** convectional foam, and **d** novel foam

Now, let us address the remaining question above that whether the significant improvement in novel foam performance is attributable to the constant pressure injection mode. We follow the manipulation above to search the lower injection pressures through pursuing injected fluid conservation for the novel foam at 36.5-day slug size with constant rate mode, saying 1548 psi for the gas well and 1553 psi for the water well. The CO_2 storage and gas utilization ratio are also listed in Table 5, which are just slightly higher than those in the constant rate mode. This is consistent with the deduction we did for the conventional foam. Thus, the improved CO_2 storage and reduced CO_2 utilization above just result from more fluid injections. In other words, for both types of foams, the injection mode is of less importance as long as a close average injection rate or pressure is fulfilled.

In the constant rate mode section, we already concluded that the novel foam was insensitive to the slug size for

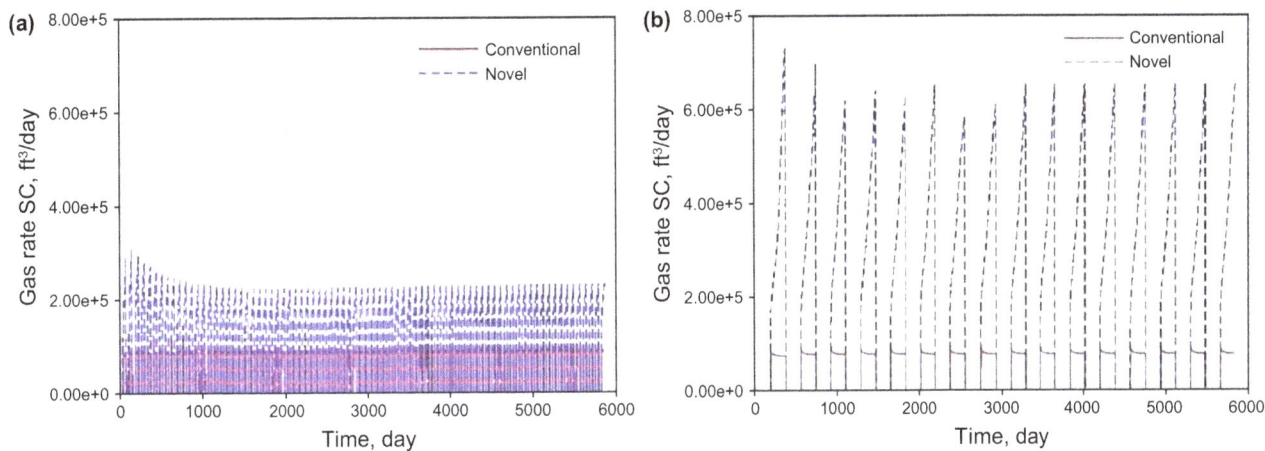

Fig. 27 Gas injection rate of alternating injection under constant pressure constraint for **a** 36.5 days and **b** 182.5 days

Table 5 Comparison of CO_2 storage and utilization ratio for constant rate and pressure injection modes for alternating injection

	CO_2 storage, 10^7 scf					CO_2 utilization ratio				
	Constant rate		Constant pressure			Constant rate		Constant pressure		
	36.5 days	182.5 days	36.5 days		182.5 days	36.5 days	182.5 days	36.5 days		182.5 days
			Gas (1581 psi) Water (1547 psi)	Gas (1548 psi) Water (1553 psi)	Gas (1581 psi) Water (1547 psi)			Gas (1581 psi) Water (1547 psi)	Gas (1548 psi) Water (1553 psi)	Gas (1581 psi) Water (1547 psi)
Novel	7.36	6.54	11.1	7.68	12.2	0.28	0.249	0.199	0.289	0.105
Conventional	1.16	2.45	1.2	–	2.28	0.044	0.0933	0.046	–	0.096

alternating injection even though smaller size held the leading position slightly. Now, we extend the above discussion to the constant pressure constraint. 1581 psi for the gas well and 1547 psi for the water well are still employed. It is observed that larger slug size increases the novel foam gas injection rate and enlarges the contrast between two types of foams (Fig. 27) since the novel foam will lower the injection pressure further with larger slug size. In turn, similar trends are expected for gas production rates (Fig. 31). The CO_2 storage and utilization ratios are listed in Table 5. For the conventional foam, it is obvious that larger slug size is preferred, indicated by the enhancement of sweep efficiency and utilization efficiency up to 90% and 100%, respectively, relative to smaller slug size. Hence, there comes the previous conclusion that the optimal injection strategy for overcoming gravity override with foam in a homogeneous reservoir is alternating injection of separate, large slugs of gas and liquid at a fixed, maximum-allowable injection pressure (Shan and Rossen 2004; Rossen and Renkema 2007). Meanwhile, if we examine the corresponding cases for two injection constraints with the same slug sizes under conservation of injection fluids, analogous improvements are present. Especially, the gas

saturation profile (Fig. 28c), surfactant concentration (Fig. 29c), and pressure distribution (Fig. 30c) are almost identical to those under constant rate constraints (Figs. 11c, 12c). Therefore, again, the injection constraint is really of less importance and the performance of the novel foam is overwhelming. With respect to sweep efficiency, we prove the novel foam is insensitive to the slug size with only 10% enhancement evidenced by the further vertical expanded override zone (Fig. 28d) and deeper extension of high pressure gradient (Fig. 30d). However, again, this costs a 47% reduction in CO_2 utilization efficiency indicated by the rocketing gas production rate (Fig. 31b) and low surfactant concentration zone in the upper layers (Fig. 29d). Therefore, for the novel foam, the sweep efficiency is a monotonic function of injection rate or pressure, but the gas utilization ratio could demonstrate a parabolic shape.

4.2.2.2 Water injection through the top part and gas into the bottom part (SINB) Similar to the manipulation above, we also looked at the injection pressures through trial and error to pursue the conservation of injected fluids for the conventional foam, 1608 psi for a gas well and 1598 psi for a water well; then, apply them to the novel

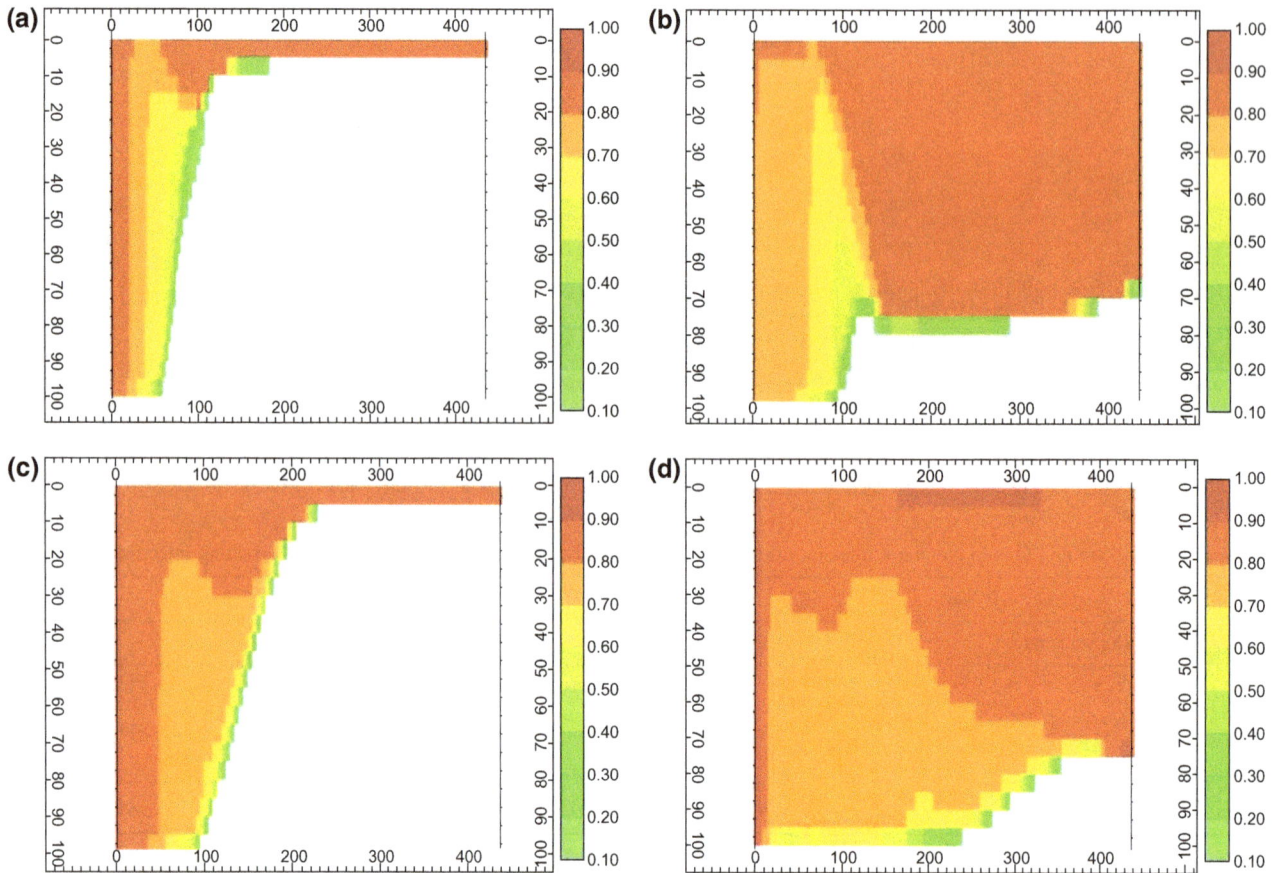

Fig. 28 Gas saturation of small slug size (36.5 days) for **a** CD1045, **b** novel surfactant, and of larger slug size (182.5 days) for **c** CD1045, **d** novel surfactant during alternating injection under constant pressure constraint

foam. The CO_2 storage and utilization ratio are listed in Table 6. A much higher gas injection rate and delayed increasing production rate (Fig. 32) indicate the higher injectivity and better efficiency for the novel foam. It is obvious that the novel foam outperforms the conventional one significantly, indicated by the extreme vertically expanded override zone (Figs. 33, 34), better surfactant transportation in the override zone (Figs. 35, 36) and more deeply extended high pressure gradient (Figs. 37, 38). A low surfactant concentration zone near the gas injector at the bottom characterizes the CO_2 partitioning ability of the novel surfactant. Similarly, the constant pressure constraint does tend to amplify the contrast between two types of foams with respect to CO_2 storage even though the CO_2 utilization ratio of the novel foam drops. It is clear that the comparison between two injection modes for the conventional foam tells us the constant pressure mode gives a little bit worse performance with close distributions of gas saturation (Figs. 24a, 33) and surfactant concentration (Figs. 25a, 35). This is not consistent with prior conclusions (Shi et al. 1998; Shan and Rossen 2004; Rossen et al.

2006; Rossen and Renkema 2007; Jamshidnezhad et al. 2008b) that regardless of co-injection or SAG, relative to the constant rate injection, the constant pressure injection can overcome gravity override better and obtain pressure distribution more evenly, i.e., most of the fixed pressure drop between wells is focused on the displacement front, with maximum suppression of the gravity effect. This discrepancy could be attributed to the different initial reservoir conditions made here. Most of the prior conclusions we mentioned up to now are based on the postulation that the reservoir is initially saturated with surfactant (Rossen et al. 1995; Shi et al. 1998; Shan and Rossen 2004), which means that these studies applied only to gravity override within the region swept by surfactant and slumping of the surfactant slug is not examined. In this study, no surfactant is present in the reservoir initially, which is more practical. Therefore, even though the constant pressure constraint tends to force the pressure drop at the displacement front (Rossen et al. 1995), it may give worse side effects for the same reason. The injection rates response (Fig. 32) gives some clues. Before enough

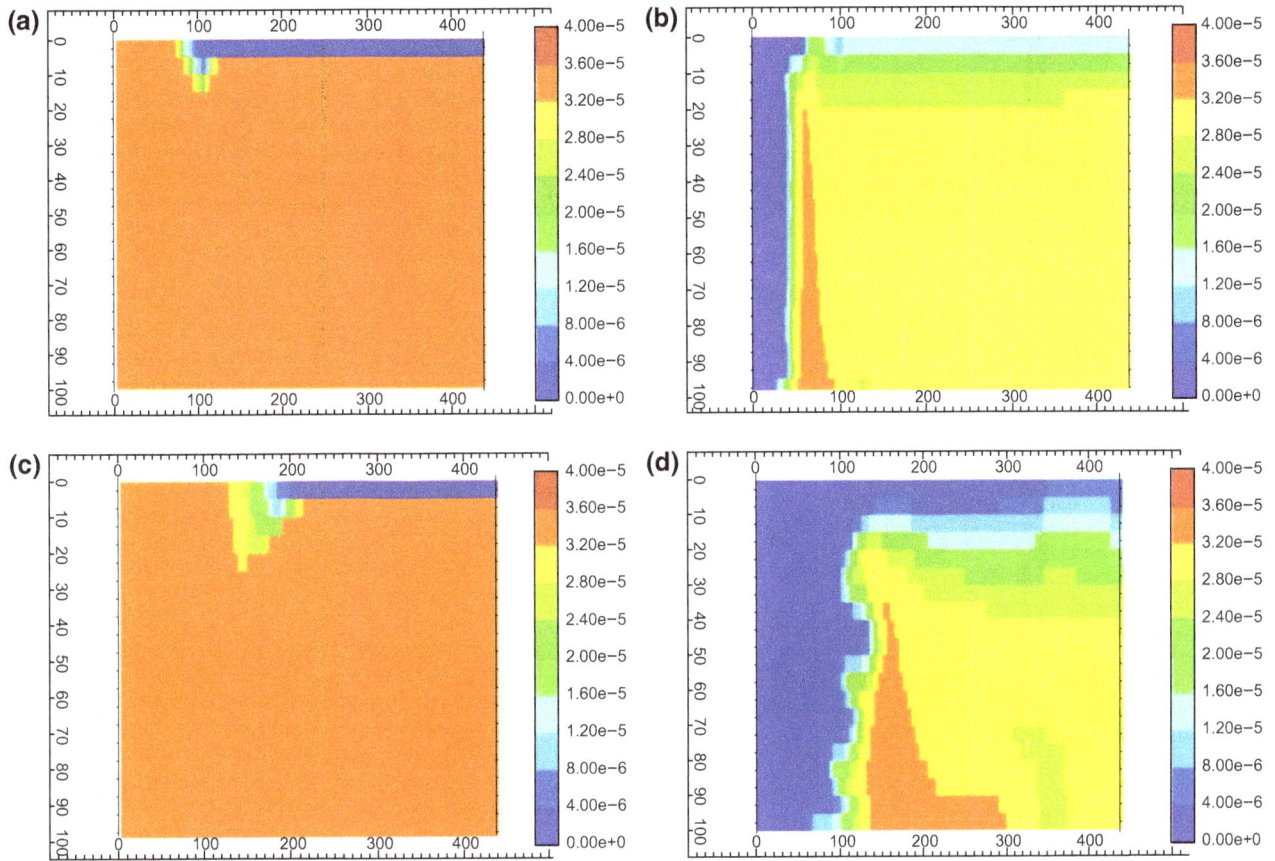

Fig. 29 Surfactant concentration of small slug size (36.5 days) for **a** CD1045, **b** novel surfactant, and of larger slug size (182.5 days) for **c** CD1045, **d** novel surfactant during alternating injection under constant pressure constraint

resistance presents ahead of the gas front, the injection rate rockets to a high level even though we employ the lower injection pressure relative to those under the constant rate constraint (Fig. 23). Then, it decreases to the similar value (90,000 scf/day) as the strong foam has been built up in the reservoir. The only advantage we can view here is the deeper extended high pressure gradient (Figs. 37, 26a).

Similar to our discussion above, with higher injection pressure, or much more fluid injected relative to the constant rate mode, CO_2 storage for the novel foam seems to significantly improve with a reduction of 58% in CO_2 utilization efficiency, even though the pressure gradient distribution is almost piston-like (Figs. 38, 26b). Therefore, to maintain conservation of injection fluids, lower injection pressures (gas well at 1585 psi and water well at 1595 psi) make both criteria comparable for the novel foam (Table 6). This supports our preliminary conclusion drawn for the conventional foam above that the injection constraint is of much less importance and it is not necessary that the constant pressure will be beneficial. The foam performance is a function of injection rate or pressure, but the most determinative factor is surfactant properties.

4.2.3 CO_2 continuous injection with dissolved CO_2-soluble surfactant

The unique injection strategy, CO_2 continuous injection with dissolved novel surfactant, was examined. For sake of comparison, analogous to the manipulation in core flooding, CO_2 injection rate is the summary of two phases in alternating and co-injection strategies under surface conditions. Accordingly, the surfactant concentration is lowered to maintain the same amount of surfactant injected. Two perforation location scenarios are investigated as well as both injection constraints, as shown in Table 2.

4.2.3.1 Constant rate injection mode
A significantly high CO_2 storage has been achieved, which is more than a 30% improvement over SINB, as shown in Table 7 and Fig. 39a. This is also implicitly indicated by the gas production rate that is far lower than the injection rate at time line (Fig. 40). It is observed that the gas "override zone" expands to the whole reservoir. The distinctions between full completion and partial completion through lower ten layers are reduced significantly indicated by the almost

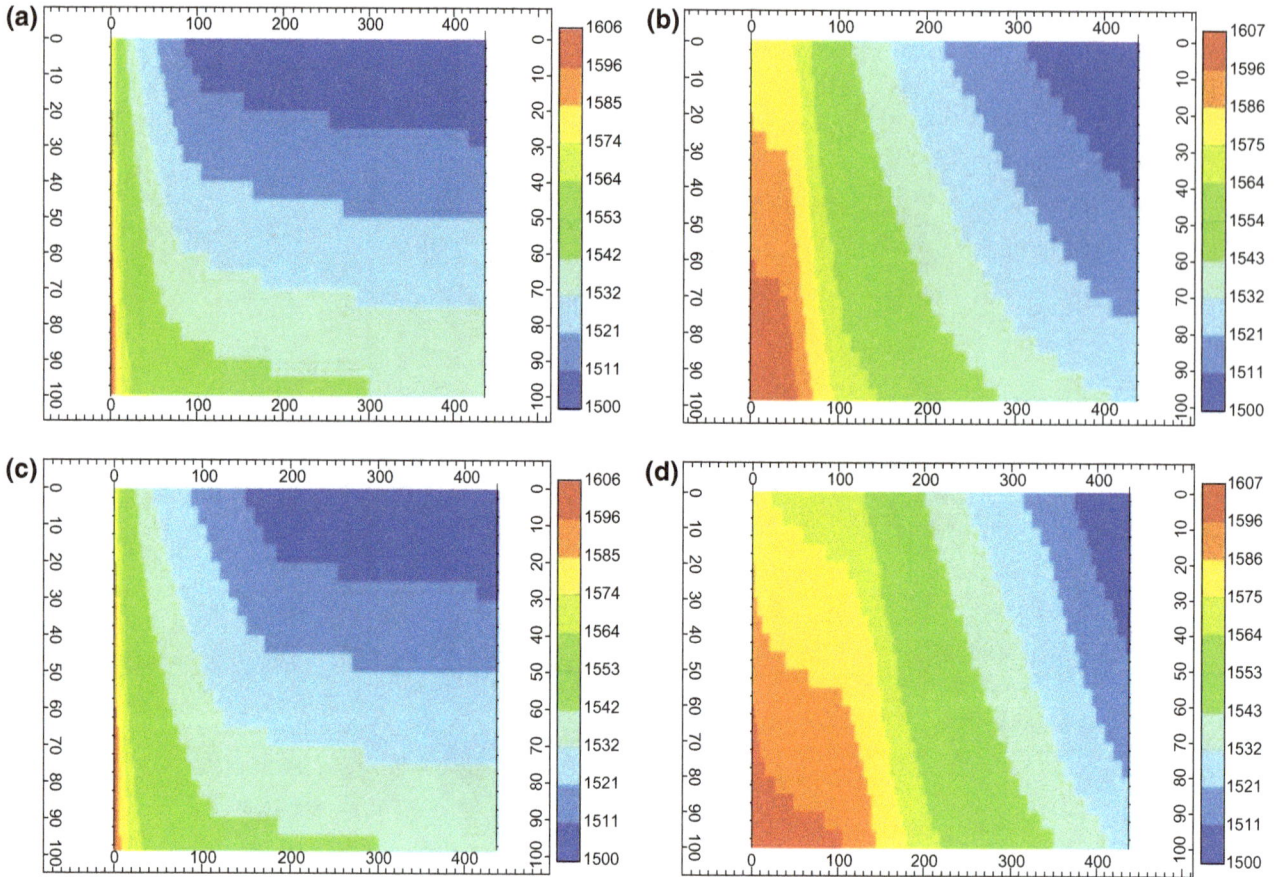

Fig. 30 Pressure distribution of small slug size (36.5 days) for **a** CD1045, **b** novel surfactant, and of larger slug size (182.5 days) for **c** CD1045, **d** novel surfactant during alternating injection under constant pressure constraint

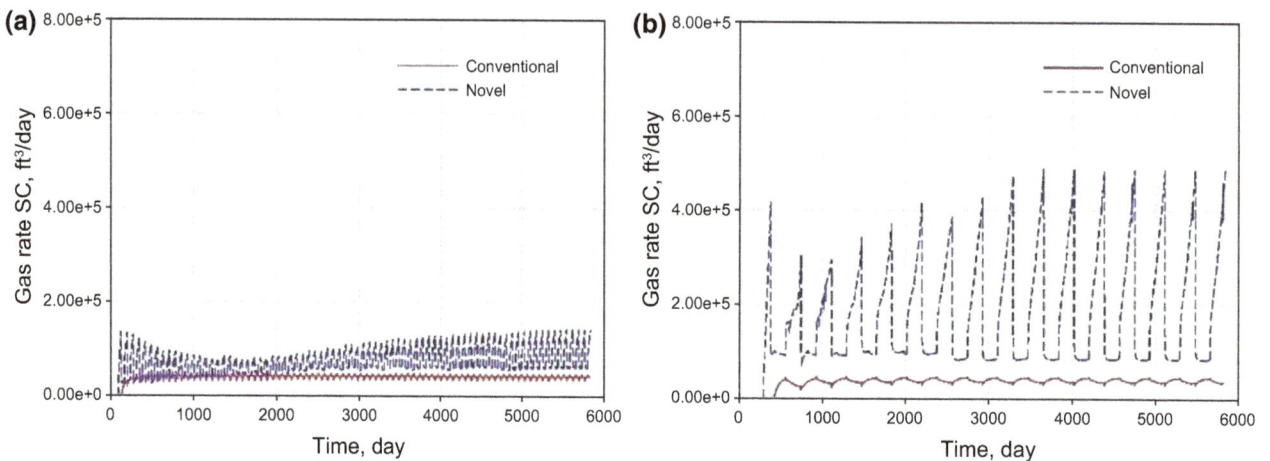

Fig. 31 Gas production rate of alternating injection under constant pressure constraint for **a** 36.5 days and **b** 182.5 days

identical gas production rate (Fig. 40), BHP in the injector (Fig. 41), gas saturation (not shown), surfactant concentration (not shown), and pressure distribution (not shown). Hence, we could take full completion for further discussion.

It may be biased to make the judgment now solely through CO_2 storage because we inject more CO_2. As shown in Table 7 and Fig. 39b, relative to SINB, this novel injection strategy improves the CO_2 utilization ratio by 3%. This can be perceived from the gas saturation profiles more

Table 6 Comparison of CO_2 storage and utilization ratio for constant rate and pressure injection modes for SINB

| | CO$_2$ storage, 10^7 scf | | | CO$_2$ utilization ratio | | |
| | Constant rate | Constant pressure | | Constant rate | Constant pressure | |
		Gas (1608 psi) Water (1598 psi)	Gas (1585 psi) Water (1595 psi)		Gas (1608 psi) Water (1598 psi)	Gas (1585 psi) Water (1595 psi)
Novel	8.9	12.3	8.38	0.338	0.142	0.33
Conventional	1.73	1.44	–	0.066	0.0554	–

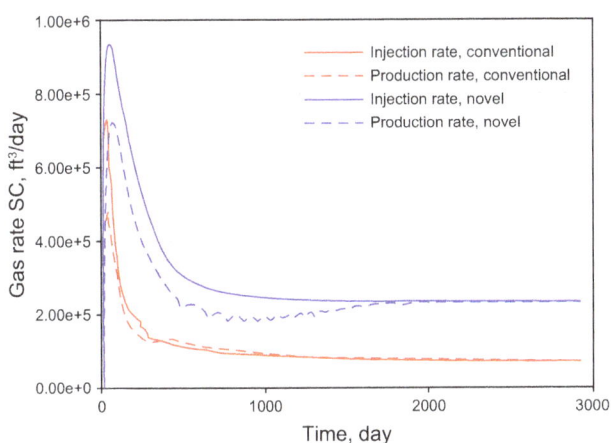

Fig. 32 Gas injection and production rates of novel and conventional foams for SINB with constant pressure constraint

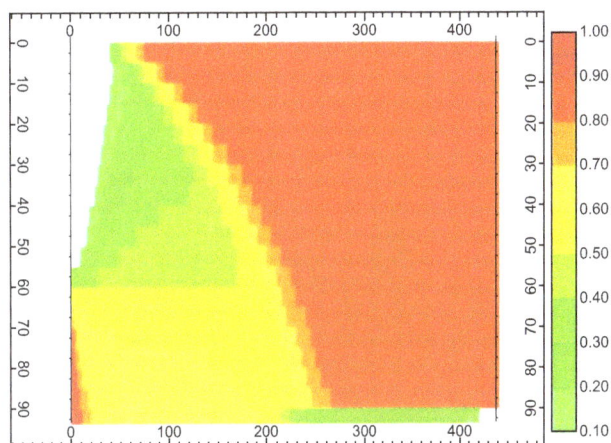

Fig. 34 Gas saturation of novel foam for SINB with constant pressure constraint

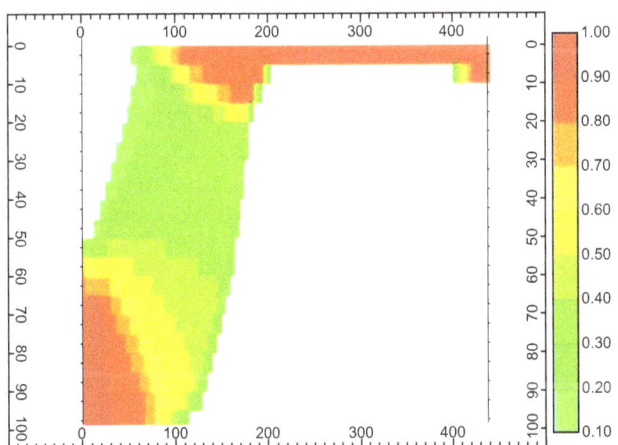

Fig. 33 Gas saturation of conventional foam for SINB with constant pressure constraint

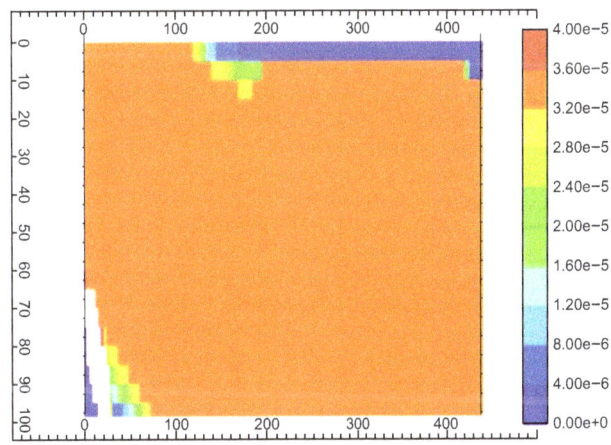

Fig. 35 Surfactant concentration of conventional foam for SINB with constant pressure constraint

directly, as shown in Fig. 42, which clearly illustrates a much more uniform gas propagation front. Furthermore, it is hard to tell the traditionally defined mixed zone which is already occupied by the so-called override zone. Surfactant concentration profiles (Fig. 43) displace a clear piston-like front of surfactant propagation. Meanwhile, it is important to note that there is a low concentration zone near the

wellbore that expands with time, which facilitates the improvement in injectivity. The reason of this phenomenon is different from that in the alternating injection with slug size increasing for the novel foam (Fig. 12b, d), which results from continuous extraction of fresh CO_2. Here, CO_2 is already saturated with surfactant and the extracted substance becomes water because the solubility of water in

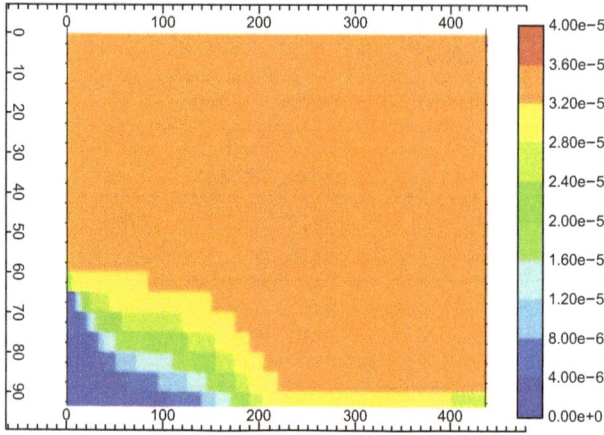

Fig. 36 Surfactant concentration of novel foam for SINB with constant pressure constraint

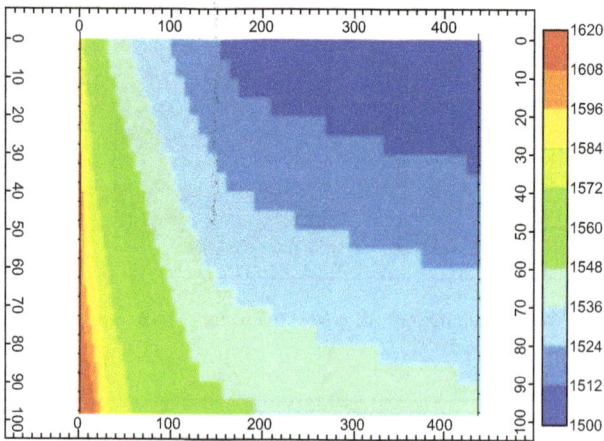

Fig. 37 Pressure distribution of conventional foam for SINB with constant pressure constraint

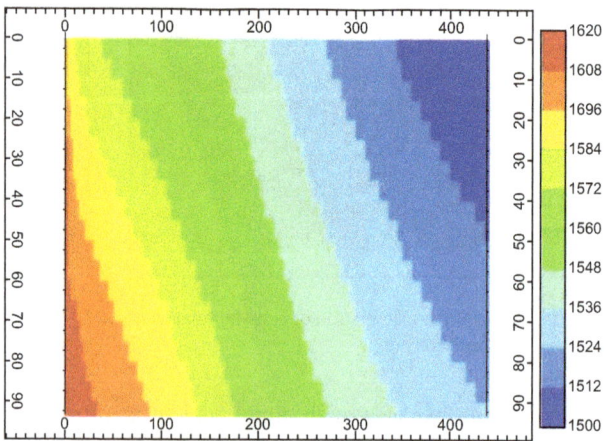

Fig. 38 Pressure distribution of novel foam for SINB with constant pressure constraint

CO_2 is not zero. With a huge amount of gas flow, bubbles will collapse when the water saturation approaches the critical value.

Here, without interference from water, the novel surfactant can be delivered much deeper into the reservoir and foam is generated in situ with formation water. It is straightforward to deduce that the injection pressure would be compellingly low among all the studied cases, as shown in Fig. 41, in which the partial completion gives a little higher value. The superiority of this novel strategy is also evidenced by the pressure distribution in the reservoir (Fig. 44). A high pressure gradient extends into the reservoir deeply, characterized by the extremely evenly distributed zones and steep contour lines, which stand for the high power utilization efficiency. In summary, this novel injection strategy is almost incomparably better with respect to saved water injection cost and highly improved sweep efficiency and gas utilization.

4.2.3.2 Constant pressure injection mode Now, we examine our conclusions for this novel injection strategy with full completion under a constant pressure injection constraint. Following the manipulations above, at first, we set an equivalent gas well injection pressure, 1585 psi, to chase the same amount of gas injection with that under the constant rate mode; then, a higher injection pressure is applied, 1610 psi, to validate our conclusions, as summarized in Table 8. It is likely that close CO_2 storage and utilization ratio are achieved with injection fluid conservation, while higher CO_2 storage and lower gas utilization efficiency occur with higher injection pressure. Again, this novel injection strategy greatly outperforms all other cases with respect to both criteria since more surfactants can be transport to the upper layers without interference of the water phase. It is noted that relative to the gradually declining injection rate in SINB (Fig. 32), the injection rate here (Fig. 45) showed the opposite trend. Meanwhile, the constant pressure constraint just improves the novel foam performance indicated by the almost identical gas saturation profiles (Figs. 46, 42), surfactant concentration profiles (Figs. 47, 43), and pressure distributions (Figs. 48, 44).

The performance of novel foam here (Table 8) supports our previous preliminary conclusions (Ren et al. 2013) that the novel foam performance is a function of injection strategy, injection rate or pressure, and partition coefficient. For certain injection strategies and novel surfactant, regardless of injection constraint, the sweep efficiency is a monotonic function of injection rate or pressure, but the gas utilization ratio demonstrates a parabolic shape. These

Table 7 Comparison of CO_2 storage and utilization ratio among different injection strategies with constant rate injection mode

	CO_2 storage, 10^7 scf			CO_2 utilization ratio		
	Alternating injection (36.5-day slug size)	SINB	CO_2 continuous injection with CO_2-soluble surfactant	Alternating injection (36.5-day slug size)	SINB	CO_2 continuous injection with CO_2-soluble surfactant
Novel	7.36	8.9	11.7	0.28	0.338	0.348
Conventional	1.16	1.73	–	0.044	0.066	–

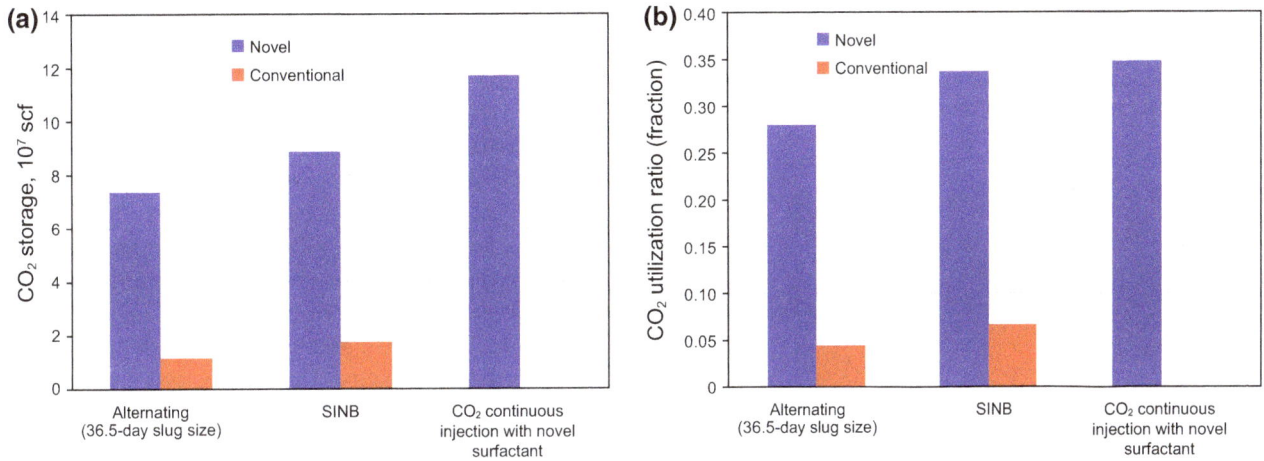

Fig. 39 Comparison of foam performances among different injection strategies. **a** CO_2 storage. **b** CO_2 utilization ratio

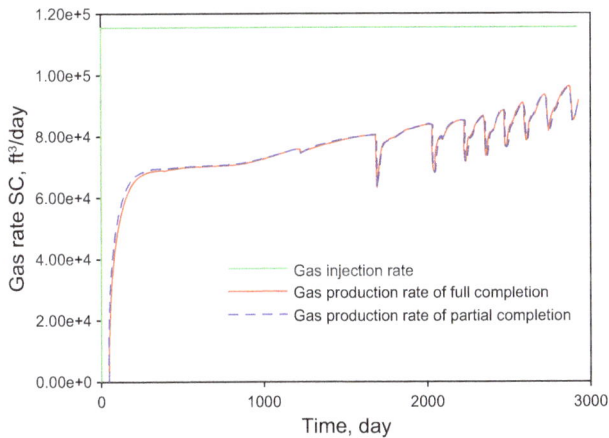

Fig. 40 Gas injection and production rates of CO_2 continuous injection with dissolved CO_2-soluble surfactant with full completion and partial completion

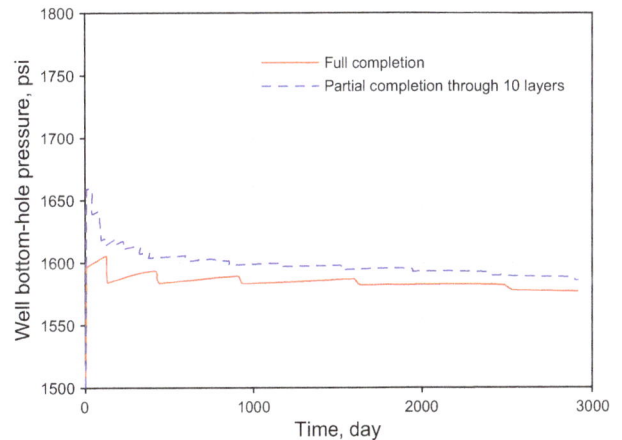

Fig. 41 Bottom-hole pressure in the gas injector of CO_2 continuous injection with dissolved CO_2-soluble surfactant with full completion and partial completion

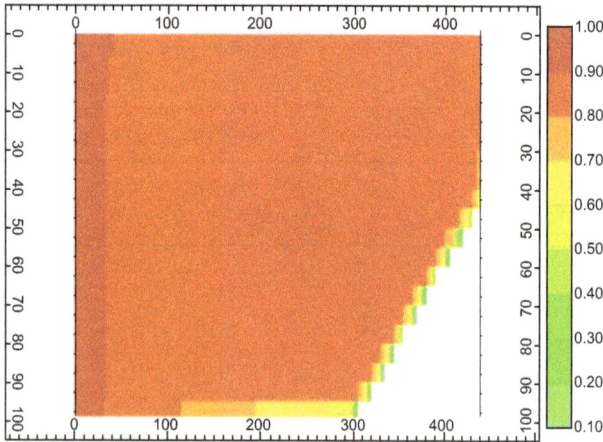

Fig. 42 Gas saturation of CO_2 continuous injection with dissolved CO_2-soluble surfactant with full completion

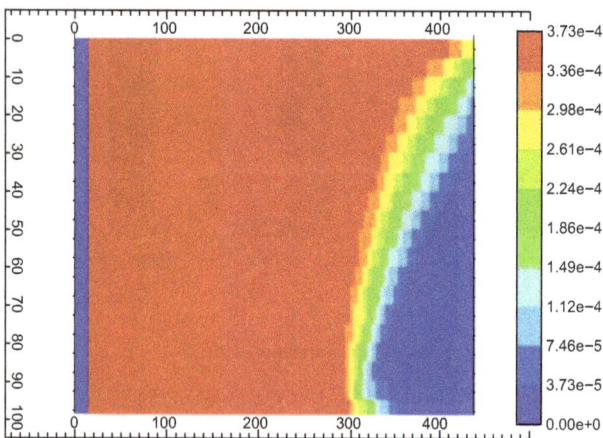

Fig. 43 Surfactant concentration of CO_2 continuous injection with dissolved CO_2-soluble surfactant with full completion

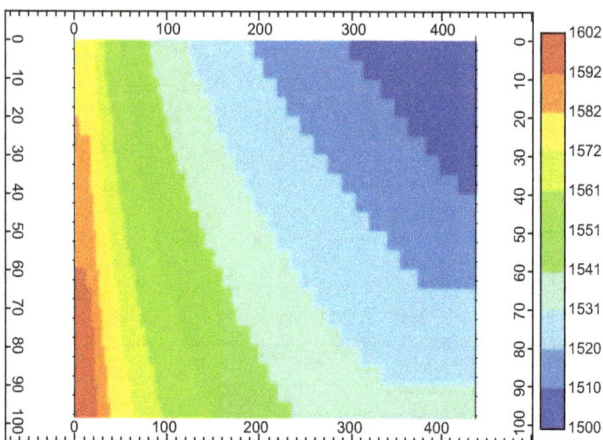

Fig. 44 Pressure distribution of CO_2 continuous injection with dissolved CO_2-soluble surfactant with full completion

conclusions highly improve the robustness of foam application according to field fluid requirement and facility availability.

5 Summary

In summary, the inherently superior properties of the novel surfactant make it outperform conventional surfactant in every injection strategy. The restriction of constant rate injection mode does not exist anymore. Injection constraint cannot solve the intrinsic problem that causes gas channeling because the constant pressure mode still arranges the pressure distribution through adjusting the injection rate (Boeije and Rossen 2014). Previous researchers (Shi and Rossen 1996, 1998; Shan and Rossen 2004; Rossen et al. 2006; Rossen and Shen 2007) have demonstrated that for conventional foam, the override zone will not expand downwards greatly and the only possible change is that the mixed zone spreads to a producer with stronger foam or higher injection pressure, as shown in Fig. 49 (black dash vs red dot dash lines), which will deteriorate injectivity. The red dot dash line indicates the cross-point of three zones could only move horizontally and this is the reason the gravity segregation length has attracted so much attention. Now, relative to the conventional foam, this novel foam tends to weaken the foam near the wellbore and strengthen it on the top layers with migration of surfactants with gas. In other words, the conflict between sweep efficiency and injectivity encountered by the conventional foam (Namdar and Rossen 2013) has been reduced significantly by the novel foam. Continuously supplying enough surfactants to the top layer is crucial for gas diversion to increase the volume of the traditionally defined override and mixed zones (Fig. 49 in blue solid line). Hence, R_g is only a criterion for fighting gravity segregation but not a sufficient condition to evaluate an injection strategy. The volume of the override zone and gravity segregation height play an important role in determining the sweep efficiency, which requires not only even pressure distribution but also steep pressure contour lines. The novel foam can perform well even with short segregation length or early gas breakthrough because they do not reflect the successive stages of gas diversion. Intrinsic superiority of the novel surfactant replaces the injection mode to dominate the foam process and gives more freedom to injection arrangement according to CO_2 acquirement.

Table 8 Comparison of CO_2 storage and utilization ratio for constant rate and pressure injection modes for the novel CO_2 injection

	CO_2 storage, 10^7 scf			CO_2 utilization ratio		
	Constant rate	Constant pressure		Constant rate	Constant pressure	
		Gas (1585 psi)	Gas (1610 psi)		Gas (1585 psi)	Gas (1610 psi)
Novel	11.7	11.9	13.8	0.348	0.351	0.298

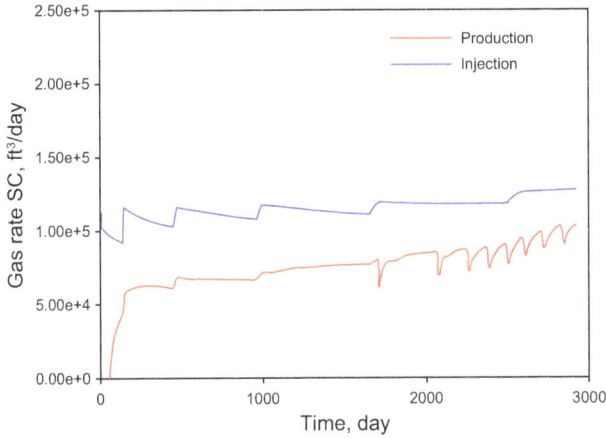

Fig. 45 Gas injection and production rates of CO_2 continuous injection with dissolved CO_2-soluble surfactant under constant pressure constraint (1585 psi)

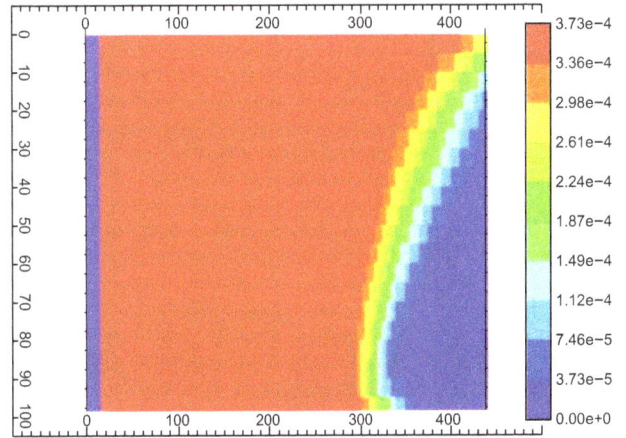

Fig. 47 Surfactant concentration of CO_2 continuous injection with dissolved CO_2-soluble surfactant under constant pressure constraint (1585 psi)

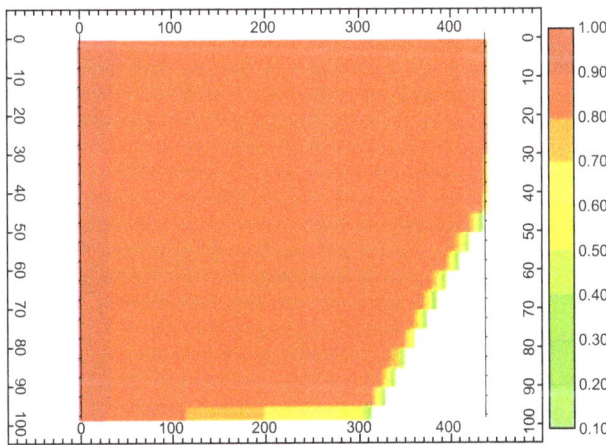

Fig. 46 Gas saturation of CO_2 continuous injection with dissolved CO_2-soluble surfactant under constant pressure constraint (1585 psi)

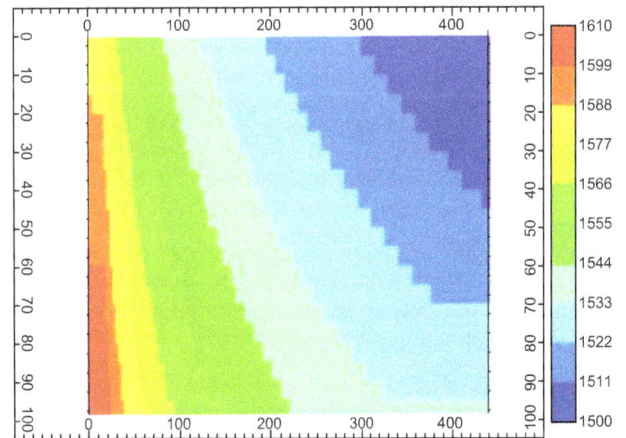

Fig. 48 Pressure distribution of CO_2 continuous injection with dissolved CO_2-soluble surfactant under constant pressure constraint (1585 psi)

6 Conclusions

(1) The novel CO_2-soluble surfactant provides better film stabilization ability than the conventional aqueous surfactant. In turn, when simultaneously injecting, the novel foam propagates faster and demonstrates higher pressure drop and sweep efficiency.

(2) Alternating injection does improve the foam propagation and injectivity regardless of surfactant type. Alternating injection also promote the superiority of the novel foam over the conventional one in quicker and stronger foam generation.

(3) It is the first time the novel injection strategy, CO_2 continuous injection with dissolved CO_2-soluble surfactant, has been tested in consolidated cores,

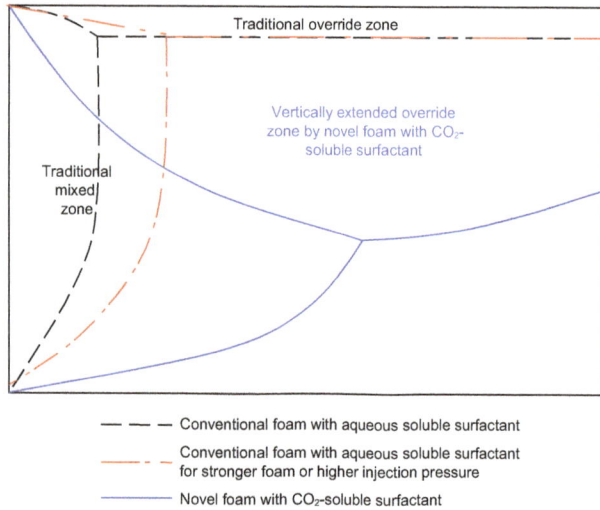

Fig. 49 Schematic plot of conventional and novel foams on fighting gravity segregation

which demonstrates superior surfactant transportation ability, in turn improving the foam propagation and displacement rate significantly.

(4) With field scale simulation, for all tested injection strategies, regardless of constant rate or pressure constraint, the novel foam significantly outperforms conventional foam in terms of much higher sweep efficiency, injectivity and much more even pressure distribution resulting from intrinsic property of the novel surfactant.

(5) The novel foam performance is a function of injection strategy, injection rate or pressure, and partition coefficient (not discussed here); for a fixed injection strategy and with the novel surfactant, regardless of injection constraint, sweep efficiency is a monotonic function of injection rate or pressure, but the gas utilization ratio demonstrates a parabolic shape.

(6) Injection constraint, i.e., constant rate or pressure, is of much less importance to both types of foams, as long as similar amount of fluids have been injected. From the point view of sweep efficiency, for alternating injection, constant pressure mode tends to amplify the superiority of the novel foam over the conventional one due to higher injectivity.

(7) Relative to conventional foam, the novel foam tends to increase the segregation height and volume of the traditionally defined override zone through gas diversion instead of solely increasing gravity segregation length and delaying gas breakthrough time. The latter two are of less importance in performance evaluation for the novel foam.

(8) For alternating injection, relative to conventional foam that is preferential to lager slug, the novel foam is not sensitive to injection fluid slug size regardless of injection constraint. The optimal slug size of novel foam with respect to gas utilization is a function of injection rate or pressure and partition coefficient.

(9) Co-injection does lower the injectivity for conventional foam relative to alternating injection while this problem has been greatly reduced with the novel foam owing to surfactant concentration deduction by gas extraction.

(10) For simultaneous injection through same sections (SIAP and SIPP), relative to full completion, partial completion lowers the injectivity and improves the sweep efficiency for both foams, while for separate injection (SINB and SIWB), the novel foam really reduces the distinction between them with respect to sweep efficiency and injectivity. The separate injection (SINB and SIWB) is able to give longer gravity segregation lengths and higher injectivity than simultaneous injection through same sections (SIAP and SIPP).

(11) The novel injection strategy, continuous CO_2 injection with dissolved surfactant, gives the best foam performance among all the tested scenarios regardless of completion sections and injection constraint. This may dramatically lower the water injection/treatment cost and improve the robustness of foam application.

Acknowledgements The authors would like to thank Hang Zhang (University of Texas at Austin, currently at Schlumberger Houston) for his dedicated help with laboratory experiments.

References

Adkins SS, Chen X, Chan I, Torino E, Nguyen QP, Sanders AW, Johnston KP. Morphology and stability of CO_2-in-water foams with nonionic hydrocarbon surfactants. Langmuir. 2010a;26(8):5335–48. doi:10.1021/la903663v.

Adkins SS, Chen X, Nguyen QP, Sanders AW, Johnston KP. Effect of branching on the interfacial properties of nonionic hydrocarbon surfactants at the air–water and carbon dioxide–water interfaces. J Colloid Interface Sci. 2010b;346(2):455–63. doi:10.1016/j.jcis. 2009.12.059.

Algharaib MK, Gharbi RB, Malallah A, Al-Ghanim W. Parametric investigations of a modified SWAG injection technique. In: SPE Middle East oil and gas show and conference, 11–14 March, Manama, Bahrain; 2007. doi:10.2118/105071-MS.

Bernard G, Holm L. Method for recovering oil from subterranean formations. U.S. Patent. No.3342256. 19 Sep 1967.

Blaker T, Celius HK, Lie T, Martinsen HA, Rasmussen L, Vassenden F. Foam for gas mobility control in the Snorre Field: the FAWAG project. In: SPE annual technical conference and exhibition, 3–6 October, Houston, Texas; 1999. doi:10.2118/56478-MS.

Boeije CS, Rossen W. Gas-injection rate needed for SAG foam processes to overcome gravity override. SPE J. 2014;20(1):49–59. doi:10.2118/166244-PA.

Bond DC, Holbrook CC. Gas drive oil recovery process. U.S. Patent. No. 2866507. Dec 1958.

Chen X, Adkins SS, Nguyen QP, Sander AS, Johnston KP. Interfacial tension and the behavior of microemulsions and macroemulsions of water and carbon dioxide with a branched hydrocarbon nonionic surfactant. J Supercrit Fluids. 2010;55(2):712–23. doi:10.1016/j.supflu.2010.08.019.

Chen Y, Elhag AS, Cui L, Worthen AJ, Reddy PP, Noguera JA, Ou AM, Ma K, Puerto M, Hirasaki GJ, Nguyen PQ, Biswal SL, Johnston KP. CO_2-in-water foam at elevated temperature and salinity stabilized with a nonionic surfactant with a high degree of ethoxylation. Ind Eng Chem Res. 2015;54(16):4252–63. doi:10.1021/ie503674m.

Chen Y, Elhag AS, Poon BM, Cui L, Ma K, Liao SY, Johnston KP. Ethoxylated cationic surfactants for CO_2 EOR in high temperature, high salinity reservoirs. In: SPE annual technical conference and exhibition, 3–6 Oct, Houston, Texas; 2012. doi:10.2118/154222-MS.

Cheng L, Reme AB, Shan D, Coombe DA, Rossen WR. Simulating foam processes at high and low foam qualities. In: SPE/DOE improved oil recovery symposium, 3–5 April, Tulsa, Oklahoma; 2000. doi:10.2118/59287-MS.

Cui L, Ma K, Abdala AA, Lu J, Tanakov IM, Biswal SL, Hirasaki GJ. Adsorption of a switchable cationic surfactant on natural carbonate minerals. SPE J. 2014;20(1):70–8. doi:10.2118/169040-PA.

Dhanuka V, Dickson J, Ryoo W, Johnston K. High internal phase CO_2-in-water emulsions stabilized with a branched nonionic hydrocarbon surfactant. J Colloid Interface Sci. 2006;298(1):406–18. doi:10.1016/j.jcis.2005.11.057.

Du D, Zitha PLJ, Uijttenhout MGH. Carbon dioxide foam rheology in porous media: a CT scan study. SPE J. 2007;12(2):245–52. doi:10.2118/97552-PA.

Eastoe J, Dupont A, Steytler DC, Thorpe M, Gurgel A, Heenan RK. Micellization of economically viable surfactants in CO_2. J Colloid Interface Sci. 2003;258(2):367–73. doi:10.1016/S0021-9797(02)00104-2.

Eastoe J, Gold S, Steytler DC. Surfactants for CO_2. Langmuir. 2006;22(24):9832–42. doi:10.1021/la060764d.

Eastoe J, Paul A, Nave S. Micellization of hydrocarbon surfactants in supercritical carbon dioxide. J Am Chem Soc. 2001;123:988–9. doi:10.1021/ja005795o.

Elhag AS, Chen Y, Chen H, Reddy PP, Cui L, Worthen AJ, Johnston KP. Switchable amine surfactants for stable CO_2/brine foams in high temperature, high salinity reservoirs. In: SPE improved oil recovery symposium, 12–16 April, Tulsa, Oklahoma, USA; 2014. doi:10.2118/169041-MS.

Fan X, Potluri VK, McLeod MC, Wang Y, Liu JC, Enick RM, Hamilton AD, Roberts CB, Johnson JK, Beckman EJ. Oxygenated hydrocarbon ionic surfactants exhibit CO_2 solubility. J Am Chem Soc. 2005;127(33):11754–62. doi:10.1021/ja052037v.

Farajzadeh R, Lotfollahi M, Eftekhari AA, Rossen WR, Hirasaki GJ. Effect of permeability on implicit-texture foam model parameters and the limiting capillary pressure. Energy Fuels. 2015;29(5):3011–8. doi:10.1021/acs.energyfuels.5b00248.

Fisher AW, Foulser RWS, Goodyear SG. Mathematical modeling of foam flooding. In: SPE/DOE enhanced oil recovery symposium, 22–25 April, Tulsa, Oklahoma; 1990. doi:10.2118/20195-MS.

Friedmann F, Chen WH, Gauglitz PA. Experimental and simulation study of high-temperature foam displacement in porous media. SPE Reserv Eng. 1991;6(1):37–45. doi:10.2118/17357-PA.

Grigg RB. Improving CO_2 efficiency for recovering oil in heterogeneous reservoirs. DOE Contract No. DE-FG26-01BC15364. 2004

Hanssen JE, Surguchev LM, Svorstol I. SAG injection in a North Sea stratified reservoir: flow experiment and simulation. In: European petroleum conference, 25–27 Oct, London, United Kingdom; 1994. doi:10.2118/28847-MS.

Heller JP. CO_2 foams in enhanced oil recovery. In: Schramm LL, editor. Foams: fundamentals and applications in the petroleum industry, vol. 242., ACS Advances in Chemistry SeriesWashington: American Chemical Society; 1994. p. 201. doi:10.1021/ba-1994-0242.ch005.

Hoefling TA, Beitle RR, Enick RM, Beckman EJ. Design and synthesis of highly CO_2-soluble surfactants and chelating agents. Fluid Phase Equilib. 1993;83:203–12. doi:10.1016/0378-3812(93)87023-T.

Hoefner ML, Evans EM, Buckles JJ, Jones TA. CO_2 foam: results from four developmental field trials. SPE Reserv Eng. 1995;10(4):273–81. doi:10.2118/27787-PA.

Jamshidnezhad M, Shen C, Kool PH, Rossen WR. Well stimulation and gravity segregation in gas improved oil recovery. In: SPE international symposium and exhibition on formation damage control, 13–15 Feb, Lafayette, Louisiana, USA; 2008a. doi:10.2118/112375-MS.

Jamshidnezhad M, van Der Bol L, Rossen WR. Injection of water above gas for improved sweep in gas IOR: performance in 3D. In: International petroleum technology conference, 3–5 Dec, Kuala Lumpur, Malaysia; 2008b. doi:10.2523/IPTC-12556-MS.

Jenkins MK. An analytical model for water/gas miscible displacements. In: SPE enhanced oil recovery symposium, 15–18 April, Tulsa, Oklahoma; 1984. doi:10.2118/12632-MS.

Kloet M, Renkema WJ, Rossen WR. Optimal design criteria for SAG foam processes in heterogeneous reservoirs. In: EUROPEC/EAGE conference and exhibition, 8–11 June, Amsterdam, The Netherlands; 2009. doi:10.2118/121581-MS.

Lake LW. Enhanced oil recovery. New York: Prentice Hall; 1989.

Lawson JB. The adsorption of non-ionic and anionic surfactants on sandstone and carbonate. In: SPE symposium on improved methods of oil recovery, 16–17 April. Tulsa, Oklahoma: Society of Petroleum Engineers; 1978. doi:10.2118/7052-MS.

Lawson JB, Reisberg J. Alternate slugs of gas and dilute surfactant for mobility control during chemical flooding. In: SPE/DOE enhanced oil recovery symposium, 20–23 April, Tulsa, Oklahoma; 1980. doi:10.2118/8839-MS.

Le VQ, Nguyen QP, Sanders A. A novel foam concept with CO_2 dissolved surfactants. In: SPE symposium on improved oil recovery, 20–23 April, Tulsa, Oklahoma, USA; 2008. doi:10.2118/113370-MS.

Lee HO, Heller JP. Carbon dioxide foam mobility measurement at high pressure. In: Annual colloid and surface science symposium, 21–24 June, Ann Arbor, Michigan; 1988. ACS Symposium Series 373: 375–386.

Leeftink TN, Latooij CA, Rossen WR. Injectivity errors in simulation of foam EOR. In: The 17th European symposium on improved oil recovery, St. Petersburg, Russia. 16–18 April 2013.

Li B, Hirasaki GJ, Miller CA. Upscaling of foam mobility control to three dimensions. In: SPE/DOE symposium on improved oil recovery, 22–26 April, Tulsa, Oklahoma, USA; 2006. doi:10.2118/99719-MS.

Li Q, Rossen WR. Injection strategies for foam generation in homogeneous and layered porous media. In: SPE annual technical conference and exhibition, 9–12 Oct, Dallas, Texas; 2005. doi:10.2118/96116-MS.

Liu J, Han B, Li G, Zhang X. Investigation of nonionic surfactant dynol-604 based reverse microemulsions formed in supercritical carbon dioxide. Langmuir. 2001;17(26):8040–3. doi:10.1021/la010743d.

Liu M, Andrianov A, Rossen WR. Sweep efficiency in CO_2 foam simulations with oil. In: SPE EUROPEC/EAGE annual conference and exhibition, 23–26 May, Vienna, Austria; 2011. doi:10.2118/142999-MS.

Liu ZT, Erkey C. Water in carbon dioxide microemulsions with fluorinated analogues of AOT. Langmuir. 2001;17:274–7. doi:10.1021/la000947e.

Ma K, Farajzadeh R, Lopez-Salinas JL, et al. Non-uniqueness, numerical artifacts, and parameter sensitivity in simulating steady-state and transient foam flow through porous media. Transp Porous Media. 2014;102:325. doi:10.1007/s11242-014-0276-9.

Ma K, Lopez-Salinas JL, Puerto MC, Miller CA, Biswal SL, Hirasaki GJ. Estimation of parameters for the simulation of foam flow through porous media. Part 1: the dry-out effect. Energy Fuels. 2013;27(5):2363–75. doi:10.1021/ef302036s.

Ma TD, Youngren GK. Performance of immiscible water-alternating-gas (IWAG) injection at Kuparuk River Unit, North Slope, Alaska. In: SPE annual technical conference and exhibition, 25–28 September, New Orleans, Louisiana; 1994. doi:10.2118/28602-MS.

Mannhardt K, Schramm LL, Novosad JJ. Effect of rock type and brine composition on adsorption of two foam-forming surfactants. SPE Adv Technol Ser. 1993;1(1):212–8. doi:10.2118/20463-PA.

McHugh MA, Krukonis VJ. Supercritical fluid extraction. 2nd ed. Boston: Butterworth; 1994.

McLendon WJ, Koronaios P, Enick RM, Biesmans G, Salazar L, Miller A, Soong Y, McLendon T, Romanov V, Crandall D. Assessment of CO_2-soluble non-ionic surfactants for mobility reduction using mobility measurements and CT imaging. J Pet Sci Eng. 2014;119:196–209. doi:10.1016/j.petrol.2014.05.010.

Mohd Shafian SR, Kamarul Bahrim R, Foo Y, Abdul Manap A, Tewari RD. Foam mobility control during WAG injection in a difficult reservoir with high temperature and high acid gas. In: SPE Asia Pacific enhanced oil recovery conference, 11–13 August, Kuala Lumpur, Malaysia; 2015. doi:10.2118/174571-MS.

Namdar Zanganeh M, Rossen W. Optimization of foam enhanced oil recovery: balancing sweep and injectivity. SPE Res Eval Eng. 2013;16(1):51–9. doi:10.2118/163109-PA.

O'Shea K, Kirmse K, Fox MA, Johnston KP. Polar and hydrogen-bonding interactions in supercritical fluids: effects on the tautomeric equilibrium of 4-(phenylazo)-1-naphthol. J Phys Chem. 1991;95:7863.

Patzek TW. Field application of foam for mobility improvement and profile control. SPE Reserv Eng. 1996;11(2):79–85. doi:10.2118/29612-PA.

Ren G, Sanders AW, Nguyen QP. New method for the determination of surfactant solubility and partitioning between CO_2 and brine. J Supercrit Fluids. 2014;91:77–83. doi:10.1016/j.supflu.2014.04.010.

Ren G, Zhang H, Nguyen Q. Effect of surfactant partitioning on mobility control during carbon-dioxide flooding. SPE J. 2013;18(4):752–65. doi:10.2118/145102-PA.

Rossen WR. Foams in enhanced oil recovery. In: Prud'homme RK, Khan SA, editors. Foams: theory, measurements and applications. New York: Marcel Dekker; 1995. p. 413–64.

Rossen W, Boeije CS. Fitting foam-simulation-model parameters to data: II. surfactant-alternating-gas foam applications. SPE Reserv Eval Eng. 2015;18(2):273–83. doi:10.2118/165282-PA.

Rossen WR, Renkema WJ. Success of foam SAG processes in heterogeneous reservoirs. In: SPE annual technical conference and exhibition, 11–14 Nov, Anaheim, California, USA; 2007. doi:10.2118/110408-MS.

Rossen WR, Shen C. Gravity segregation in gas-injection IOR. In: EUROPEC/EAGE conference and exhibition, 11–14 June, London, UK; 2007. doi:10.2118/107262-MS.

Rossen WR, van Duijn CJ. Gravity segregation in steady-state horizontal flow in homogenous reservoirs. J Pet Sci Eng. 2004;43(1–2):99–111. doi:10.1016/j.petrol.2004.01.004.

Rossen WR, Kibodeaux KR, Shi JX, Zeilinge SC, Lim MT. Injectivity and gravity override in surfactant-alternating-gas foam processes. In: SPE annual technical conference and exhibition, 22–25 Oct, Dallas, Texas; 1995a. doi:10.2118/30753-MS.

Rossen WR, van Duijn CJ, Nguyen QP, Vikingstad AK. Injection strategies to overcome gravity segregation in simultaneous gas and liquid injection in homogeneous reservoirs. In: SPE/DOE symposium on improved oil recovery, 22–26 April, Tulsa, Oklahoma, USA; 2006. doi:10.2118/99794-MS.

Rossen WR, Zeilinger SC, Shi J, Lim MT. Mechanistic simulation of foam processes in porous media. In: SPE annual technical conference and exhibition, 25–28 Sep, New Orleans, Louisiana; 1994. doi:10.2118/28940-MS.

Rossen WR, Zhou ZH, Mamun CK. Modeling foam mobility in porous media. SPE Adv Technol Ser. 1995b;3(1):146–53. doi:10.2118/22627-PA.

Sagir M, Tan IM, Mushtaq M, Ismail L, Nadeem M, Azam MR. Synthesis of a new CO_2 philic surfactant for enhanced oil recovery applications. J Dispers Sci Technol. 2014a;35(5):647–54. doi:10.1080/01932691.2013.803253.

Sagir M, Tan IM, Mushtaq M, et al. Novel surfactant for the reduction of CO_2/brine interfacial tension. J Dispers Sci Technol. 2014b;35(3):463–70. doi:10.1080/01932691.2013.794111.

Sagir M, Tan IM, Mushtaq M, Nadeem M. CO_2 mobility and CO_2/brine interfacial tension reduction by using a new surfactant for EOR applications. J Dispers Sci Technol. 2014c;35(11):1512. doi:10.1080/01932691.2013.859087.

Sagir M, Tan IM, Mushtaq M, Talebian SH. FAWAG using CO_2-philic surfactants for CO_2 mobility control for enhanced oil recovery applications. In: SPE Saudi Arabia section technical symposium and exhibition, 21–24 April, Al-Khobar, Saudi Arabia; 2014d. doi:10.2118/172189-MS.

Sanchez NL. Management of water alternating gas (WAG) injection projects. In: Latin American and Caribbean petroleum engineering conference, 21–23 April, Caracas, Venezuela; 1999. doi:10.2118/53714-MS.

Sanders A, Jones RM, Rabie A, Putra E, Linroth MA, Nguyen QP. Implementation of a CO_2 foam pilot study in the SACROC Field: performance evaluation. In: SPE annual technical conference and exhibition, 8–10 Oct, San Antonio, Texas, USA; 2012. doi:10.2118/160016-MS.

Sanders A, Nguyen QP, Nguyen N, Adkins S, Johnston KP. Twin-tailed surfactants for creating CO_2-in-water macroemulsions for sweep enhancement in CO_2-EOR. In: Abu Dhabi international petroleum exhibition and conference, 1–4 Nov, Abu Dhabi, UAE; 2010. doi:10.2118/137689-MS.

Shan D, Rossen WR. Optimal injection strategies for foam IOR. SPE J. 2004;9(2):132–50. doi:10.2118/88811-PA.

Shi JX, Rossen WR. Simulation and dimensional analysis of foam processes in porous media. In: Permian Basin oil and gas recovery conference, 27–29 March, Midland, Texas; 1996. doi:10.2118/35166-MS.

Shi JX, Rossen WR. Improved surfactant-alternating-gas foam process to control gravity override. In: SPE/DOE improved oil recovery symposium, 19–22 April, Tulsa, Oklahoma; 1998. doi:10.2118/39653-MS.

Stalkup FI Jr. Miscible displacement. SPE Monograph No. 8. New York: Society of Petroleum Engineers of AIME; 1983.

Soong Y, Xing D, Wei B, Enick R M, Eastoe J, et al. CO_2-soluble surfactants for enhanced oil recovery mobility control via thickening or in-situ foam generation. In: Proc., 2009 AIChE Annual Meeting, Nashville, TN, USA; 8–13 Nov 2009.

Stevens JE. CO_2 foam field verification pilot test at EVGSAU: phase IIIB–project operations and performance review. SPE Reserv Eng. 1995;10(4):266–72. doi:10.2118/27786-PA.

Stone HL. A simultaneous water and gas flood design with extraordinary vertical gas sweep. In: SPE international petroleum conference in Mexico, 7–9 Nov, Puebla Pue., Mexico; 2004a. doi:10.2118/91724-MS.

Stone HL. Method for improved vertical sweep of oil reservoirs. US Patent No. 7303006 (PCT/US2004/014519). 2004b.

Stone HL. Vertical, conformance in an alternating water-miscible gas flood. In: SPE annual technical conference and exhibition, 26–29 Sep, New Orleans, Louisiana; 1982. doi:10.2118/11130-MS.

Svorstol I, Vassenden F, Mannhardt K. Laboratory studies for design of a foam pilot in the Snorre Field. In: SPE/DOE improved oil recovery symposium, 21–24 April, Tulsa, Oklahoma; 1996. doi:10.2118/35400-MS.

Tan B, Cooper AI. Functional oligo(vinyl acetate) CO_2-philes for solubilization and emulsification. J Am Chem Soc. 2005;127(25):8938–9. doi:10.1021/ja052508d.

Xing D, Wei B, McLendon WJ, Enick RM, McNulty S, Trickett K, Soong Y. CO_2-soluble, nonionic, water-soluble surfactants that stabilize CO_2-in-brine foams. SPE J. 2012;17(4):1172–85. doi:10.2118/129907-PA.

Xu Q, Rossen WR. Experimental study of gas injection in surfactant-alternating-gas foam process. In: SPE annual technical conference and exhibition, 5–8 Oct, Denver, Colorado; 2003. doi:10.2118/84183-MS.

Zeng YC, Ma K, Farajzadeh R, Puerto M, Biswal SL, Hirasaki GJ. Effect of surfactant partitioning between gaseous phase and aqueous phase on CO_2 foam transport for enhanced oil recovery. Transp Porous Media. 2016;114:777–93. doi:10.1007/s11242-016-0743-6.

A new mathematical model for horizontal wells with variable density perforation completion in bottom water reservoirs

Dian-Fa Du[1] · Yan-Yan Wang[2] · Yan-Wu Zhao[1] · Pu-Sen Sui[3] · Xi Xia[1]

Abstract Horizontal wells are commonly used in bottom water reservoirs, which can increase contact area between wellbores and reservoirs. There are many completion methods used to control cresting, among which variable density perforation is an effective one. It is difficult to evaluate well productivity and to analyze inflow profiles of horizontal wells with quantities of unevenly distributed perforations, which are characterized by different parameters. In this paper, fluid flow in each wellbore perforation, as well as the reservoir, was analyzed. A comprehensive model, coupling the fluid flow in the reservoir and the wellbore pressure drawdown, was developed based on potential functions and solved using the numerical discrete method. Then, a bottom water cresting model was established on the basis of the piston-like displacement principle. Finally, bottom water cresting parameters and factors influencing inflow profile were analyzed. A more systematic optimization method was proposed by introducing the concept of cumulative free-water production, which could maintain a balance (or then a balance is achieved) between stabilizing oil production and controlling bottom water cresting. Results show that the inflow profile is affected by the perforation distribution. Wells with denser perforation density at the toe end and thinner density at the heel end may obtain low production, but the water breakthrough time is delayed. Taking cumulative free-water production as a parameter to evaluate perforation strategies is advisable in bottom water reservoirs.

Keywords Bottom water reservoirs · Variable density perforation completion · Inflow profile · Cresting model · Cumulative free-water production

1 Introduction

Bottom water reservoirs are widely distributed on earth and hold a large proportion of oil reserves (Islam 1993). Taking China for example, there exist a large number of bottom water reservoirs, most of which are developed using horizontal wells. Compared with vertical wells, the producing sections of horizontal wells have direct contact with oil reservoirs, which not only reduces the producing pressure drawdown, but also ensures bottom water flowing into the wellbore more smoothly in a form of "pushing upward" (Besson and Aquitaine 1990; Dou et al. 1999; Permadi et al. 1996; Zhao et al. 2006). Owing to these advantages, it can effectively control bottom water cresting. The need of economic and effective development of bottom water reservoirs leads to the appearance of many types of completion methods, such as barefoot well completion, slotted screen well completion and perforation completion (Ouyang and Huang 2005). Recently, partial completion, variable density perforation completion and other new completion methods have been put forward to further control bottom water cresting (Goode and Wilkinson 1991; Sognesand et al. 1994). By accurately finding out the water

✉ Dian-Fa Du
dudf@upc.edu.cn

✉ Yan-Yan Wang
514375721@qq.com

[1] School of Petroleum Engineering, China University of Petroleum, Qingdao 266580, Shandong, China

[2] Sinopec Petroleum Exploration and Production Research Institute, Sinopec, Beijing 100083, China

[3] Shengli Production Plant, Shengli Oilfield Branch Company, Sinopec, Dongying 257051, Shandong, China

Edited by Yan-Hua Sun

production interval of horizontal wells, adopting plugging strategies or properly adjusting the bottom water inflow profile, these techniques can effectively prolong the life of production wells. And among all these techniques, perforation completion, including variable density perforation and selectively perforated completion, plays a critical role in alleviating water cresting (Pang et al. 2012).

Previously, scholars put more emphasis on the productivity evaluation of horizontal wells (Dikken 1990; Novy 1995; Penmatcha et al. 1998) and bottom water cresting (Permadi et al. 1995; Wibowo et al. 2004; Chaperon 1986). The published papers mostly focused on horizontal wells with open-hole completion. There is little research into horizontal wells with variable density perforation completion, and the ones that exist turned out to be very problematic: (1) The method for open-hole completion horizontal wells was used ignoring the fluid flow in perforation tunnels in these studies (Landman and Goldthorpe 1991; Yuan et al. 1996; Zhou et al. 2002). Then, a model describing the damage zone must be introduced to characterize the influence of perforation holes (Umnuayponwiwat and Ozkan 2000; Muskat and Wycokoff 2013). Some scholars utilized the numerical simulation method to discuss the impact of selective perforation on the productivity of horizontal wells and built a single-phase flow variable density perforation model for horizontal wells by two filtration zones (Li et al. 2010). Since the seepage resistance needs to be considered more precisely, especially in the middle and later periods of the oilfield development, the existing results are somewhat inaccurate. (2) Conventional simplified models cannot analyze formation pressure thoroughly and predict bottom water cresting. Furthermore, a non-uniformly distributed bottom water inflow profile along the wellbore was obtained without considering the wellbore pressure drop (Guo et al. 1992). (3) In order to optimize completion parameters for horizontal wells, oil production is usually viewed as the only objective function. It is reasonable for horizontal wells in conventional reservoirs. However, it is not accurate for horizontal wells located in bottom water reservoirs because of ignoring bottom water cresting, which decreases the effective production period of wells (Luo et al. 2015).

In this paper, based on the precise consideration of the fluid flow in each perforation, the flow behavior in perforations, wellbores, as well as reservoirs, was analyzed. Coupling the fluid flow in reservoirs and wellbore pressure drop, a comprehensive model, which can be used to evaluate productivity of horizontal wells, was developed based on potential functions and solved using the numerical discrete method. Then, a model describing bottom water cresting was established on the basis of the piston-like displacement principle. Finally, both the bottom water cresting behavior and the factors influencing inflow profile

were analyzed using the developed model, and a more systematic optimization method was proposed by introducing the concept of the cumulative free-water production, which could realize a balance between stabilizing oil production and controlling bottom water coning.

2 Productivity analysis for horizontal wells with variable density perforation completion

For horizontal wells with perforation completed in bottom water reservoirs, formation fluids firstly flow into perforation holes before converging in the horizontal wellbore. Under this circumstance, the effect of perforation holes is similar to that of short producing branches in inclined horizontal wells (Holmes et al. 1998). The real producing part for horizontal wells should be quantities of perforation holes. Therefore, the perforation holes can be regarded as "source-sink" term, and this kind of problem can be solved with a source function.

2.1 Analysis of fluid flow near wellbore regions

A model was built to describe the flow of formation fluid near horizontal wellbores, and the assumptions for this model are as follows:

(1) The formation is homogeneous with a uniform thickness;

(2) The horizontal permeability meets the following basic relationship: $K_x = K_y = K_h$, and the vertical permeability is $K_z = K_v$. The target reservoir is infinite in the horizontal plane;

(3) The single-phase fluid flowing in the reservoir is incompressible and the fluid flow obeys Darcy's law;

(4) The wellbore is horizontal, in which the perforations are unevenly distributed;

(5) The perforating direction is perpendicular to the wellbore, and the lengths as well as radius of all the perforation tunnels are the same.

Acoordinate system is established as shown in Fig. 1, in which the heel end of the wellbore is M_0 (x_0, y_0, z_0). The horizontal part of the wellbore (total length L, m) is divided into N segments. Therefore, the length of each segment is L/N.

Different segments have different characteristic parameters: the perforation density, $n_p(i)$; the perforation depth, l_p; bore diameter, D_p; phase angle ω; and the initial perforation angle ω_0. The heel end is selected as the origin of this coordinate, and the x direction is parallel with the wellbore. The coordinates of any point (x, y, z) in the jth perforation tunnel in the ith segment are as follows:

$$
\begin{cases}
x_p(i,j,t) = x_0 + \dfrac{L}{N}\left(\displaystyle\sum_{k=1}^{i-1}\sin\theta_k\cos\alpha_k + \dfrac{j}{n_p(i)}\sin\theta_i\cos\alpha_i + t\cdot l_p\sin\gamma_{ij}\cos\chi_{ij}\right) \\[2ex]
y_p(i,j,t) = y_0 + \dfrac{L}{N}\left(\displaystyle\sum_{k=1}^{i-1}\sin\theta_k\sin\alpha_k + \dfrac{j}{n_p(i)}\sin\theta_i\sin\alpha_i + t\cdot l_p\sin\gamma_{ij}\sin\chi_{ij}\right) \quad (0\le t\le 1) \\[2ex]
z_p(i,j,t) = z_0 + \dfrac{L}{N}\displaystyle\sum_{k=1}^{i-1}\cos\theta_k + \dfrac{j}{n_p(i)}\cos\theta_i + t\cdot l_p\cos\gamma_{ij}
\end{cases}
\tag{1}
$$

where θ_k is the angle between the kth segment and the vertical direction, $k = 1, 2, \ldots, i-1$; α_k is the angle between the x-axis and the projection of the kth segment in the horizontal plane, $k = 1, 2, \ldots, i-1$. θ_i is the angle between the ith segment and the vertical direction; α_i is the angle between the x-axis and the projection of the ith segment in the horizontal plane; γ_{ij} is the angle between the jth perforation in the ith producing part of the horizontal well and the z-axis; χ_{ij} is the angle between the x-axis and the projection in the xy plane of the jth perforation in the ith segment of the horizontal well; in this paper, when all the perforation is perpendicular to the horizontal wellbore, then $\chi_{i,j} = \pi/2$.

According to the classical theory of flow in porous media (Cheng 2011), the fluid flow at any point in an infinite reservoir obeys the Laplace equation. So we have:

$$
\frac{\partial^2 p}{\partial x^2} + \frac{\partial^2 p}{\partial y^2} + \frac{K_v}{K_h}\frac{\partial^2 p}{\partial z^2} = 0
\tag{2}
$$

where p is the pressure, MPa; K_v is the vertical permeability, μm^2; and K_h is the horizontal permeability, μm^2.

Substituting linear transformation $z' = \beta z$, where $\beta = \sqrt{K_h/K_v}$, into Eq. (2) gives:

$$
\frac{\partial^2 \Phi}{\partial x^2} + \frac{\partial^2 \Phi}{\partial y^2} + \frac{\partial^2 \Phi}{\partial z'^2} = 0
\tag{3}
$$

where $\Phi = \sqrt{K_h K_v}/\mu_o p$, m^2/s; μ_o is the oil viscosity, mPa s; and also the corresponding parameters of the perforation become:

$$
l'_p = l_p\sqrt{\beta^2\cos\gamma_{ij}^2 + \frac{1}{\beta^2}\sin\gamma_{ij}^2};
$$

$$
\sin\gamma'_{ij} = \frac{\beta\sin\gamma_{ij}}{\sqrt{\beta^2\cos\gamma_{ij}^2 + \frac{1}{\beta^2}\sin\gamma_{ij}^2}};
$$

$$
\cos\gamma'_{ij} = \frac{\cos\gamma_{ij}}{\beta\sqrt{\beta^2\cos\gamma_{ij}^2 + \frac{1}{\beta^2}\sin\gamma_{ij}^2}}.
$$

In order to simplify the solution procedure, one assumption that the flow rates of all perforations in each segment are equal, is proposed (Wang et al. 2006):

$$
q_p = \frac{q_{ra}(i)}{\Delta x\, n_p(i)}
\tag{4}
$$

where q_p is the flow rate of each perforation at the ith segment, m^3/s; $q_{ra}(i)$ is the flow rate of the ith segment, m^3/s; Δx is the length of the ith segment, which is equal to L/N, m; and $n_p(i)$ is the perforation density of the ith segment, holes/cm.

Compared with the length of the horizontal wellbore, the perforation is rather short. Therefore, it can be regarded as an infinitesimal line source. After integrating the point sink solution over the perforation direction, the pressure response of any perforation hole in the formation is obtained. For example, the potential at point $M(x, y, z')$ caused by the jth perforation in the ith segment can be described as:

$$
\Phi_{ij}(x,y,z') = \int_0^{l_p} -\frac{q_r(i,j)}{4\pi l_p r}\,ds + C_{ij}
$$

$$
= \int_0^{l_p} -\frac{Nq_{ra}(i)}{4\pi l_p L n_p(i)\sqrt{(x_p - x)^2 + (y_p - y)^2 + (z'_p - z')^2}}\,ds + C_{ij}
\tag{5}
$$

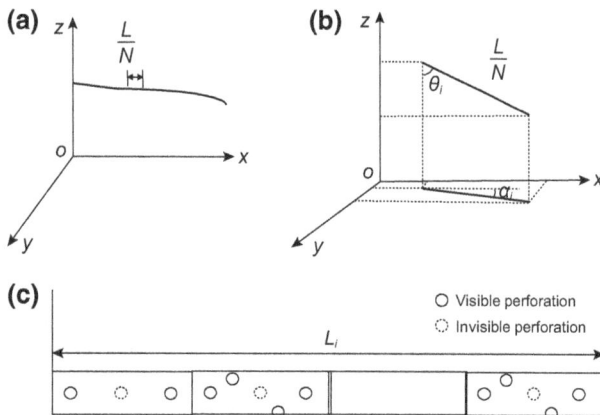

Fig. 1 Schematic diagram near the horizontal wellbore in a bottom water reservoir

The integration of Eq. (5) is as follows:

$$\Phi_{ij}(x, y, z') = -\frac{Nq_{ra}(i)}{4\pi l_p L n_p(i)} \ln \frac{r_{1ij} + r_{2ij} + l'_p}{r_{1ij} + r_{2ij} - l'_p} + C_{ij} \quad (6)$$

where $q_r(i, j)$ is the flow rate of the jth perforation tunnel in the ith segment, m³/s; r is the distance between the source point $M_p(x, y, z')$ and the target point $M(x, y, z')$; C_{ij} is an integration constant; r_{1ij} is the distance between the heel end and the target point; r_{2ij} is the distance between the toe end and the target point, and they observe the following expressions, respectively:

$$r_{1ij} = \sqrt{[x_p(i,j,0) - x]^2 + [y_p(i,j,0) - y]^2 + [z'_p(i,j,0) - z']^2} \quad (7)$$

$$r_{2ij} = \sqrt{[x_p(i,j,1) - x]^2 + [y_p(i,j,1) - y]^2 + [z'_p(i,j,1) - z']^2} \quad (8)$$

where $x_p(i, j, 0)$, $y_p(i, j, 0)$, $z_p(i, j, 0)$ and $x_p(i, j, 1)$, $y_p(i, j, 1)$, $z_p(i, j, 1)$ are the coordinates of the left and right ends of the jth perforation in the ith producing part.

Based on the mirror image reflection and superposition principle, the potential of the jth perforation tunnel for the ith production segment at point $M_p(x, y, z')$ is obtained:

$$\Phi(x, y, z') = -\frac{Nq_{ra}(i)}{4\pi l_p L n_p(i)} \sum_{n=-\infty}^{+\infty} \{\xi_{ij}(2h' + 4nh' - z'_{pij}(i,j,0),$$
$$2h' + 4nh' - z'_{pij}(i,j,1), x, y, z')$$
$$+ \xi_{ij}(4nh' + z'_{pij}(i,j,0), 4nh' + z'_{pij}(i,j,1), x, y, z')$$
$$- \xi_{ij}(-2h' + 4nh' + z'_{pij}(i,j,0), -2h'$$
$$+ 4nh' + z'_{pij}(i,j,1), x, y, z')$$
$$- \xi_{ij}(4nh' - z'_{pij}(i,j,0), 4nh'$$
$$- z'_{pij}(i,j,1), x, y, z') + C'_i\} \quad (9)$$

with

$$\xi_{ij}(\varepsilon_0, \varepsilon_1, x, y, z') = \ln \frac{r_{1ij} + r_{2ij} + l'_p}{r_{1i} + r_{2i} - l'_p} \quad (10)$$

$$r_{1i} = \sqrt{[x(i,j,0) - x]^2 + [y(i,j,0) - y]^2 + [\varepsilon_0 - z']^2} \quad (11)$$

$$r_{2i} = \sqrt{[x(i,j,1) - x]^2 + [y(i,j,1) - y]^2 + [\varepsilon_1 - z']^2} \quad (12)$$

According to the superposition principle, the potential at any point of the infinite formation created by all the perforation tunnels of the horizontal well is:

$$\Phi_{ij}(x, y, z') = -\frac{N}{4\pi l_p L} \sum_{i=1}^{N} \frac{q_{ra}(i)}{n_p(i)} \sum_{j=1}^{\frac{L}{N}n_p(i)} \ln \phi_{ij} + C \quad (13)$$

with

$$\phi_{ij} = \sum_{n=-\infty}^{+\infty} \left\{ \xi_{ij}\left(2h' + 4nh' - z'_{pij}(i,j,0), 2h' + 4nh' \right. \right.$$
$$\left. - z'_{pij}(i,j,1), x, y, z'\right) + \xi_{ij}\left(4nh' + z'_{pij}(i,j,0), 4nh' \right.$$
$$\left. + z'_{pij}(i,j,1), x, y, z'\right)$$
$$- \xi_{ij}\left(-2h' + 4nh' + z'_{pij}(i,j,0), -2h' \right.$$
$$\left. + 4nh' + z'_{pij}(i,j,1), x, y, z'\right)$$
$$\left. - \xi_{ij}\left(4nh' - z'_{pij}(i,j,0), 4nh' - z'_{pij}(i,j,1), x, y, z'\right) + \frac{2l_p}{nh'} \right\} C'_i \quad (14)$$

The boundary pressure at the oil–water interface is assumed to be constant. Therefore, the potential at any point of the formation may be expressed as:

$$\Phi_{ij}(x, y, z') = \Phi_e - \frac{N}{4\pi l_p L} \sum_{i=1}^{N} \frac{q_{ra}(i)}{n_p(i)} \sum_{j=1}^{\frac{L}{N}n_p(i)} \ln \phi_{ij} \quad (15)$$

Combining the definition of potential, the pressure at any point of the formation can be given as follows:

$$p_{wf,ij}(x, y, z') = p_e - \frac{N\mu_o}{4\pi l_p L \sqrt{K_h K_v}} \sum_{i=1}^{N} \frac{q_{ra}(i)}{n_p(i)} \sum_{j=1}^{\frac{L}{N}n_p(i)} \ln \phi_{ij} \quad (16)$$

where p_e is the boundary pressure, MPa.

For perforated completion, the perforation tunnels directly contact the formation. Therefore, the flow potential of some point, which is just located in the perforation, can be obtained using Eq. (15). In this case, there are two points involved: the target point and the source point (perforation point). To get rid of singularity phenomenon, the central hole in the wall is chosen as the jth perforation's target point when calculating the distance between two perforation points.

Only calculating the pressure of one central point for the segment and regarding it as the pressure of the whole segment will result in deviation when analyzing the segment's pressure of a horizontal wellbore. In this paper, taking the average over all the perforations' pressure in the same segment and using the average value as the representative pressure of that segment, the pressure of the ith segment is as follows:

$$p_{wf}(i) = \frac{1}{\Delta x \, n_p(i)} \sum_{j=1}^{\Delta x \, n_p(i)} p_{wf}(i,j) \quad (17)$$

where $p_{wf}(i)$ is the flow pressure for all the perforations in the ith segment, MPa.

There are $2N$ variables required to be calculated by analyzing pressure distribution along the wellbore: $p_{wf}(i)$ and $q_{ra}(i)$, $(i = 1, 2, \ldots, N)$. However, it only contains N equations in the flow model [Eq. (15)]. Therefore, one more model is needed to describe pressure drop along the wellbore (Li et al. 1996, 2006).

2.2 Wellbore pressure drop model

For perforated horizontal wells, the main idea to develop a wellbore pressure drop model is to divide the horizontal wellbore into several segments, and each segment is subdivided into several smaller parts that only include one perforation (Su and Gudmundsson 1994).

According to the analysis of the pressure drop in a wellbore, the total pressure loss can be written as:

$$\mathrm{d}p_{wf}(i) = \mathrm{d}p_{fric}(i) + \mathrm{d}p_{acc}(i) + \mathrm{d}p_{mix}(i) + \mathrm{d}p_{G}(i) \quad (18)$$

where $\mathrm{d}p_{fric}(i)$ is the friction loss of the ith segment, MPa; $\mathrm{d}p_{acc}(i)$ is the acceleration loss of the ith segment, MPa; $\mathrm{d}p_{mix}(i)$ is the mixing loss of the ith segment, MPa; and $\mathrm{d}p_{G}(i)$ is the gravity loss of the ith segment, MPa.

It should be noted that, in the previous section, a mechanical field described by the flow model is established based on the potential function. Therefore, when the potential function is used to deal with flow problems, the pressure loss caused by viscous force has been considered. As we all know, the mathematical expression of fluid potential is Kp/μ, where K is the formation permeability, p represents pressure, while μ means the fluid viscosity, and it is used to describe the viscous force, which will lead to pressure loss along the perforation. So it means that the viscous force has been taken into consideration in the first model. In other words, when building wellbore pressure here, only four kinds of pressure loss should be calculated.

The calculation method of frictional loss and acceleration loss between two perforation tunnels is expressed, respectively.

$$\Delta p_{fric} = 1.34 \times 10^{-13} f_{fric}(i,j) \frac{\Delta l}{D} \frac{\rho \bar{v}_s^2(i,j)}{2}$$
$$= 1.0862 \times 10^{-13} f_{fric}(i,j) \frac{\rho q_L^2(i,j)}{D^5 n_p(i)} \quad (19)$$

$$\Delta p_{acc}(i,j) = \rho \left[\bar{v}_s^2(i,j) - \bar{v}_s^2(i,j+1) \right]$$
$$= 3.5215 \times 10^{-13} \frac{\rho}{D^5} \left[q_L^2(i,j) - q_L^2(i,j+1) \right] \quad (20)$$

where ρ is the liquid density, kg/m^3; D is the wellbore diameter, m; $q_L(i,j)$ is the flow rate along the wellbore, m^3/d; $\bar{v}_s(i,j)$ is the average velocity of the jth perforation in the ith segment, m/s; and $f_{fric}(i,j)$ is the friction coefficient of the jth perforation in the ith segment.

In Eq. (19), one parameter, called the friction coefficient, is introduced. The calculation of the friction coefficient is dependent on the Reynolds number. If the Reynolds number is less than or equal to 2000, the flow is laminar; otherwise, it is turbulent flow. And the expression for the Reynolds number is as follows:

$$Re = 7.3682844 \times 10^{-3} \frac{Q\rho}{r\mu}$$

where Re is the Reynolds number; Q is the axial flow rate along the wellbore, m^3/d; μ is the viscosity of the flowing fluid, mPa s; and r is the wellbore radius, m.

In addition, we also introduced a criterion when calculating the mixing loss between two perforation holes. The specific calculation method for frictional loss and mixing losses is listed in Table 1. In Table 1, q_c is the critical rate, m^3/d; ε is the surface roughness, mm.

When calculating the mixing loss, one parameter, Δp_{per}, is introduced. It is the frictional loss showing up after perforating and can be figured out using Eq. (19).

The friction loss, acceleration loss and the mixing loss of the ith segment are listed below:

$$\mathrm{d}p_{fric}(i) = 1.0862 \times 10^{-13} \frac{\rho}{D^5} \sum_{j=1}^{\Delta x n_p(i)} f_{fric}(i,j) \frac{q_L^2(i,j)}{n_p(i)} \quad (21)$$

$$\mathrm{d}p_{acc}(i) = \sum_{j=1}^{\Delta x n_p(i)} \Delta p_{acc}(i,j)$$
$$= 3.5215 \times 10^{-13} \frac{\rho}{D^5} \sum_{j=1}^{\Delta x n_p(i)} \left[q_L^2(i,j) - q_L^2(i,j+1) \right] \quad (22)$$

$$\mathrm{d}p_{mix}(i) = \sum_{j=1}^{n_p(i)} \Delta p_{mix}(i,j) \quad (23)$$

Table 1 Calculation method for friction coefficient and mixing loss between two perforation holes

Flow pattern	Discriminant standard	Calculation of friction coefficient $f_{fric}(i,j)$ (Cheng 2011)	Calculation of mixing losses $\Delta p_{mix}(i,j)$ (Cheng 2011)
Laminar flow	$q_L(i,j) < q_c$	$\frac{64}{Re(i,j)}$	$\Delta p_{per}(i,j) - 0.31 \times 10^7 Re(i,j) \frac{q_{ra}(i)}{\Delta x n_p(i) q_L(i,j)}$
Turbulent flow	$q_L(i,j) > q_c$	$\left[-1.8 \lg\left[\frac{6.9}{Re(i,j)} + \left(\frac{\varepsilon}{3.7D}\right)^{1.11} \right] \right]^{-2}$	$0.76 \times 10^3 Re(i,j) \frac{q_{ra}(i)}{\Delta x n_p(i) q_L(i,j)}$

If the horizontal well is inclined, the gravity loss is non-ignorable, and the wellbore pressure drop model becomes:

$$\Delta p_{wf}(i,j) = \begin{cases} 1.0862 \times 10^{-13} \dfrac{\rho}{D^5} \sum\limits_{j=1}^{\Delta x n_p(i)} f_{fric}(i,j) \dfrac{q_L^2(i,j)}{n_p(i)} + 3.5215 \times 10^{-13} \dfrac{\rho}{D^5} \sum\limits_{j=1}^{\Delta x n_p(i)} [q_L^2(i,j) - q_L^2(i,j+1)] \\ + \rho g \Delta x \cos \theta_i + \sum\limits_{j=1}^{n_p(i)} \Delta p_{mix}(i,j) \qquad q_{ra}(i) \neq 0 \\ 1.0862 \times 10^{-13} \dfrac{\rho}{D^5} \sum\limits_{j=1}^{\Delta x n_p(i)} f_{fric}(i,j) \dfrac{q_L^2(i,j)}{n_p(i)} + \rho g \Delta x \cos \theta_i \qquad q_{ra}(i) = 0 \end{cases} \tag{24}$$

2.3 Coupling model

When a horizontal well begins to produce reservoir fluids, the fluids in the perforations connect the wellbore and oil reservoir together. Therefore, perforation is considered as an infinitesimal linear sink, which directly contacts reservoir and wellbore. Meanwhile, pressure responses are generated in the whole reservoir. The generated pressure responses near the perforations are associated with oil inflow in the radial direction of the horizontal wellbore, which can be calculated by utilizing the reservoir flow model. Since the pressure in perforation holes is relevant to the wellbore pressure, a coupling relationship exists between the reservoir flow model and the wellbore pressure drop model.

According to the reservoir flow model, a model is developed to calculate steady-state productivity of the horizontal well with variable density perforation, in which the pressure drop along the horizontal wellbore is considered:

3 Analysis of inflow profiles in bottom water reservoirs

In this paper, an iterative method is used to solve the above-mentioned model, and the iterative process is shown in Fig. 2.

The bottom water rises fastest in the vertical plane of the horizontal wellbore (Cheng et al. 1994). In other words, the bottom water will firstly break through into the wellbore in this plane due to the highest pressure gradients in this profile. Therefore, a complicated 3-D problem can be turned into a 2-D problem in the xz profile where the wellbore lies. The formation between the wellbore and the oil–water interface is discretized according to the division of the horizontal wellbore in the reservoir flow model. And the total number of grids in the vertical direction is n_z, which is shown in Fig. 3.

The rise of bottom water is treated as a piston-like flooding process. That is to say, there is an obvious interface between the oil zone and the water zone. The oil–water contact moves upward to the horizontal wellbore in the vertical direction. Once it reaches any point of the wellbore, water breakthrough occurs there. According to the material balance theory (Xiong et al. 2013), we have:

$$\begin{cases} p_{wf}(i) = p_e - \dfrac{1}{l_p} \dfrac{\mu}{4\pi k \Delta x} \sum\limits_{k=1}^{N} \dfrac{q_{ra}(k)}{n_p(k)} \sum\limits_{j=1}^{\frac{L}{\Delta x} n_p(k)} \phi_{kj} \qquad k = 1, 2, \ldots, N; \quad i = 1, 2, \ldots, \dfrac{L_k}{\Delta x} \\ p_{wf}(i) = p_{wf}(i-1) + 0.5[\Delta p_{wf}(i-1) + \Delta p_{wf}(i)] \end{cases} \tag{25}$$

where $p_{wf}(i)$ is the wellbore pressure in the ith segment, MPa; N is the number of divided segments along the wellbore; and ϕ_{kj} is a function corresponding to the horizontal wellbore as well as the oil–water interface.

$$[S_w(i, k+1) - S_{wc}]\varphi dxdydz = v_w dxdydt \tag{26}$$

Based on Darcy's law, the water rise velocity is as follows:

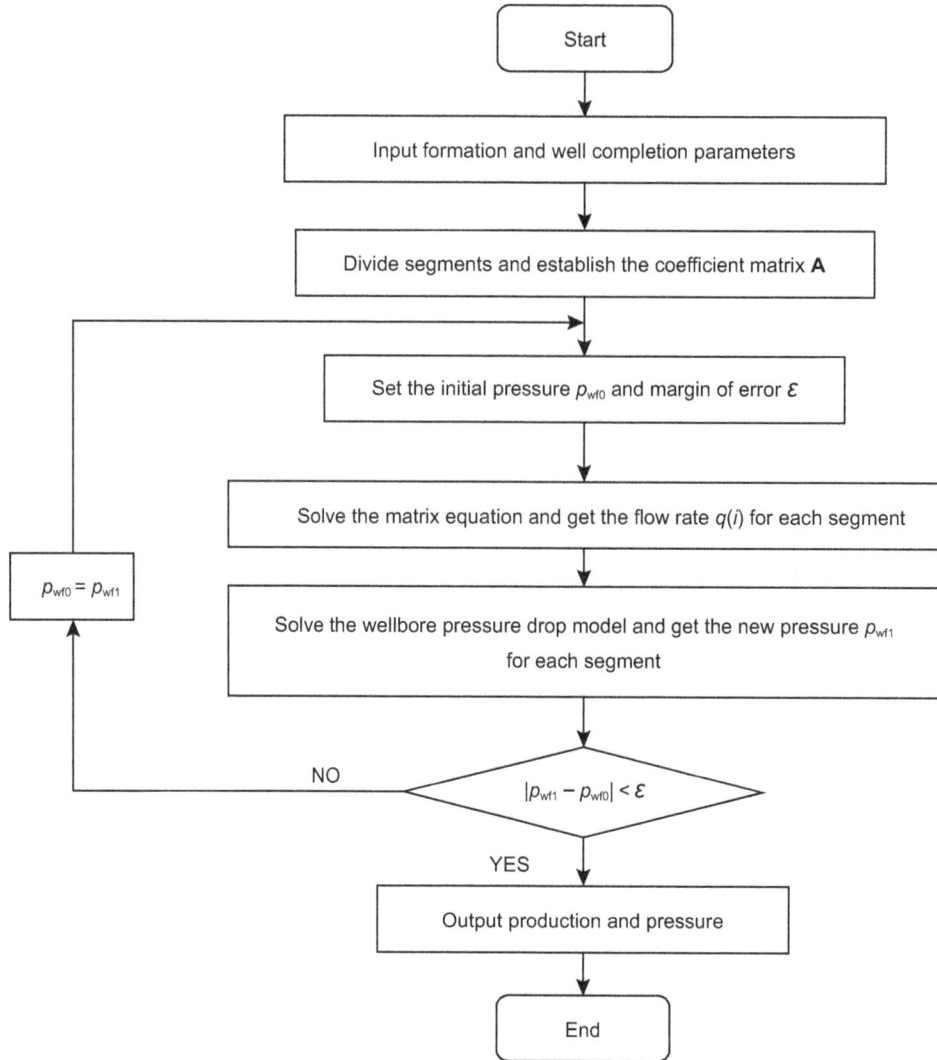

Fig. 2 Flowchart for solving the coupling model

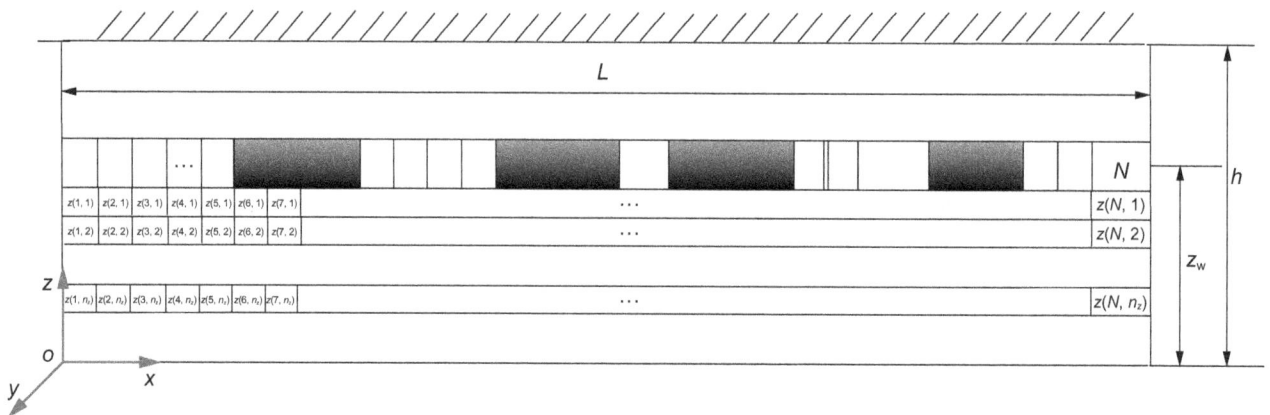

Fig. 3 Schematic diagram of the physical model for bottom water cresting

$$v_{\mathrm{w}} = \frac{K_{\mathrm{rw}} K_{\mathrm{v}}}{\mu_{\mathrm{w}}} \frac{\partial p(i,k)}{\partial z} \tag{27}$$

where $S_{\mathrm{w}}(i, k+1)$ is the water saturation of the $(i, k+1)$ grid at time t in the longitudinal profile; S_{wc} is the connate water saturation; K_{rw} is the relative permeability to water; and μ_{w} is the water viscosity, mPa s.

According to the results of grid discretization, the vertical pressure gradients between any two contiguous grids are as follows:

$$\begin{cases} \Delta p(i,k) = p(i, k+1) - p(i,k) & k = 1, 2, \ldots, n_z - 1 \\ \Delta p(i,k) = p_{\mathrm{wf}}(i) - p(i,k) & k = n_z \end{cases} \tag{28}$$

Substituting Eqs. (28) into (27) gives the rise velocity of bottom water:

$$v_{\mathrm{w}}(i,k) = \frac{K_{\mathrm{rw}} K_{\mathrm{v}}}{\mu_{\mathrm{w}}} \frac{\Delta p(i,k)}{\Delta z} \tag{29}$$

According to the established mathematical model, the pressure at any grid between the oil–water contact surface and the horizontal wellbore can be written as:

$$p_{\mathrm{wf}}(i,k) = p_{\mathrm{e}} - \frac{N \mu_{\mathrm{o}}}{4 \pi L \sqrt{K_{\mathrm{h}} K_{\mathrm{v}}}} \sum_{i=1}^{N} q(i,k) \phi_{ik} \tag{30}$$

Combining with Eq. (26), the time required for the water rising from the $(k+1)$th grid to the kth grid is obtained:

$$t(i,k) = \frac{\varphi \mu_{\mathrm{w}}}{K_{\mathrm{rw}} K_{v}} \frac{[S_{\mathrm{w}}(i,k) - S_{\mathrm{wc}}](\Delta z)^2}{\Delta p(i,k)} \tag{31}$$

The breakthrough time at the ith segment is:

$$t_{\mathrm{a}} = \sum_{k=1}^{n_z} t(i,k) \tag{32}$$

where φ is the porosity; n_z is the number of meshes in the longitudinal direction between the wellbore and the oil–water interface.

4 Case study

Using the developed model, the well productivity and water breakthrough for a horizontal well in a bottom water drive reservoir were evaluated. Table 2 lists the bottom water drive reservoir properties and its drilling and completion parameters.

The steady-state productivity of the horizontal well with variable density perforation completion was evaluated, and also the bottom water inflow profile was calculated. A set of basic variable density perforation cases are designed and are shown in Fig. 4. For simplicity, the average perforation

density of each case is 2 shots/m. In Case 1, the perforation is uniformly distributed. In Cases 2 and 3, the perforation density at the heel end of the horizontal well is larger than that at the toe end of the horizontal well, while the perforation density is denser at the toe end than that at the heel end for Cases 4–6.

The simulation results for all the six cases are shown in Figs. 5, 6, 7, 8 and 9. Figure 5 shows the pressure distribution along the horizontal wellbore. In fact, there exists a pressure drop along the perforation hole. However, it has little relationship with our research object; thus, the relevant calculation was not carried out in this work. In order to better identify their characteristics, the pressure distribution curves of only three cases (Cases 1, 2 and 5) are plotted together in Fig. 5. It can be seen that the pressure distribution curves are steep near the heel end of the horizontal wellbore, while relatively flat at the toe end. In addition, the greater the pressure drop is, the denser the perforations will be. Near the toe end, the pressure of Case 5 is higher than those of Case 1 and Case 2, indicating a denser perforation and a greater pressure drop in this location, while near the heel end, the difference of these three curves is smaller. Figure 6 shows the friction and acceleration losses (Case 1) along the wellbore, respectively. At any point of the horizontal wellbore, the friction loss is greater than the acceleration loss, and the former is nearly six times as much as the latter, which means the friction loss plays a leading role.

Figure 7 gives the flow rate distribution along the horizontal wellbore. For the horizontal well with uniformly distributed perforation, the flow rate at the toe end is lower than that at the heel end, due to the influence of the wellbore pressure drop. These six cases have the same number of perforations, and thus, the production is also similar, especially for Cases 1–3. For Case 3, although the difference of the flow rate between the heel end and the toe end is larger compared with the other five cases (Fig. 7), the production rate of the horizontal well is slightly larger (Fig. 8). Therefore, in order to maximize the production of the horizontal well, a perforation scheme with a larger perforation density at the toe end should be adopted. It should be noted that a larger perforation density at the toe end does not necessarily result in a higher production rate. The reason is that this kind of perforation scheme will lead to much lower flow rate at the toe end. In addition, Fig. 7 also shows that the flow rate distribution of Case 6 is more uniform compared with the other five cases. Due to the influence of the pressure drop along the wellbore and the reduced end effect, the flow rate distribution of Case 6 has a more uniform distribution of cresting height, which will delay the occurrence of bottom water breakthrough. Simulation results indicate that reducing the perforation density at the heel end is helpful for obtaining an evenly advancing

Table 2 Parameters for the bottom water drive reservoir and its drilling and completion parameters

Parameter	Value	Parameter	Value
Reservoir thickness, m	28	Bottom hole pressure, MPa	20
Pressure at the oil–water interface, MPa	25	Horizontal well length, m	600
Horizontal permeability, μm^2	0.2	Wellbore diameter, cm	17.45
Vertical permeability, μm^2	0.05	Water avoidance height, m	21
Oil density, kg/m^3	845	Relative wellbore roughness	0.0001
Oil viscosity, mPa s	15.4	Initial perforation phase angle	$\pi/2$
Perforation density, shots/m	2	Angle between the wellbore and the x-axis	0

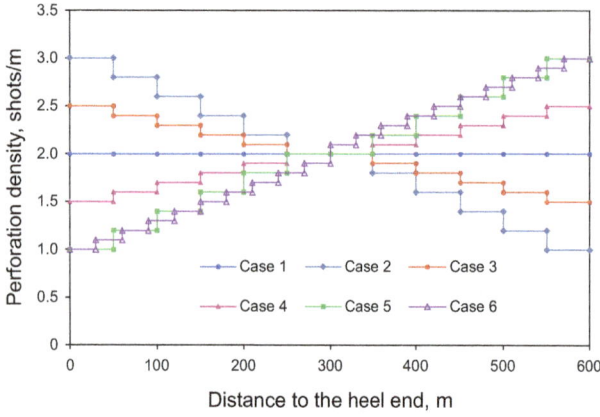

Fig. 4 Cases for variable density perforation

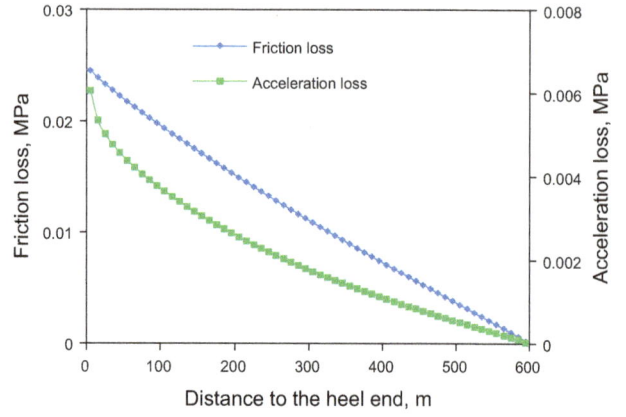

Fig. 6 Friction loss and acceleration loss along the horizontal wellbore for Case 1

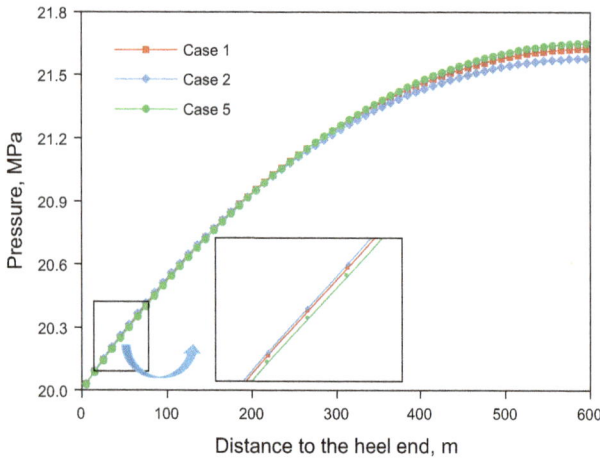

Fig. 5 Pressure distribution along the horizontal wellbore

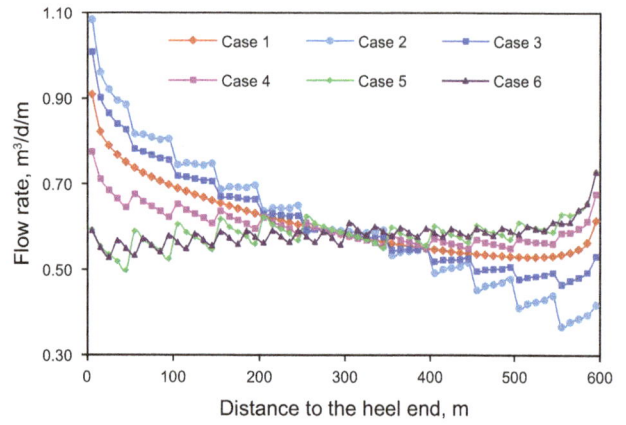

Fig. 7 Flow rate distribution along the horizontal wellbore for different cases

water profile on the vertical plane of the horizontal wellbore.

The detailed distribution of bottom water breakthrough time along the wellbore is shown in Fig. 9. Compared with Case 1 (evenly distributed perforation), the redistribution of perforating density changes both the breakthrough location and breakthrough time simultaneously. It is obvious that the larger perforation density means a shorter breakthrough time. Meanwhile, the bottom water

breakthrough location and time of the whole wellbore can be calculated by the developed model, as shown in Table 3.

Figure 10 illustrates the distribution of the bottom water rising height along the wellbore for three different cases, which is in good agreement with the results in Fig. 9. The bottom water cresting height of Case 2 is shown in Fig. 11. It shows that the effect of the perforation density on bottom water cresting height becomes more serious with the increase of time. The deformation of the water ridge is

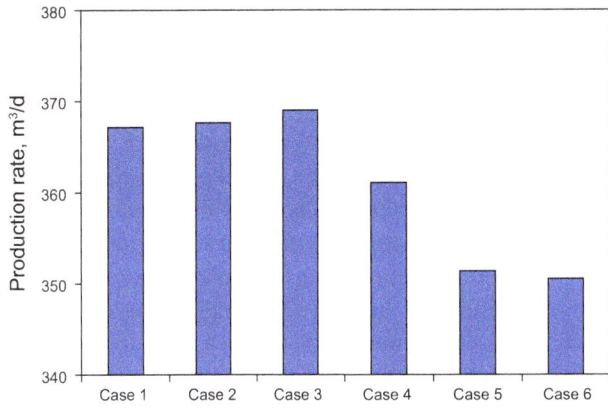

Fig. 8 Production rate of the horizontal well for different cases

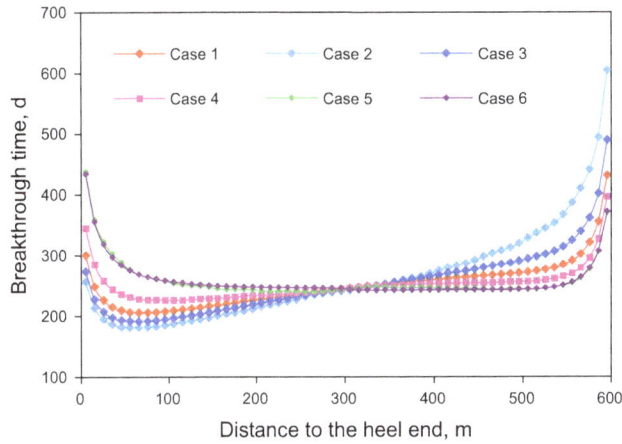

Fig. 9 Distribution of breakthrough time for different cases

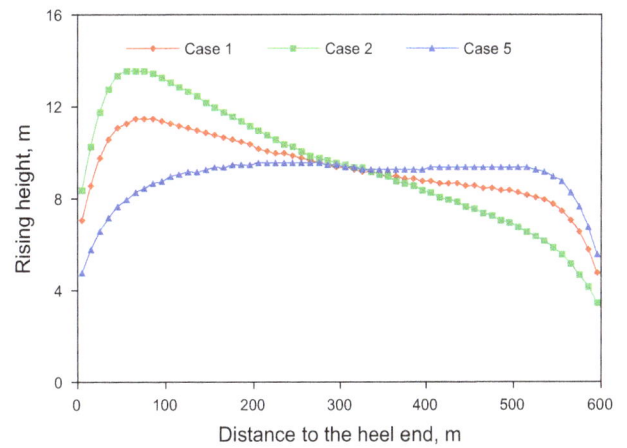

Fig. 10 Bottom water rising height for different cases (production time, 150 days)

horizontal well with a higher production rate will have a shorter water-free production period. So there exists a contradiction between increasing the well production and delaying the breakthrough time of bottom water. In order to comprehensively evaluate the effect of both controlling bottom water cresting and stabilizing oil production, a new parameter, called the cumulative free-water production, is defined as follows:

$$Q_o = q \times T_o \tag{33}$$

where Q_o is the cumulative oil production without water, m^3; q is the rate when the horizontal well produces steadily, m^3/d; and T_o is the production time without water, d.

Obviously, the cumulative oil production without water of the horizontal well could consider both the well production and effective production time.

According to the calculation results (Table 4), the Q_o of Cases 4 and 5 is higher than those of other Cases owing to the more uniform inflow profile. Therefore, for the horizontal well studied in this paper, the perforation scheme with denser perforation hole at the toe end and sparser perforation hole at the heel end is a more appropriate choice.

obvious, which reflects on the degree of water crest asymmetry.

According to the above results, the perforation scheme with higher production (Cases 2 and 3) will advance the time of the bottom water breakthrough and thus have a negative effect on the development of bottom water reservoirs using horizontal wells. In other words, the

Table 3 Bottom water breakthrough location and time

Case	Distance between the breakthrough point and the heel end, m	Breakthrough time at the first breakthrough point, d	Breakthrough time at the point 100 m away from the heel, d	Breakthrough time at the point 500 m away from the heel, d
1	65	205.3	208.5	271.9
2	55	180.8	187.1	328.5
3	60	190.9	196.4	294.2
4	105	226.0	226.2	256.5
5	265	239.1	253.4	243.6
6	335	242.1	256.0	284.9

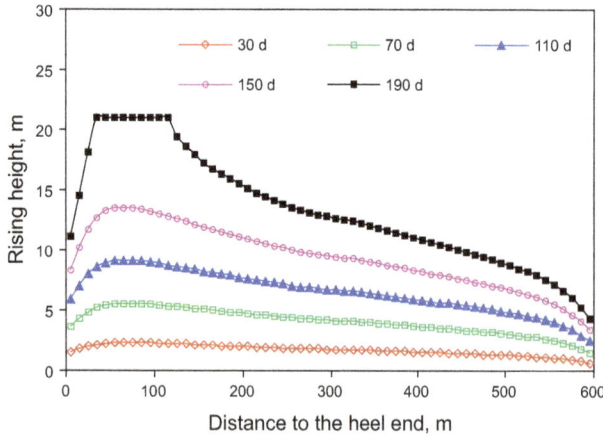

Fig. 11 Rising height of bottom water at different production times (Case 2)

Table 4 Parameters required for calculating Q_o of different cases

Case	Breakthrough location, m	T_o, d	q, m³/d	Q_o, 10^4 m³
1	65	205.3	367.1	7.5
2	55	180.8	367.6	6.6
3	65	190.9	369.0	7.0
4	105	226.0	361.0	8.2
5	265	239.1	351.4	8.4

5 Conclusions

In this paper, a model coupling fluid flow in reservoirs and pressure drop along the wellbore was developed to obtain the distribution of flow rate along the horizontal wellbore and to determine the horizontal well productivity. In this model, the fluid flow at each perforation hole was taken into account, and the perforation tunnels were treated as many infinitesimal sections. After that, a bottom water cresting model was developed based on the piston-like displacement principle, which could be used to calculate the bottom water breakthrough location, breakthrough time and the shape of water cresting at different times.

For a horizontal well with uniform density perforation completed in a bottom water reservoir, the pressure drop along the wellbore may significantly affect the water breakthrough mode. Specifically, the bottom water will firstly break through at the heel end of the wellbore. Neglecting the end effects, water breakthrough occurs in proper sequence from heel end to toe end. The bottom water breakthrough at a specific position will happen earlier if the perforation density is increased. When a horizontal well with denser perforation density at the toe end and sparser perforation density at the heel end, the lowest

point of the distribution curve will move upward, which means the water breakthrough time is delayed. Under these circumstances, the horizontal well obtains a longer effective production period.

It is meaningful to take cumulative oil production without water as a parameter to evaluate the perforation strategies for horizontal wells. With this parameter, both the effects of productivity of the horizontal well and bottom water breakthrough time can be considered comprehensively. However, in order to simplify the calculation process, some assumptions have been introduced in this model, some of which are ideal ones and different from the actual reservoirs. Further work needs to be done in order to extend the application of this model.

References

Besson J, Aquitaine E. Performance of slanted and horizontal wells on an anisotropic medium. In: SPE European petroleum conference, 21–24 October. The Hague, Netherlands; 1990. doi:10.2118/20965-MS.

Chaperon I. A theoretical study of coning toward horizontal and vertical wells in anisotropic formations: subcritical and critical rates. In: SPE annual technical conference and exhibition, 5–8 October. New Orleans, Louisiana; 1986. doi:10.2118/15377-MS.

Cheng LS. Higher seepage mechanics. Beijing: Petroleum Industry Press; 2011 (**in Chinese**).

Cheng LS, Lang ZX, Zhang LH. Reservoir engineering problem of horizontal wells coning in bottom-water drive reservoir. J Univ Pet China. 1994;18(4):43–7 (**in Chinese**).

Dikken BJ. Pressure drop in horizontal wells and its effect on production performance. In: SPE India oil and gas conference and exhibition, 17–19 February. New Delhi, India; 1990. doi:10.2118/39521-MS.

Dou H, Lian S, Guan C. The experimental studies of physical simulation of bottom water reservoirs with barrier and permeable interbed on horizontal well. In: SPE western regional meeting, 26–27 May. Anchorage, Alaska; 1999. doi:10.2118/55995-MS.

Goode PA, Wilkinson DJ. Inflow performance of partially open horizontal wells. J Pet Technol. 1991;43(8):983–7. doi:10.2118/19341-PA.

Guo BY, Molinard JE, Lee RL. A general solution of gas/water coning problem for horizontal wells. In: SPE European petroleum conference, 16–18 November. Cannes, France; 1992. doi:10.2118/25050-MS.

Holmes JA, Barkve T, Lund O. Application of a multi-segment well model to simulate flow in advanced wells. In: European petroleum conference, 20–22 October. The Hague, Netherlands; 1998. doi:10.2118/50646-MS.

Islam MR. Oil recovery from bottom-water reservoirs. J Pet Technol. 1993;45(6):514–6. doi:10.2118/25394-PA.

Landman MJ, Goldthorpe WH. Optimization of perforation distribution for horizontal wells. In: SPE Asia-Pacific conference, 4–7 November. Perth, Australia; 1991. doi:10.2118/23005-MS.

Li XG, Wang QH, Li Y. Numerical simulation model of perforated well completions. J Univ Pet China. 1996;20(2):48–53 (**in Chinese**).

Li GS, Song J, Xiong W, et al. Simulation model and calculation of seepage flow field for high pressure waterjet perforated wells. Pet Explor Dev. 2006;32(6):97–100 (**in Chinese**).

Li H, Chen DC, Meng HX. Optimized models of variable density perforation in the horizontal well. Pet Explor Dev. 2010;37(3): 363–8 (in Chinese).

Luo X, Jiang L, Su Y, Huang K. The productivity calculation model of perforated horizontal well and optimization of inflow profile. Petroleum. 2015;1(2):154–7. doi:10.1016/j.petlm.2015.04.002.

Muskat M, Wycokoff RD. An approximate theory of water-coning in oil production. Trans AIME. 2013;114(1):144–63. doi:10.2118/935144-G.

Novy RA. Pressure drops in horizontal wells: when can they be ignored. SPE Reserv Eng. 1995;10(1):29–35. doi:10.2118/24941-PA.

Ouyang LB, Huang B. A comprehensive evaluation of well-completion impacts on the performance of horizontal and multilateral wells. In: SPE annual technical conference and exhibition, 9—12 October. Dallas, Texas; 2005. doi:10.2118/96530-MS.

Pang W, Chen DC, Zhang ZP, et al. Segmentally variable density perforation optimization model for horizontal wells in heterogeneous reservoirs. Pet Explor Dev. 2012;39(2):230–8. doi:10.1016/S1876-3804(12)60036-6.

Penmatcha VR, Arbabi S, Aziz K. A comprehensive reservoir/wellbore model for horizontal well. In: SPE India oil and gas conference and exhibition, 17–19 February. New Delhi, India; 1998. doi:10.2118/39521-MS.

Permadi P, Lee RL, Kartoatmodjo RST. Behavior of water cresting under horizontal wells. In: SPE annual technical conference and exhibition, 22–25 October. Dallas, Texas; 1995. doi:10.2118/30743-MS.

Permadi P, Gustiawan E, Abdassah D. Water cresting and oil recovery by horizontal wells in the presence of impermeable streaks. In: SPE/DOE improved oil recovery symposium, 21–24 April. Tulsa, Oklahoma; 1996. doi:10.2118/35440-MS.

Sognesand S, Skotner P, Hauge J. Use of partial perforations in Oseberg horizontal wells. In: SPE annual technical conference and exhibition, 25–28 September. New Orleans, Louisiana; 1994. doi:10.2118/28569-MS

Su Z, Gudmundsson JS. Pressure drop in perforated pipes: experiments and analysis. In: SPE Asia Pacific oil and gas conference, 7–10 November. Melbourne, Australia; 1994. doi:10.2118/28800-MS.

Umnuayponwiwat S, Ozkan E. Water and gas coning toward finite-conductivity horizontal wells: cone buildup and breakthrough. In: SPE rocky mountain regional/low-permeability reservoirs symposium and exhibition, 12–15 March, Denver, Colorado; 2000. doi:10.2118/60308-MS.

Wang RH, Zhang YZ, Bu YH, et al. A segmentally numerical calculation method for estimating the productivity of perforated horizontal wells. Pet Explor Dev. 2006;33(5):630–3 (in Chinese).

Wibowo W, Permadi P, Mardisewojo P, et al. Behavior of water cresting and production performance of horizontal well in bottom water drive reservoir: a scaled model study. In: SPE Asia Pacific conference on integrated modelling for asset management, 29–30 March, Kuala Lumpur, Malaysia; 2004. doi:10.2118/87046-MS.

Xiong J, He HP, Xiong YM, et al. The effect of partial completion parameters in horizontal well on water coning. Nat Gas Geosci. 2013;24(6):1232–7 (in Chinese).

Yuan H, Sarica C, Brill JP. Effect of perforation density on single phase liquid flow behavior in horizontal wells. In: SPE international conference on horizontal well technology, 18–20 November, Calgary, Alberta, Canada; 1996. doi:10.2118/37109-MS.

Zhao G, Zhou J, Liu X. An insight into the development of bottom water reservoirs. J Can Pet Technol. 2006;45(4):22–30. doi:10.2118/06-04-CS.

Zhou ST, Ma DQ, Liu M. Optimization of perforation tunnel distribution in perforated horizontal wells. J China Univ Pet (Ed Nat Sci). 2002;26(3):52–4 (in Chinese).

8

Naturally fractured hydrocarbon reservoir simulation by elastic fracture modeling

Mehrdad Soleimani[1]

Abstract Accurate fluid flow simulation in geologically complex reservoirs is of particular importance in construction of reservoir simulators. General approaches in naturally fractured reservoir simulation involve use of unstructured grids or a structured grid coupled with locally unstructured grids and discrete fracture models. These methods suffer from drawbacks such as lack of flexibility and of ease of updating. In this study, I combined fracture modeling by elastic gridding which improves flexibility, especially in complex reservoirs. The proposed model revises conventional modeling fractures by hard rigid planes that do not change through production. This is a dubious assumption, especially in reservoirs with a high production rate in the beginning. The proposed elastic fracture modeling considers changes in fracture properties, shape and aperture through the simulation. This strategy is only reliable for naturally fractured reservoirs with high fracture permeability and less permeable matrix and parallel fractures with less cross-connections. Comparison of elastic fracture modeling results with conventional modeling showed that these assumptions will cause production pressure to enlarge fracture apertures and change fracture shapes, which consequently results in lower production compared with what was previously assumed. It is concluded that an elastic gridded model could better simulate reservoir performance.

Keywords Reservoir performance · Discrete fracture model · Naturally fractured reservoir · History matching · Elastic gridding

1 Introduction

To handle the complexity of reservoir heterogeneity which comes from natural fractures, the nature of the reservoir should be determined in advance (Agar and Hampson 2014). The complexity in reservoir modeling either comes from reservoir characterization (e.g., heterogeneity and anisotropy in permeability) or from the process of oil recovery (e.g., capillarity, gravity and phase behavior), or from both (Kresse et al. 2013). To resolve this complexity, a dual-porosity model (Nie et al. 2012) and recently a triple-porosity model (Huang et al. 2015; Sang et al. 2016) have been introduced to simulate fractured reservoirs. Despite many useful features of the dual-porosity model, it cannot provide reliable results in reservoirs in which fractures do not intersect (Karimi-Fard and Firoozabadi 2003). Applicability of the classic double-porosity model with a constant shape factor to low-permeability reservoir simulation is questionable (Cai et al. 2015). It also does not describe discrete fractures, which are the main source of challenge in naturally fractured reservoirs (Chen et al. 2008; Presho et al. 2011; Soleimani 2016a). History matching of naturally fractured reservoirs becomes more challenging particularly when these models are represented using discrete fracture network (DFN) models in carbonate rocks (Bahrainian et al. 2015). DFN models were also represented using generally unstructured grids to achieve high degrees of geological realism. Karimi-Fard et al. (2004) introduced an efficient discrete fracture model applicable for general-purpose reservoir simulators. Accurate representation of each individual fracture requires use of unstructured grids which can

✉ Mehrdad Soleimani
  msoleimani@shahroodut.ac.ir

[1] Faculty of Mining, Petroleum and Geophysics, Shahrood University of Technology, Shahrood, Iran

Edited by Jie Hao

consider effects of the fracture aperture (Mi et al. 2014). Jiang and Younis (2015) implemented a lower-dimensional DFN model based on unstructured gridding for handling complex fracture geometry in simulated formation. Although DFN models have several advantages, they cannot be used directly with standard history matching techniques for strongly heterogeneous reservoirs with complex fracture geometry (Yan et al. 2016). Figure 1a shows a real structure and shape of different porosities in a limestone cube. Figure 1b, c shows conventional dual-porosity and DFN approaches that model the realistic fractures in a limestone fractured reservoir. In such models, if in the course of history matching, more fractures are added, moved or deleted, the model must be re-gridded from the beginning (Salimi and Bruining 2010).

In this study, we try to increase realism of the fracture model by considering shape of the fracture and the variation of its properties through production and pressure regime change in fluid flow procedure. Fracture characteristics which change through time of production are included in elastic properties of fractures. The elastic gridding scheme was introduced in this study into the fracture modeling scheme to handle heterogeneity of a complex reservoir. This approach was then applied on a fractured limestone reservoir in southwest Iran. Results of the application of the proposed strategy for fracture modeling in the study field show production of more water in comparison with conventional fracture modeling result.

2 Fracture simulation methods for carbonate rocks

The intersection relationships of fractures are probably very complex in a realistic DFN model (Zhang 2015). Bisdom et al. (2016) investigated impact of in situ stress and outcrop fracture geometry on hydraulic apertures in reservoirs. They stated that each fracture network containing fractures is created of at least one fracture set, but is not necessarily limited to it. By proposing a fracture propagation model using multiple planar fractures with a mixed model, Jang et al. (2016) stated that there is a large discrepancy in reservoir volume stimulation, because of a number of intersections of fracture connectivity. Heffer and King (2006) introduced a spatial correlation function of fractures as displacement strain vectors using renormalization techniques in representation of stochastic tensor fields for strain modeling. Masihi and King (2007) applied this method to generate fracture networks based on the assumption that the elastic energy in the fractured media follows a Boltzmann distribution. Koike et al. (2012) used geostatistical fracture distribution and fracture orientation (strike and dip) in simulation of the fracture system to estimate the hydraulic conductivity. Bisdom et al. (2016) also proved that the fracture orientation and the associated hydraulic aperture distribution have stronger impact on equivalent permeability than length or spacing. Thus, spatial correlation of fractures is the most important parameter in any fracture networks model or gridding scheme. To make this correlation between fractures, various methods are introduced for assigning precise values from fracture characteristics to the model of fractures. Among them, various approaches of using outcrop fracture characteristics such as fracture spatial distribution, length, height, orientation, spacing and aperture are widely used for model regularization (Wilson et al. 2011; Hooker et al. 2012). Lapponi et al. (2011) used outcrop data to construct a 3D model in a dolomitized carbonate reservoir rock from the Zagros Mountains, southwest Iran. Lee et al. (2011) studied the spatial fracture intensity effect on hydraulic flow in fractured rock. They used outcrop for simulation of spatial fracture intensity

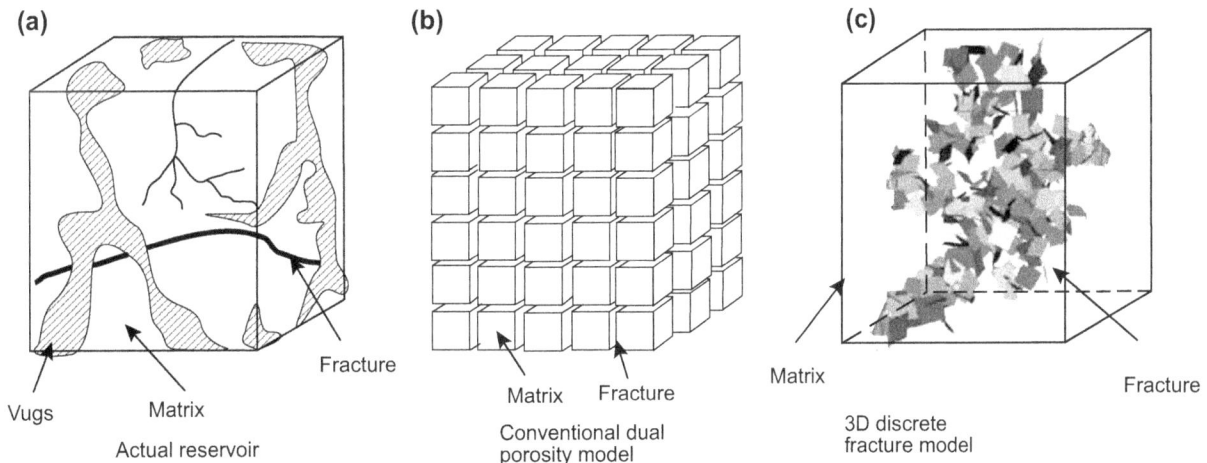

Fig. 1 Dual-porosity conventional model **a** actual porosity model, **b** sugar cube representation and **c** discrete fracture model for a fractured reservoir (Biryukov and Kuchuk 2012)

distribution. They have built three spatial fracture intensity distribution models and showed that flow vectors are strongly affected by spatial fracture intensity. They also proved that the higher the fracture intensity, the higher flow velocity. Boro et al. (2014) presented a workflow to construct an upscaled fracture model based on outcrop studies in a carbonate platform. It is important to note that fractures in the outcrop might have been affected by surface processes like weathering and stress release. The overall analysis, however, can help to constrain possible scenarios on fracture populations that may be relevant to the subsurface reservoir. Malinouskaya et al. (2014) illustrated a method to rapidly estimate permeability of a fracture network, using fracture data from outcrops of a Jurassic carbonate ramp. The method proposed by Maffucci et al. (2015) used outcrop data combined with a discrete fracture network (DFN) model to increase the reliability of fracture system characterization in the case of limited data for carbonate rocks. Connectivity of fracture networks in carbonate rock is dependent on orientation, size distribution and densities of the different fracture sets. These parameters also define size of the blocks enveloped by fractures (i.e., the matrix block size), which is generally used to model transfer of fluids between the matrix and fractures (Wennberg et al. 2016). To incorporate interaction between the matrix and fractured media, Huang et al. (2014) divided the fractured porous medium into two non-overlapping subdomains. One domain has a continuum model in the rock matrix, and the other, in deep fractured and fissure zones, is described by a DFN model. Then, they coupled these domains to simulate groundwater flow in their case study. However, Boro et al. (2014) stated that in general, fracture intensities, apertures and their intrinsic permeability would have more significant impact on the permeability of the field. Fracture shape and orientations are more important in affecting connectivity. Bisdom et al. (2016) also showed the strong importance of fracture orientation and associated hydraulic aperture distribution on equivalent permeability. In general, it is widely believed that fluid flow is affected by heterogeneities at all scales, from millimeter scale (porosity) to kilometer scale (Shekhar et al. 2014). In the present work, field evidence of different solutions and fractures in limestone is used to certify the nature of elasticity of fractures through time of production.

The proposed strategy also exposes incorrect assumptions in conventional fracture modeling, which has a great impact on history matching studies. Afterward, the concept of allocating each fracture to a fracture set and subsequently to a fracture network was considered by comparing the formation outcrop and considering the elasticity nature of the fractures, which comes from formation fluid pressure and/or regional stress in modeling.

3 Elastic fracture modeling

Dennis et al. (2010) stated that the physical structure properties and complexity of fracture characterization both have a significant effect on fluid flow in fractured rock. Thus, they proposed that the fracture zone should be characterized fully before simulation. Wang et al. (2016) studied the flow stress damage and reservoir responses to injection rate under different DFN-connected configuration states. Their results proved significant influence of the hydraulic pressure flow on the properties of hydraulic fractures. Generally, Gan and Elsworth (2016) stated that in any fracture modeling for complex reservoirs, the upcoming assumptions should not be neglected: Fractures initiate from flaws and the process is controlled by the elastic stress around them, the material surrounding the flaws can be viewed as continuum media, and individual flaws are spaced widely enough so that stress anomalies associated with each do not overlap. These assumptions are necessary for analytical formulation and therefore will be retained in the analysis, although a slight degree of plasticity is allowed (Wang and Shahvali 2016). However, in any elastic gridding, it should be kept in mind that a fracture is initiated when the maximum stress concentration occurring on the critical flaw boundary reaches the strength of the material which surrounds the flaw. Fracture extension also occurs from the tensile and not the compressive stress concentrations under both tensile and compressive loading. Fan et al. (2012) stated that microcracks induced by the excess oil/gas pressure may propagate and form an interconnected fracture network. This indicates also that during production, different pressure regimes could change characteristics of fractures. Not only might they be closed in the case of pressure drop or reservoir depletion, but they also might be widened due to high production rates and excessive pressure of fluid flow to the walls of fractures or cracks. It also might create fractures which make connections between vugs, while cracks could be widened and/or become fractures. Fan et al. (2012) investigated mechanism of fracture propagation by a linear elastic model. They have shown that critical crack propagation takes place if the intensity of the induced stress reaches the fracture toughness of the reservoir rock. On the other hand, subcritical crack propagation occurs in the rock when the stress intensity has not reached the fracture toughness of the reservoir rock, but exceeds a threshold value, which is usually a fraction (e.g., 20%–50%) of fracture toughness of the reservoir rock. This conclusion states how important it is to consider fracture–matrix interaction and/or boundary condition of fractures in accurate flow simulation. Subsequently, Hassanzadeh and Pooladi-Darvish (2006) considered the time variability of the fracture boundary condition by the Laplace domain analytical solutions of the diffusivity equation for different geometries of fractures in constant fracture pressure through a large number of pressure steps. Guerriero et al. (2013) proposed an analytical model that could pave the way to a full numerical model allowing one to calculate the pore pressure within fractures, at several scales of observation, in a reasonable time. They also suggested that the model also allows one to obtain a better understanding of the hydraulic behavior of fractured porous rock. However, not only the pressure regime change, but other factors affecting fracture shape and apertures could be accounted for by considering elastic behavior of fractures in the model construction during gridding and throughout production history matching investigation.

Bisdom et al. (2016) stated relationships between the fracture geometrical parameters and some other parameters such as the stress applied in the medium. However, for this specific case, accurate relationship between degree of elasticity and fracture's aperture would be defined only by core analyses in a wide range of applied stresses from pore fluid to the walls of fractures. However, previous studies have shown that this relationship is linear in a narrow range of applied stress (Bisdom et al. 2016).

Elastic gridding is a variant of the grid optimization-based technique on a length functional with a non-Euclidean metric tensor. In creation of the grid, it is considered to be a system of springs connecting neighboring grid vertices along the grid lines. On the other hand, the problem of grid optimization is thereby reduced to a problem of elasticity and the problem of translating grid optimization criteria into criteria for assigning spring constants to grid lines. After having assigned values to the spring constants of the elastic grid, the grid vertex positions can be found by solving the equilibrium equation for the elastic system. In case of reservoirs with parallel flow channels, strong pressure regime change will cause fracture apertures to become wider to some extent. However, not all the fractures in carbonate rocks, regardless of their sizes, are responsible for fluid flow. Observations of fractures in core and outcrop indicate that flow in open fractures in carbonate rock tends to be channeled rather than through fissures. Most of the flow takes place along a few dominating channels in the fracture plane, whereas most of the fracture plane is not effective for fluid flow. Wennberg et al. (2016) stated that the effect of channeled flow should be taken into account during evaluation of fractured carbonate reservoirs and building dynamic flow models. However, the consequence of the aperture variation is that the fluid flow in fractured carbonate reservoirs will tend to be channeled instead of that of fissure-type flow, which is the assumption in most flow simulators (Wennberg et al. 2016). Figure 2 shows different solutions and fractures of the Sarcheshmeh

limestone formation outcrops in the Shahrood area, Iran. Major fractures shown by black lines (which are considered as the pathway of fluid) are oriented normal to the maximum stress direction, while red lines show minor fractures crosscut the major ones. The nature of major fractures is that these are the main fluid flow channels, while minor fractures are not necessarily important for that purpose. This is a typical fracture pattern in limestone reservoirs in Iran. This pattern would make only those parallel fractures, which are the only pathway of fluid flow, undergo effects of pressure regime change through high rate production. This effect will cause aperture widening, which changes the shape of the fracture, the parameter that is going to be considered in fracture modeling in this study. Another high production rate effect in such reservoirs is connecting vugs by propagation and/or widening of fractures. Lapponi et al. (2011) showed that vuggy porosity seems to increase porosity only locally and to a limited extent, developing a non-connected pore network. However, fracture propagation under pressure or high production rate would make connections between vugs. This is not an important effect in production, since these connecting fractures show small apertures, not appropriate for fluid flow. Figure 3 shows an outcrop of the same formation in the same location and an example of vugs connected by fracture propagation under pressure.

However, in this study reservoir, the thermal fracturing is not planned in the master development plan of the field and the production history of the reservoir also shows some degree of overpressure fluid in the first periods of initial production, when accurate data were not available. Thus, neither the thermal fluid injection nor the fluid overpressure was considered here. The majority of fractures that we had in the study reservoir were of the type of fractures shown in Fig. 2.

As was previously mentioned, some fractures connect vugs, which results in fracture propagation and/or fracture shape. All the main fractures responsible for fluid flow are modeled as flat planes in fracture modeling and are fixed through the production simulation procedure. However, in the proposed strategy, these fractures will change in shape through history matching. This is what we called the elastic gridding modeling. Figure 4 shows an example of an elastic grid and the elastic nature of fractures in modeling. As it was previously mentioned, fracture properties might change through high pressure regime change.

The more the pressure regime changes in a short time, the more this will cause more change in the shape of fractures. Figure 4a shows a conventional model of fractures. Figure 4b, c illustrates shape change in the same fracture after high pressure regime change, modeled by elastic fracture modeling. Figure 4d also shows change in fracture aperture modeled by elastic fracture modeling.

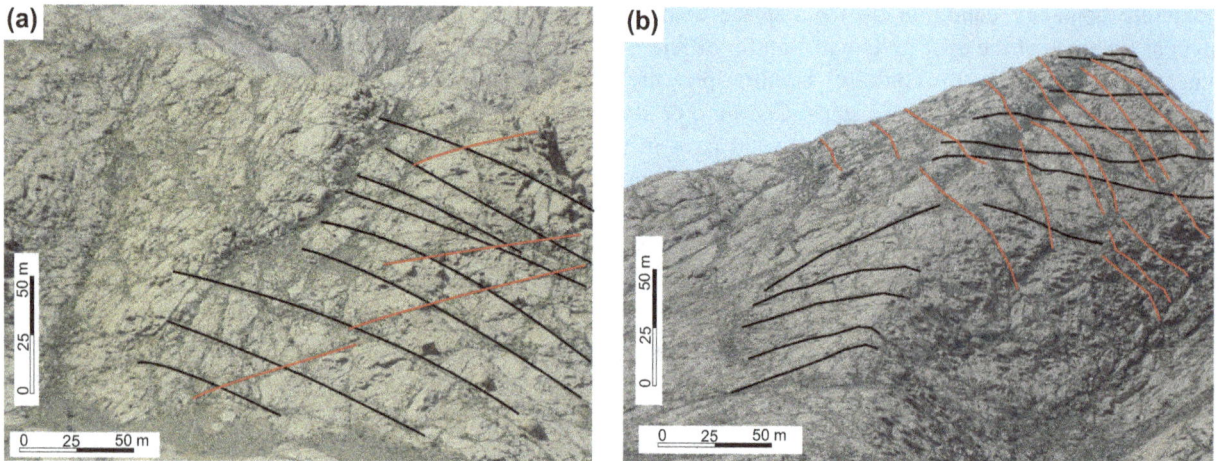

Fig. 2 **a** Outcrop of parallel fractures in the limestone Sarcheshmeh Formation, Iran. Only fractures shown by *black lines* are responsible for fluid flow. Fractures defined by *red lines* do not make flow paths. **b** The same formation in another part of the study area

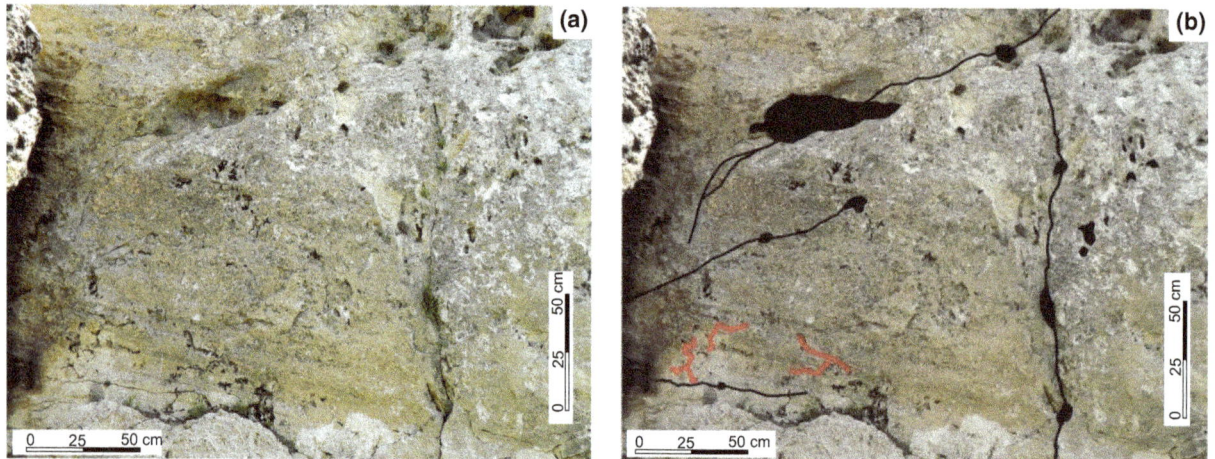

Fig. 3 Outcrop of the Sarcheshmeh Formation, same location as in Fig. 2, Iran. **a** Various porosity types in the outcrop, **b** connected and non-connected vugs and worm channels (*red lines*)

However, the most important parameter that changes the permeability of the reservoir rock is the dilation of fractures. Unlike other fracture parameters, the dilation degree of a fracture under stress cannot be described from core sample tests. Clearly, long fractures cause the core to fall apart during the experiment. Core recovery is also very poor in intensely fractured intervals, and the stress release when the core is taken to the surface will affect the observed apertures (Wennberg et al. 2016). Use of electrical image logs also suffers from uncertainty if the absolute aperture is large. This problem would be boosted if we want to accurately measure the dilation of the fracture. Therefore, it necessitates that the relationship between aperture dilation and applied stress is defined by numerical analysis and other permeability tests in various conditions.

Taron et al. (2014) tested fracture dilation in a geothermal system.

However, Min et al. (2004) and Farahmand et al. (2015) derived relationships between aperture dilation and applied stress. Min et al. (2004) studied completely all cases and we used the results that they derived in their complete study. Min et al. (2004) have stated that the exact extent of shear dilations of fractures can only be identified through numerical experiments. Their experiments showed that on the one hand, equivalent permeability decreases with increase in stresses, when the differential stress is not large enough to cause shear dilation of fractures. On the other hand, the equivalent permeability increases with the increase in differential stresses, when the stress ratio was large enough to cause continued shear dilation of fractures.

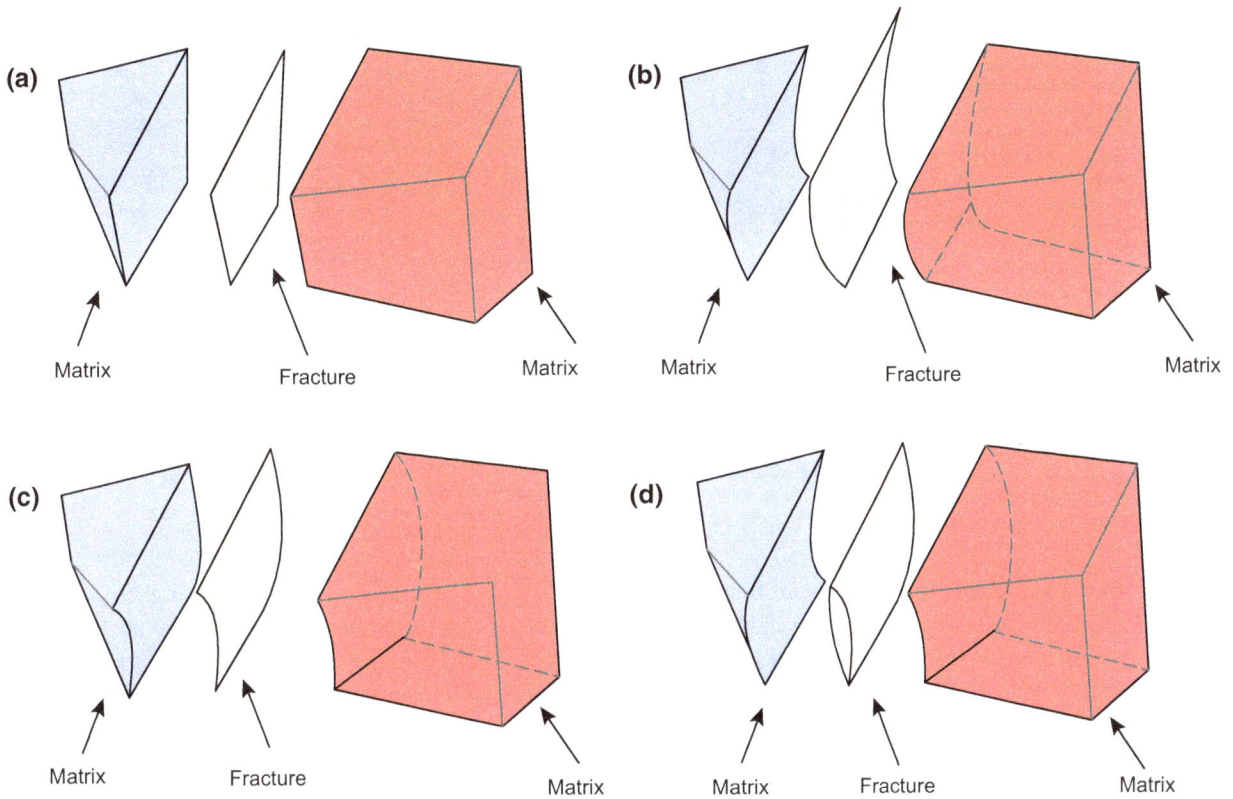

Fig. 4 **a** Conventional model of matrix and fractures (Karimi-Fard 2013). **b** an example of change in shape of a fracture modeled by elastic fracture modeling, **c** another change in shape and **d** change in fracture aperture modeled by elastic fracture modeling

In this case, shear dilation is the dominating mechanism in characterizing the stress-dependent permeability. They have proved that the maximum contribution of dilation is more than one order of magnitude in permeability. Thus, we have used this role in our study for fracture dilation.

Since the elastic system is attached to the boundary of the model, it would not collapse. For both convex and concave domains, the elastic method is also very flexible in fracture geometry and robust to system collapse (Fig. 5). Figure 5 shows an example of a conventional grid, and Fig. 5b illustrates only a schematic of the same system but with elastic grids. Since we do not exactly know how the grids would differ during simulation, the applied pressure and elastic properties of rock would define that thus Fig. 5b is an example of any shape of grids with the unique geometry. Figure 6a depicts a simple case of conventional fracture modeling in a medium. Figure 6b shows example of a low allowed degree of elasticity and/or low pressure regime change, and Fig. 6c exhibits a sample of high degree of elasticity allowed in the model due to possible high production rates with high fluctuation in the pressure regime. Every parameter that is going to model the elastic behavior of a fracture should be defined by numerous experimental tests on core samples.

The pore fluid effects needed to calculate elastic properties of fluid-saturated rock could be obtained also in each study from the simulation case. However, in this study, we did not have enough data to derive an implicit model for elastic behavior simulation of a fracture. However, it is not only the matter of data, but it is about the matter of accurate and implicit relationship between the applied stress and strain of the medium and elastic properties and behavior of fractures. Thus, in this study, we used an explicit relationship between the stress applied to a fracture and the allowed degree of elasticity of the fracture. Consequently, we should define a maximum degree of change in shape that we consider for a fracture and the value of changes in curvature of the fracture. The former is defined based on the Bulk and/or Young's modulus of the rock. For hard rocks, the lowest grade of change in shape is allowed and it is vice versa for soft rocks. The latter needs more explanation. At first, it should define whether curvature has any effect on permeability change or it is only a fracture shape change, without effect on permeability. According to Fig. 4, it changes the permeability only if both sides of the fracture experience convex curvature, which increases the permeability. In case of same convexity or concavity of fracture walls, no changes in permeability would happen.

(a)

(b)

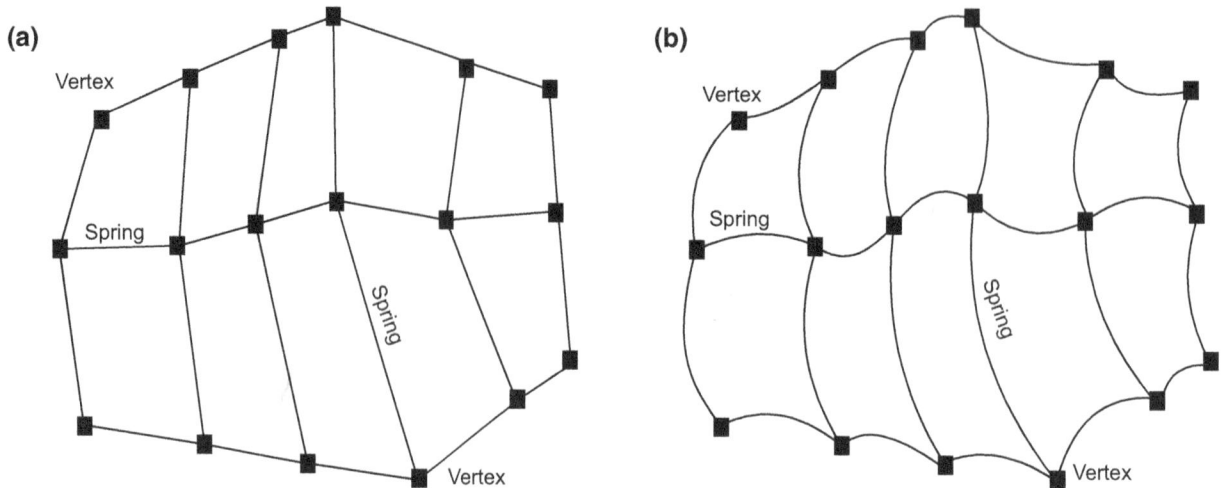

Fig. 5 **a** Example of a conventional grid and **b** one out of thousands of examples of an elastic grid

In this study, we assumed the first case for our fractures, as shown in Fig. 4d. The degree of curvature is defined based on the toughness of rock. Although the exact degree of convexity and curvature of the fracture wall should be defined by core sample test, it could be defined as a linear function of applied stress for medium to rocks.

4 The study reservoir

The study reservoir is an extensively faulted anticline 74 km long and 6–8 km wide, located in the Dezful Embayment, southwest Iran (Fig. 7a). The study anticline is an asymmetrical anticline with a NW–SE trend and a sinuous axis. The target formation is a prominent carbonate unit of Oligocene and early Miocene age called the Asmari Formation. This formation in the study field contains limestone, dolomite and minor marl and shale (Abdollahie Fard et al. 2006). The study field was divided into five sectors based on geological and engineering data. Most of the boundaries of these sectors are in agreement with faults. These sectors are depicted in the underground contour (UGC) map of the target formation shown in Fig. 7b. It should be mentioned that the UGC map in Fig. 7b was obtained from time–depth conversation of 3D migrated seismic data with well top adjustment supervision. However, an advanced migration algorithm is needed for imaging in such complex media (Soleimani 2016b). Limited lateral communication, the presence of faults acting as barriers, hydrodynamic system and imbalanced offtake have caused the Asmari Formation to be divided into several compartments (Sherkati and Letouzey 2004). The oil production from the study reservoir has been very imbalanced mainly because early development took place in the central and northwestern part of the field, where oil

column was thicker and the wells had higher productivity. In spite of extensive fracturing of the Asmari Formation, a pressure differential has developed in the gas zone across the field due to imbalanced oil and gas production.

4.1 Petrophysical data

Limited cores were cut from the Asmari Formation for routine and special analyses. Total length of cores cut was 550 m out of which 421 m was recovered. The mean porosity of plugs cut from 421 m of cores in six wells was 9.8%, while 21% of samples have porosities less than 4% and only 3% of samples have porosity more than 20%, implying that the Asmari Formation is a low porosity reservoir (Hoseinzadeh et al. 2015). The median permeability of cores was calculated to be 0.43 mD with 60% of samples having permeability of less than one milli-darcy, indicating a very low permeable carbonate rock matrix. Production logging tools (PLT) logs were recorded in 15 oil wells and in 5 gas wells. Reviewing of the PLT log results indicates that distance between flowing intervals varies from 1 m to the maximum of 44 m. This indicates an active mechanism of fracture production in this field.

5 Zonation and fracture study

Porosity distribution in the Asmari Formation is very diverse. Therefore, division of this formation into several zones and subzones is unavoidable. Based on petrophysical and petrographic characteristics, the target formation was divided into seven different oil-bearing zones with the water column as zone 8. Consequently, zones 1, 2, 6 and 7 were divided into two subzones (Table 3 of Appendix). From the study of cores, some parameters were extracted

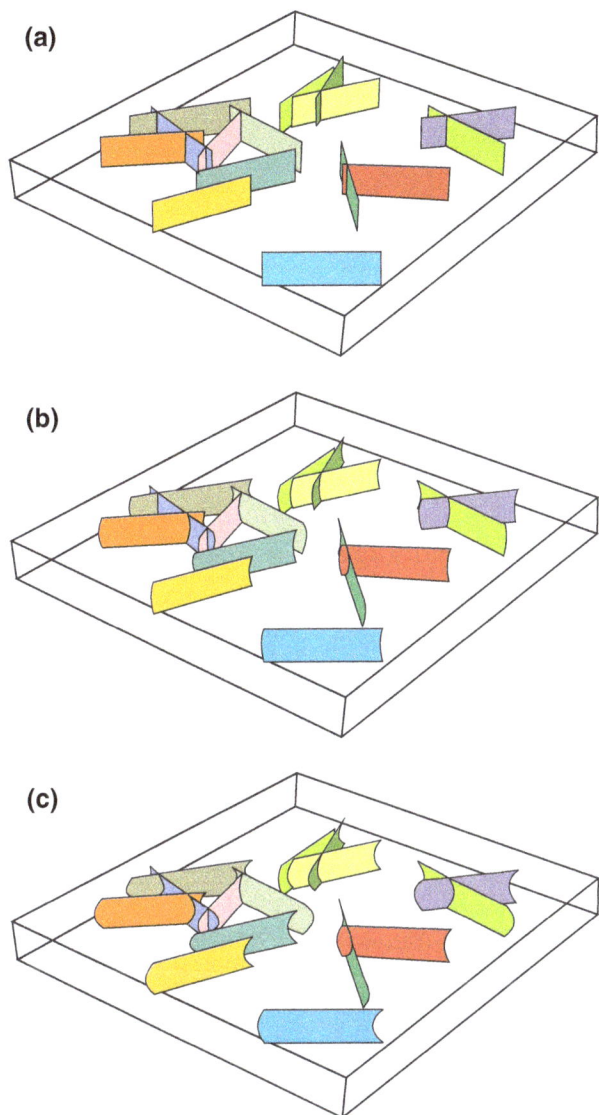

Fig. 6 **a** Conventional modeling in which fractures were modeled by *flat planes* with no change in shape through production (Karimi-Fard 2013), **b** Low degree of elasticity allowed, and **c** high degree of elasticity allowed in elastic fracture modeling in the proposed strategy

such as type of fracture; spacing (number of fractures per meter); dip of fractures (relative to the core axis in the core description tables and relative to horizontal in the interpretation); width of fractures measured in laboratory conditions; and filling mineral type. Among these parameters, type of fracture and filling mineral type are classified and coded as shown in Tables 1 and 2, respectively. Occurrence of anomalous losses of mud during drilling is often the first indirect indication of the existence of a naturally fractured interval. Comparison of the mud loss rate in different wells may give the intensity of fracturing in different parts of the reservoir (Xia et al. 2015).

In spite of the fact that there are many parameters that can affect the mud loss rate and inherent errors in the results, such as mud weight, weight on bit, stroke per minute of pumps and rock type, the mud loss rate is one of the most important parameters in indirect observations of fractures. Other information that could be useful in fracture studies includes productivity index (PI), PLT data and maximum daily flow rate. Production rate is usually dependent on the extent of fracturing of the formation in this low-matrix permeability reservoir (Xia et al. 2015a). Therefore, higher production rates correspond to a more extensive fracturing of the strata in the well location (Fig. 8). To determine the fractured zones within the reservoir, a radius of curvature method (Yoshida et al. 2016) was used. The 3D curvature model was not shown here, but it was obtained from 3D migrated seismic data and application of maximum and minimum curvature attributes. After applying this method to the target formation, we have concluded that the most common types of fracture are type 1 (cross-axial tensional and conjugate shear fractures) and type 2 (axial tensional and conjugate shear fractures). Interpretation of mud losses, daily flow rate, PI, PLT and other data suggest that sector 1N and partially sector 4 are highly fractured. Fracturing in sector 3 is relatively intensive and not extensive in sector 1S. Figure 9a shows the fracture quality map, and Fig. 9b illustrates strain distribution map on top of the target formation. The strain regime here is of shortening type. The mud loss, PI and PLT integration was performed by finding the relationship between fracture quality index and these three parameters in GIS media.

6 Elastic fracture model generation

All the required data for elastic fracture and geological model construction were collected. Petrophysical data contain two series of net-to-gross (NTG) ratio, porosity and water saturation related to each horizon. Monte Carlo statistical simulation was used to obtain a 3D distribution of fracture spacing density and, consequently, matrix block size. Input probabilistic frequencies were scaled in such a way as to have the mode of matrix block height distributions changing from approximately 3 m at the top to approximately 6 m at the bottom of the Asmari Formation, respectively. Matrix–fracture communication was defined by considering elasticity behavior of fractures. At that stage, matrix and fracture properties were considered as first approximations to be modified during history matching. After construction of the model, a geological model containing $633 \times 451 \times 12$ (3,425,796 cells) mesh cells was prepared including all faults. Configuration of the elastic gridding model was based on fault traces and a contour line of -2600 m from the structure map on top of the Asmari Formation. The elastic grid has 133 blocks in

(a)

(b)

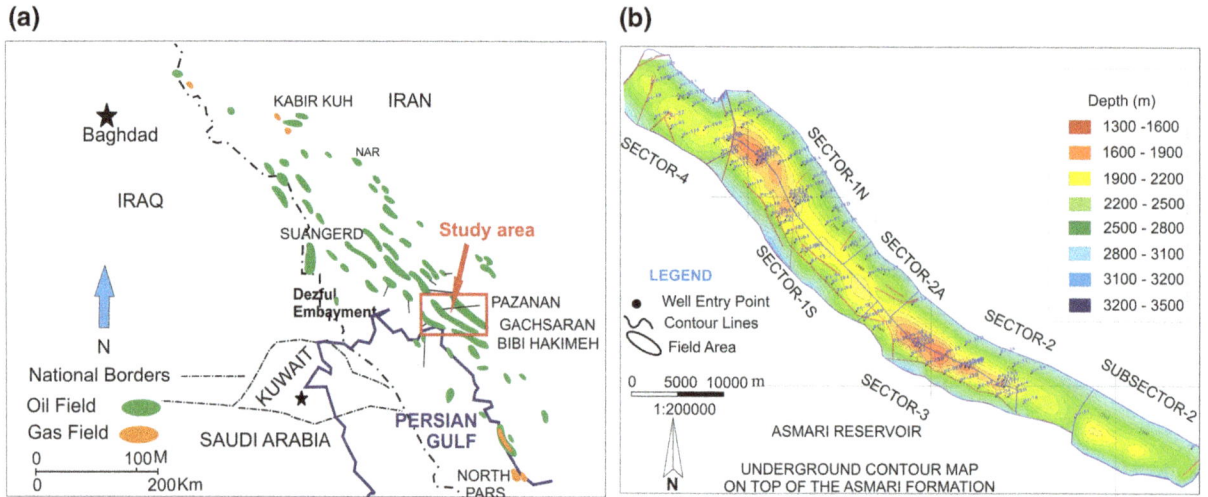

Fig. 7 a Location of the study field is shown by the *rectangle* (Soleimani and Jodeiri-Shokri 2015). **b** Zonation of the study field and underground contour map (UGC) of top of the Asmari Formation

Table 1 Classifying different types of fracture in the cores

Type of fracture	Open	Partly filled	Filled	Closed or hairline
Score	1	2	3	4

Table 2 Classifying different types of filling minerals in the cores

Filling mineral	Calcite	Dolomite	Anhydrite	Clay minerals
Code	1	2	3	4

one direction and 17 blocks in the other direction with 30 layers. The elastic grid structural elevations were obtained from underground contour maps and verified against well picks. Gross thickness of zones (subzones) was based on isopach and isochore maps. Porosity versus normalized depth profiles were used as guides to split zones (subzones) into 30 layers. Figure 10 displays a map view of the elastic grids on the top of the target formation. Zone (subzone) matrix porosity and NTG ratio were obtained from geologic maps and then downscaled to layers based on the porosity–normalized depth profiles.

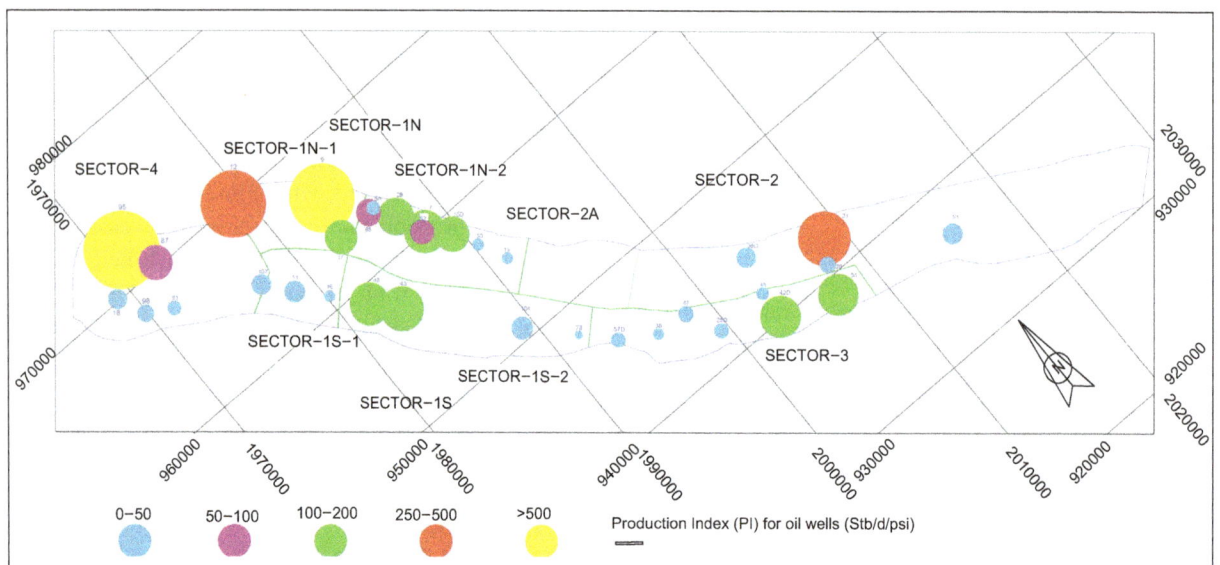

Fig. 8 Production index based on production of the wells in each zone

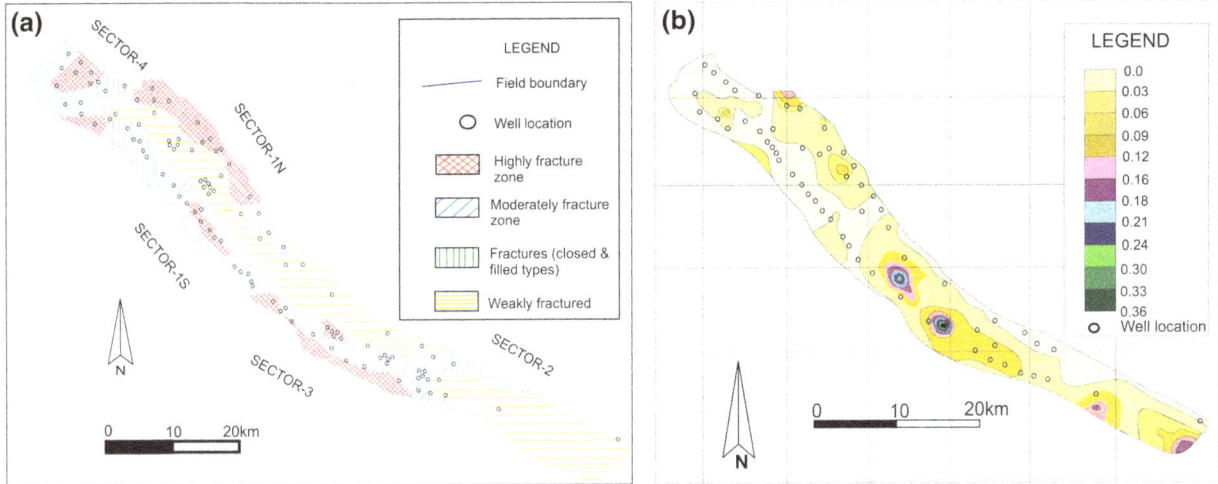

Fig. 9 **a** Fracture quality map based on mud loss, PI and PLT and **b** strain distribution map of the study field. Small symbols show well location on both maps

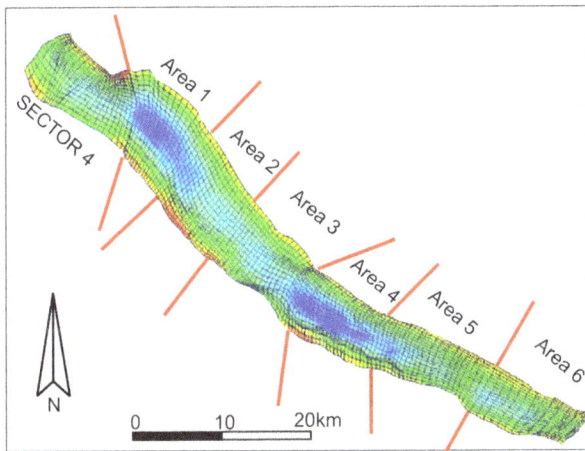

Fig. 10 Map view of the grids on the top of the target formation

Matrix permeability was based on matrix porosity and permeability–porosity correlations. A fracture porosity distribution was generated based on the fracture porosity map. This map was horizontally scaled according to the qualitative map of fracture intensity. It was vertically scaled also in accordance with empirical correlations reflecting reduction in fracture intensity from top to bottom of the Asmari Formation (Fig. 11). Well trajectories and perforation, production, and static pressure histories were incorporated in the model. Sporadic information about salt and/or water production was only available for a limited number of oil wells. Individual well gas/oil ratios were found to have resulted from prorating rather than from measurements; on this basis, they were excluded from history matching.

6.1 Model initialization

Providing the model with initial distributions of water saturation, reservoir pressure and hydrocarbon components were the subject of this part of the study. To obtain distribution models of these characteristics, we took seven available drainage capillary pressure curves (not shown here) and scaled them in accordance with the following equation:

$$\phi S_{\mathrm{wi}} = \left(\frac{R_{\mathrm{w}}}{R_{\mathrm{t}}}\right)^{1/w} \tag{1}$$

where R_{w} is the formation water resistivity, R_{t} is the formation resistivity, and $w = f(R_{\mathrm{w}}, R, \emptyset)$. The right side of Eq. (1) was derived from well logs and averaged for different regions and zones (subzones). To have a reasonable value for the right side of equation, a primary blocking step of the target formation was performed for averaging in each block. Subsequently, a weighted averaging based on the thickness of each zone/subzone was performed. Due to the large number of obtained maps with different blocking approaches and based on different interpretation of the averaged map result with the help of geological and hydrological data, the maps are not shown here. Then, S_{wi} was estimated from Eq. (1) for each simulation model elastic grid block based on its matrix porosity. The resulting S_{wi} values were introduced in the model as connate water saturation. To obtain different oil–water contacts (OWC), two aquifers and six equilibration regions were introduced into the model (Fig. 12). Two analytical aquifers were introduced instead of one aquifer, while the northwest part of the reservoir was completely disconnected from any analytical aquifer. Aquifer 2 is stronger

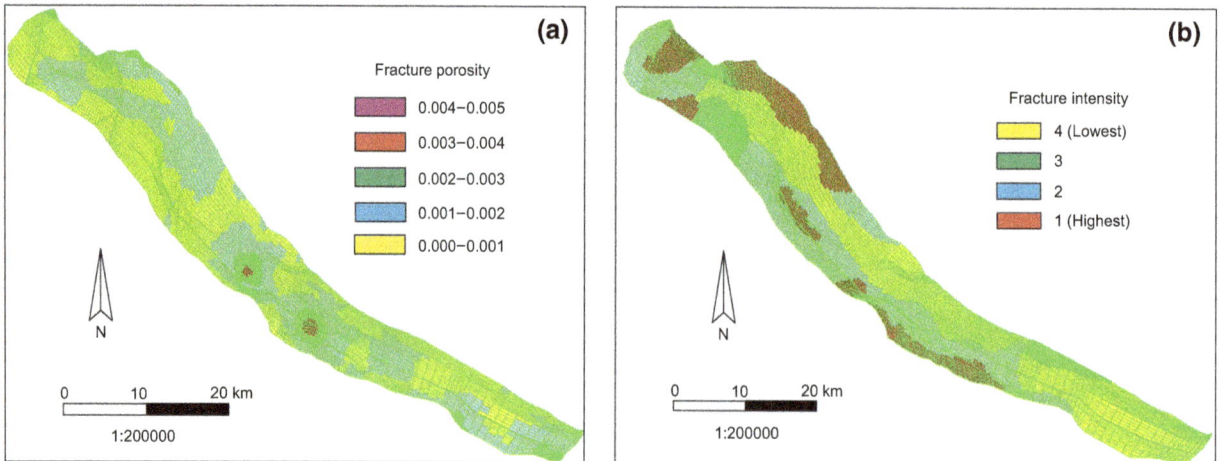

Fig. 11 **a** Fracture porosity and **b** fracture intensity map of top of the target formation

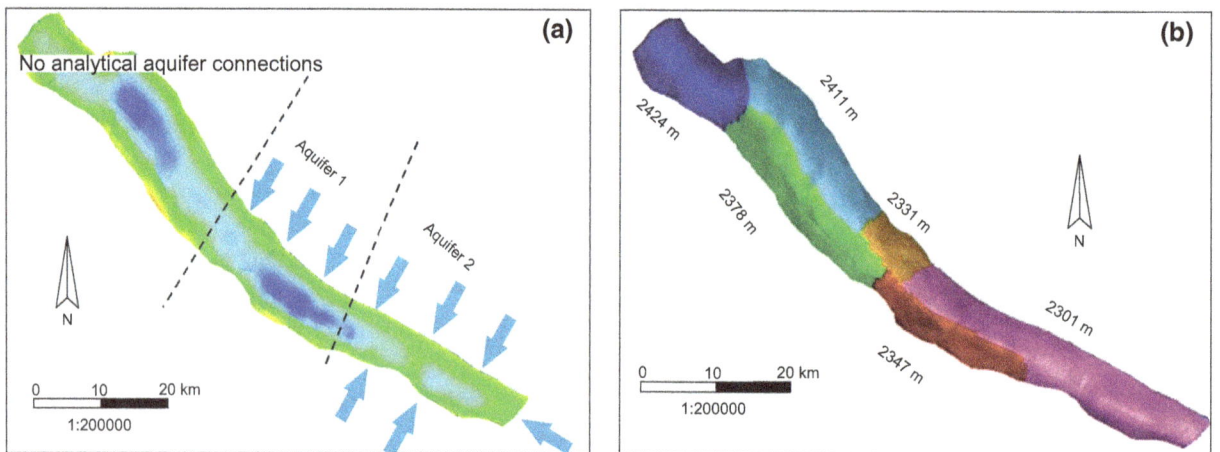

Fig. 12 **a** Two different aquifers in the field and **b** six different OWC zones, all introduced into the model

than the other aquifers. The aquifers are separated by faults, while six OWC zones are controlled by the pressure regime of the field. The water saturation in the matrix also could affect the OWC in different zones.

Thus for model initiation, the pressure regime and the matrix water saturation (besides the fracture water saturation) should be introduced as the initial condition for model running. Figure 13 shows the pressure regime and matrix water saturation in the model.

6.2 Preliminary simulation

The study field's production history is complicated, and the aquifer strength varies from the southeast to the northwest of the field. Thus, introducing the elastic behavior of fractures into the model was proposed here to realize the history matching of the reservoir production.

However, running time was a significant issue right from the beginning of the matching process. Only using advanced hardware (a total of 12 CPUs of 3 GHz each) and implementing extensive debugging allowed the reduction of the running time for a history match to nearly 75 h. In the first step, the conventional history matching objectives in the traditional gridded model showed the entire field's solution of gas production and water production only in the oil wells located southwest of the field. To introduce the elastic behavior of fractures into the model, initial global modifications involved elevating the aquifer permeability, elevating fracture permeability in x direction by a factor of 10 and reducing fracture permeability in z direction by a factor of 10 to compensate for the elasticity of microfracturing observed in the cores. Examination of elastic model runs showed that oil flow from matrix to fractures should be increased to match water and gas production from the oil wells. Elevating of this

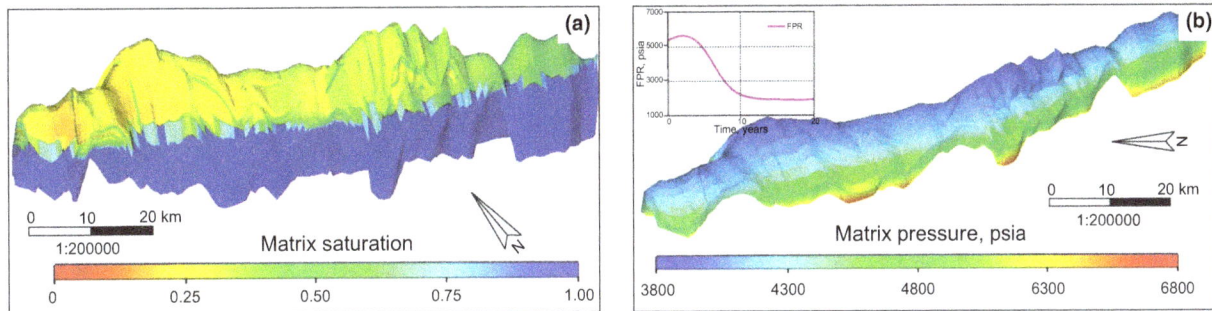

Fig. 13 a Matrix water saturation and **b** the pressure regime both introduced into the model for initiation

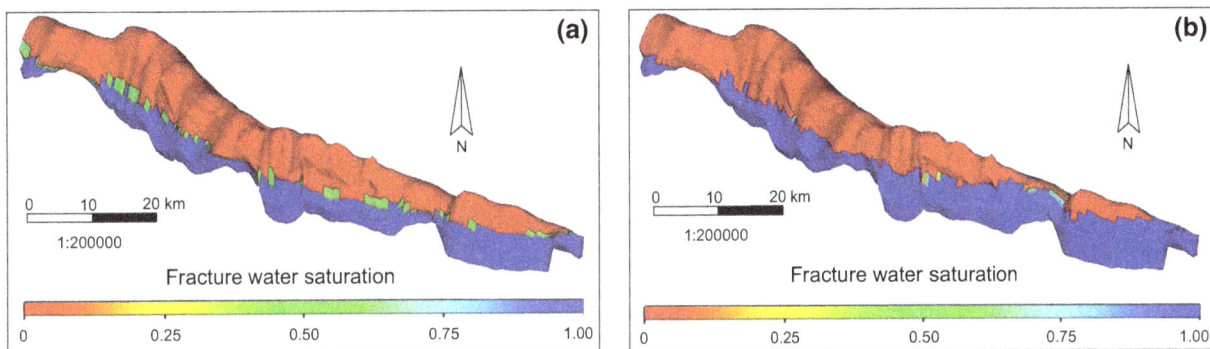

Fig. 14 Fracture water saturation in **a** conventional fractures and **b** elastic fracture modeling

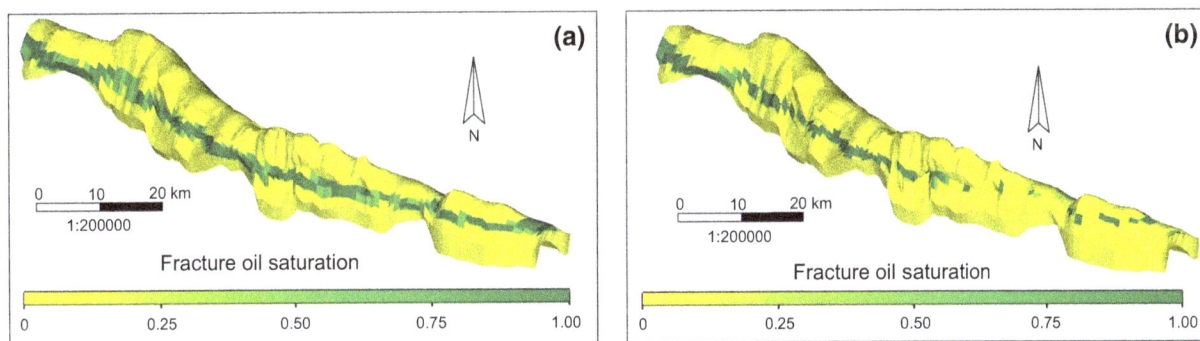

Fig. 15 Fracture oil saturation in **a** conventional fracture and **b** elastic fracture modeling

oil transfer was done by modifying the imbibition capillary pressure. Individual oil well modifications included changing capillary pressure curves assigned to the matrix grid blocks surrounding the well, varying the height of matrix blocks, modification of the pore volume of initially oil- and gas-saturated grid blocks and modification of the fracture permeability in x, y and z directions. Figure 14 shows the conventional and elastic fracture water saturation model. As it can be seen on the elastic fracture model, wells located in the southeast (SE) of the field may produce more water. This is

due to the increase in cracks and/or fracture apertures, which means previous small cracks were filled with water. Figure 15 shows elastic and conventional models of fracture oil saturation. Again, the southeast (SE) part of the field does not produce more oil after a while according to the elastic modeling results. This phenomenon is less observed for gas saturation. Figure 16 shows result of elastic and conventional fracture modeling. As a result, change in shape of fractures, (increasing aperture) would replace oil by gas in the vicinity of oil-gas contact. This is obvious in Fig. 16.

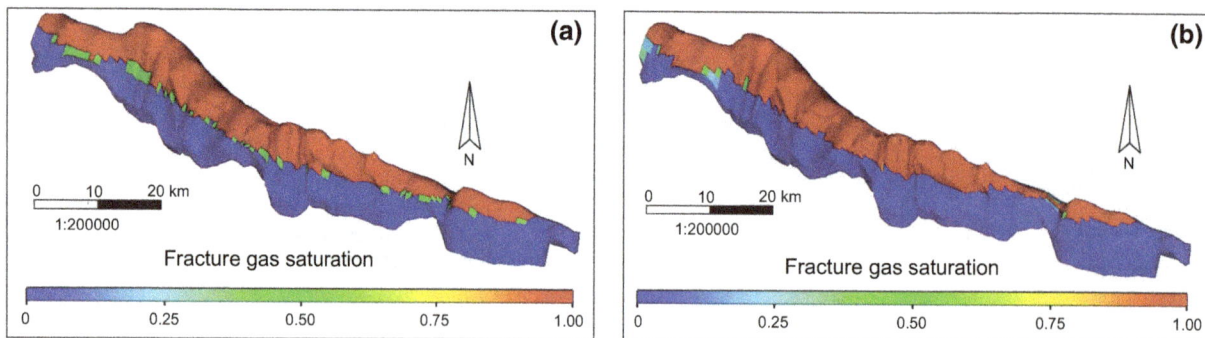

Fig. 16 Fracture gas saturation in **a** conventional fracture and **b** elastic fracture modeling

7 Conclusions

There are many problems in fractured reservoirs that may require new approaches in fracture modeling based on advanced concepts. Some problems cannot be simplified beyond a certain limit. The complexity, scale and uncertainty of natural systems are the primary reasons for most of the modeling difficulties which have to be handled.

The objective of this work was to develop a prototype workflow for history matching in naturally fractured reservoirs. This concern is also with the problem of generating more realistic computational grids for reservoir simulations. Here we face the particular problems connected to the complexity of the reservoir. The proposed strategy combines the accuracy of fracture modeling with the efficiency of elastic gridding. Elastic gridding is very simple, and in principle, the method should be well suited for weighing between mutually conflicting optimization criteria with highly nonlinear and discontinuous cost functions. The strategy was applied on a complex reservoir from southwest Iran. The strategy was observed to be reasonably effective in achieving field-level agreement in total oil and water production rates. Aperture increase in

cracks will mean they will be filled by water, while it replaces gas by oil in the vicinity of the gas oil contact. Thus, elastic fracture modeling shows that more water will be produced compared to what was assumed by conventional fracture models, which do not take into account fracture properties change through production.

Appendix

Table 3 shows zonation of the target formation. As was mentioned in the text, the target formation is divided into seven zones with the aquifer as zone 8. This zonation is based on the geological and petrophysical information. In each zone and subzone, lithology porosity, permeability and NTG of the zone are fully described. This zonation was used as the base of the reservoir modeling and simulation.

Table 3 Zonation of the target formation in the study reservoir

Zonation		Lithology	Porosity and permeability	NTG
Zone 1	Zone 1–1	Average thickness 36 m Mainly composed of limestone and dolomite	Most of pore volume is filled with cements. The average porosity is 10%. Water saturation increases from 10% in the southern flank to 30% in northern flank as well as from west to east	The NTG is about 90%–95% in the western half and is variable in the eastern half; it varies from 80% to 100% in the southern limb, from 60% to 42% in the northern flank
	Zone 1–2	It is 28 m thick. Mainly consists of dolomite, limestone, sandy dolomite, thin layer of shale and sandstone	Average porosity is 8% in sector 4 and varies from 10% to 14% in both flanks. In eastern portion, it ranges from 11% in the southern limb to 7% in the northern flank. Water saturation varies from 20% to 25% in the southern flank to 55%–60% in the northern flank	The calculated NTG ratio of the subzone is about 90% in the central portion of the field, 60% in the eastern portion of the northern flank and 45% in the sector 4

Table 3 continued

Zonation		Lithology	Porosity and permeability	NTG
Zone 2	Zone 2–1	This subzone predominantly contains dolomite and limestone. On average, it is 41 m thick	In most of the wells throughout the field, porosity varies from 8% to 10%. Water saturation in this subzone is distributed similarly to the previous subzones with the minimum saturation of 20% at the southern flank and the maximum saturation of 60% at the northern flank.	Average NTG ranges from 85 to 95% in the southern flank with exception of a few wells having NTG about 65%. In the northern flank of the central and eastern portions, NTG averages 65% and 35% accordingly
	Zone 2–2	Its thickness is 27 m. Composed of limestone, dolomitic limestone, calcareous dolomite and partly thin shaly layers.	Porosity decreases from 12% in the southern flank to 7%–10% in the northern limb. In sector 4, porosity decreases to 5%. Water saturation increases from 15% in the south to 55% in the western part and to 80% in the eastern part of the northern flank	NTG decreases from 90% to 95% in the southern flank to 25%–50% in the northern flank. In sector 4, NTG value averages 80%
Zone 3		Thickness of 40 m. Contains alternation of limestone, dolomitic limestone and dolomite	Porosity varies form 15% in the southern limb to 9%–11% in the northern flank. Water saturation increases from 10–20% in the southern flank to 50% in the western part of the northern flank	Since rock quality is relatively good, NTG ratio is almost 100% in the southern flank and decreases to 85% in the northern flank
Zone 4		18 m thick and is composed of dense limestone and partly of dolomite. A thin shaly bed is observed.	The average porosity of the zone ranges from 8% to 14%. In the central and eastern portions of the field water, saturation increases uniformly from 20% in the southern limb to about 70%–80% in the northern flank	Except sector 4, NTG ratio decreases from 75% to 95% in the southern flank to 25% and to 35% in the western and eastern portions of the northern flank, respectively
Zone 5		Thickness is 60 m. Composed of limestone, partly of dolomite and of thin shaly layers	Average porosity varies between 7 and 12%. Average water saturation ranges between 25% and 35%. A general trend shows a south to north increase in the water saturation	The NTG ratio varies between 80% and 95% throughout the field
Zone 6	Zone 6–1	18.5 m thick and composed of dolomite and intercalation of argillaceous layers	The average porosity in the eastern part is 7%. It is 16% in the southern and 8% in the northern flank The water saturation is 40% in the south flank and 80% in the northern limb	NTG ratio ranges between 13 and 54%. It is 95% in the south flank and 60% in the north flank of the central part of the field
	Zone 6–2	Thickness of 33 m and composed of limestone and partly of dolomite. Dies out at the western part	The average calculated porosity of the subzone reduces from 16% in the southern limb to 9%–10% in the northern flank. Water saturation changes from 25% in the southern flank to 50% in the northern flank	NTG ratio ranges from 90 to 95% in sector 1, between 50 and 60% in sectors 2 and 3, and from 66% to 100% in the southern flank of sector 4
Zone 7	Zone 7–1	37 meters thick and consists of limestone and partly of dolomite and of a few thin shaly layers	The average porosity is 12%. In the southern flank, the porosity is 13% decreasing to 10% in the northern limb. As a result, the water saturation value of 75% was reached	NTG ratio varies from 80% to 100% decreasing to 34%. In the eastern part of the field, the average NTG ratio ranges between 75 and 80%
	Zone 7–2	Thickness 38 m. Composed of limestone, marly limestone and shale	In sector 1, porosity is 12% in the southern flank and 7% in the northern flank. Average porosity values of 8% to 10% were assumed In sector 4, water saturation varies from 45% in the southern flank to 90% in the northern limb	The NTG ratio is about 95% in the southern flank decreasing to 50% in the northern limb

Table 3 continued

Zonation	Lithology	Porosity and permeability	NTG
Zone 8	Thickness 72 m and composed of shale, marl and marly limestone	Porosity of 9%–10% was estimated. Water saturation in the southern and northern flanks was evaluated to be 40% and 55%, respectively	The NTG ratio in sector 4 (37%) and in sector 1 (53%), decreasing from about 60% in the southern flank to about 35% in the northern flank

References

Abdollahie Fard I, Braathen A, Mokhtari M, Alavi A. Interaction of the Zagros fold–thrust belt and the Arabian-type, deep-seated folds in the Abadan plain and the Dezful embayment SW Iran. Pet Geosci. 2006. doi:10.1144/1354-079305-706.

Agar SM, Hampson GJ. Fundamental controls on flow in carbonates: an introduction. Pet Geosci. 2014. doi:10.1144/petgeo2013-090.

Bahrainian SS, Daneh Dezfuli A, Noghrehabadi A. Unstructured grid generation in porous domains for flow simulations with discrete-fracture network model. Transport Porous Media. 2015. doi:10.1007/s11242-015-0544-3.

Biryukov D, Kuchuk FJ. Transient pressure behavior of reservoirs with discrete conductive faults and fractures. Transport Porous Media. 2012. doi:10.1007/s11242-012-0041-x.

Bisdom K, Bertott G, Nick HM. The impact of in situ stress and outcrop-based fracture geometry on hydraulic aperture and upscaled permeability in fractured reservoirs. Techtonophysics. 2016. doi:10.1016/j.tecto.2016.04.006.

Boro H, Rosero E, Bertotti G. Fracture-network analysis of the Latemar Platform (northern Italy): integrating outcrop studies to constrain the hydraulic properties of fractures in reservoir models. Pet Geosci. 2014. doi:10.1144/petgeo2013-007.

Cai L, Ding DY, Wang C, Wu YS. Accurate and efficient simulation of fracture–matrix interaction in shale gas reservoirs. Transport Porous Media. 2015. doi:10.1007/s11242-014-0437-x.

Chen Y, Cai D, Fan Z, Li K, Ni J. 3D geological modeling of dual porosity carbonate reservoirs: a case from the Kenkiyak pre-salt oilfield Kazakhstan. Pet Explor Dev. 2008. doi:10.1016/S1876-3804(08)60097-X.

Dennis I, Pretorius J, Steyl G. Effect of fracture zone on DNAPL transport and dispersion: a numerical approach. Environ Earth Sci. 2010. doi:10.1007/s12665-010-0468-8.

Fan ZQ, Jin ZH, Johnson SE. Gas-driven subcritical crack propagation during the conversion of oil to gas. Pet Geosci. 2012. doi:10.1144/1354-079311-030.

Farahmand K, Baghbanan A, Shahriar K, Diederichs MS. Effect of fracture dilation angle on stress-dependent permeability tensor of fractured rock. In: 49th U.S. Rock mechanics/geomechanics symposium, San Francisco, 2015, ARMA-2015-542.

Gan Q, Elsworth D. A continuum model for coupled stress and fluid flow in discrete fracture networks. Geomech Geophy Geo Energy Geo Resour. 2016. doi:10.1007/s40948-015-0020-0.

Guerriero V, Mazzoli S, Iannace A, Vitale S, Carravetta A, Strauss C. A permeability model for naturally fractured carbonate reservoirs. Mar Pet Geol. 2013. doi:10.1016/j.marpetgeo.2012.11.002.

Hassanzadeh H, Pooladi-Darvish M. Effects of fracture boundary conditions on matrix-fracture transfer shape factor. Transport Porous Media. 2006. doi:10.1007/s11242-005-1398-x.

Heffer KJ, King PR. Spatial scaling of effective modulus and correlation of deformation near the critical point of fracturing. Pure Appl Geophys. 2006. doi:10.1007/978-3-7643-8124-010.

Hoseinzadeh M, Daneshian J, Moallemi SA, Solgi A. Facies analysis and depositional environment of the Oligocene-Miocene Asmari Formation, Bandar Abbas hinterland Iran. Open J Geol. 2015. doi:10.4236/ojg.2015.54016.

Hooker JN, Gomez LA, Laubach SE, Gale JFW, Marrett R. Effects of diagenesis (cement precipitation) during fracture opening on fracture aperture-size scaling in carbonate rocks. Geol Soc Lond Spec Publ. 2012. doi:10.1144/SP370.9.

Huang T, Guo X, Chen F. Modeling transient flow behavior of a multiscale triple porosity model for shale gas reservoirs. J Nat Gas Sci Eng. 2015. doi:10.1016/j.jngse.01.022.

Huang Y, Zhou Z, Wang J, Dou Z. Simulation of groundwater flow in fractured rocks using a coupled model based on the method of domain decomposition. Environ Earth Sci. 2014. doi:10.1007/s12665-014-3184-y.

Jang A, Kim J, Ertekin T, Sung W. Fracture propagation model using multiple planar fracture with mixed mode in naturally fractured reservoir. J Pet Sci Eng. 2016. doi:10.1016/j.petrol.02.015.

Jiang J, Younis RM. Numerical study of complex fracture geometries for unconventional gas reservoirs using a discrete fracture-matrix model. J Nat Gas Sci Eng. 2015. doi:10.1016/j.jngse.2015.08.013.

Karimi-Fard M, Firoozabadi A. Numerical simulation of water injection in fractured media using discrete-fracture model and the Galerkin method. SPE Reserv Eval Eng. 2003. doi:10.2118/83633-PA.

Karimi-Fard M, Durlofsky LJ, Aziz K. An efficient discrete fracture model applicable for general-purpose reservoir simulators. SPE J. 2004. doi:10.2118/88812-PA.

Karimi-Fard M. Modeling tools for fractured systems: gridding, discretization, and upscaling. Stanford University. 2013. http://cees.stanford.edu/docs/KarimiFard13.

Koike K, Liu C, Sanga T. Incorporation of fracture directions into 3D geostatistical methods for a rock fracture system. Environ Earth Sci. 2012. doi:10.1007/s12665-011-1350-z.

Kresse O, Weng XW, Gu HR, Wu RT. Numerical modeling of hydraulic fractures interaction in complex naturally fractured formations. Rock Mech Rock Eng. 2013. doi:10.1007/s00603-012-0359-2.

Lapponi F, Casini G, Sharp I, Blendinger W, Fernández N, Romaire I, Hunt D. From outcrop to 3D modelling: a case study of a dolomitized carbonate reservoir, Zagros Mountains Iran. Pet Geosci. 2011. doi:10.1144/1354-079310-040.

Lee CC, Lee CH, Yeh HF, Lin HI. Modeling spatial fracture intensity as a control on flow in fractured rock. Environ Earth Sci. 2011. doi:10.1007/s12665-010-0794-x.

Maffucci R, Bigi S, Corrado S, Chiodi A, Di Paolo L, Giordano G, Invernizzi C. Quality assessment of reservoirs by means of outcrop data and discrete fracture network models: the case history of Rosario de La Frontera (NW Argentina) geothermal system. Tectonophysics. 2015. doi:10.1016/j.tecto.2015.02.016.

Malinouskaya I, Thovert JF, Mourzenko VV, Adler PM, Shekhar R, Agar S, Rosero E, Tsenn M. Fracture analysis in the Amellago outcrop and permeability predictions. Pet Geosci. 2014. doi:10.1144/petgeo2012-094.

Masihi M, King PR. A correlated fracture network: modeling and percolation properties. Water Resour Res. 2007. doi:10.1029/2006WR005331.

Mi L, Jiang H, Li J, Li T, Tian Y. The investigation of fracture aperture effect on shale gas transport using discrete fracture model. J Nat Gas Sci Eng. 2014. doi:10.1016/j.jngse.09.029.

Min KB, Rutqvist J, Tsang CF, Jing L. Stress-dependent permeability of fractured rock masses: a numerical study. Int J Rock Mech Min Sci. 2004. doi:10.1016/j.ijrmms.2004.05.005.

Nie RS, Meng YF, Jia YL, Zhang FX, Yang XT, Niu XN. Dual porosity and dual permeability modeling of horizontal well in naturally fractured reservoir. Transport Porous Media. 2012. doi:10.1007/s11242-011-9898-3.

Presho M, Woc S, Ginting V. Calibrated dual porosity, dual permeability modeling of fractured reservoirs. J Pet Sci Eng. 2011. doi:10.1016/j.petrol.2011.04.007.

Salimi H, Bruining H. Upscaling in vertically fractured oil reservoirs using homogenization. Transport Porous Media. 2010. doi:10.1007/s11242-009-9483-1.

Sang G, Elsworth D, Miao X, Mao X, Wang J. Numerical study of a stress dependent triple porosity model for shale gas reservoirs accommodating gas diffusion in kerogen. J Nat Gas Sci Eng. 2016. doi:10.1016/j.jngse.2016.04.044.

Shekhar R, Sahni I, Benson G, et al. Modelling and simulation of a Jurassic carbonate ramp outcrop, Amellago, High Atlas Mountains. Morocco. Pet Geosci. 2014. doi:10.1144/petgeo2013-010.

Sherkati S, Letouzey J. Variation of structural style and basin evolution in the central Zagros (Izeh zone and Dezful Embayment) Iran. Mar Pet Geol. 2004. doi:10.1016/j.marpetgeo.01.007.

Soleimani M, Jodeiri-Shokri B. 3D static reservoir modeling by geostatistical techniques used for reservoir characterization and data integration. Environ Earth Sci. 2015. doi:10.1007/s12665-015-4130-3.

Soleimani M. Seismic imaging by 3D partial CDS method in complex media. J Pet Sci Eng. 2016a. doi:10.1016/j.petrol.2016.02.019.

Soleimani M. Seismic image enhancement of mud volcano bearing complex structure by the CDS method, a case study in SE of the Caspian Sea shoreline. Russ Geol Geophs. 2016b. doi:10.1016/j.rgg.2016.01.020.

Taron J, Hickman S, Ingebritsen SE, Williams C. Using a fully coupled, open-source THM simulator to examine the role of thermal stresses in shear stimulation of enhanced geothermal systems. In: 48th US Rock mechanics/geomechanics symposium held in Minneapolis, 2014, ARMA 14-7525.

Wang Y, Shahvali M. Discrete fracture modeling using Centroidal Voronoi grid for simulation of shale gas plays with coupled nonlinear physics. Fuel. 2016. doi:10.1016/j.fuel.2015.09.038.

Wang Y, Li X, Tang CA. Effect of injection rate on hydraulic fracturing in naturally fractured shale formations: a numerical study. Environ Earth Sci. 2016. doi:10.1007/s12665-016-5308-z.

Wennberg OP, Casini G, Jonoud S, et al. The characteristics of open fractures in carbonate reservoirs and their impact on fluid flow: a discussion. Pet Geosci. 2016. doi:10.1144/petgeo2015-003.

Wilson CE, Aydin A, Karimi-Fard M, et al. From outcrop to flow simulation: constructing discrete fracture models from a LIDAR survey. AAPG Bull. 2011. doi:10.1306/03241108148.

Xia Y, Jin Y, Chen M. Comprehensive methodology for detecting fracture aperture in naturally fractured formations using mud loss data. J Pet Sci Eng. 2015a. doi:10.1016/j.petrol.10.017.

Xia Y, Jin Y, Chen M, et al. Hydrodynamic modeling of mud loss controlled by the coupling of discrete fracture and matrix. J Pet Sci Eng. 2015b. doi:10.1016/j.petrol.2014.07.026.

Yan X, Huang Z, Yao J, Li Y, Fan D. An efficient embedded discrete fracture model based on mimetic finite difference method. J Pet Sci Eng. 2016. doi:10.1016/j.petrol.2016.03.013.

Yoshida N, Levine JS, Stauffer PH. Investigation of uncertainty in CO2 reservoir models: a sensitivity analysis of relative permeability parameter values. Int J Greenh Gas Control. 2016. doi:10.1016/j.ijggc.2016.03.008.

Zhang QH. Finite element generation of arbitrary 3-D fracture networks for flow analysis in complicated discrete fracture networks. J Hydrol. 2015. doi:10.1016/j.jhydrol.2015.08.065.

9

Impact of formation water on the generation of H_2S in condensate reservoirs: a case study from the deep Ordovician in the Tazhong Uplift of the Tarim Basin, NW China

Jin Su[1,2] · Yu Wang[1,2] · Xiao-Mei Wang[1,2] · Kun He[1,2] · Hai-Jun Yang[3] · Hui-Tong Wang[1,2] · Hua-Jian Wang[1,2] · Bin Zhang[1,2] · Ling Huang[1,2] · Na Weng[1,2] · Li-Na Bi[1,2] · Zhi-Hua Xiao[4]

Abstract A number of condensate reservoirs with high concentrations of H_2S have been discovered in the deep dolomite reservoirs of the lower Ordovician Yingshan Formation (O_1y) in the Tazhong Uplift, where the formation water has a high pH value. In the O_1y reservoir, the concentrations of Mg^{2+} and SO_4^{2-} in the formation water are higher than those in the upper Ordovician formation. The concentration of H_2S in the condensate reservoirs and the concentration of Mg^{2+} in the formation water correlate well in the O_1y reservoirs of the Tazhong Uplift, which indicates a presumed thermochemical sulfate reduction (TSR) origin of H_2S according to the oxidation theory of contact ion-pairs (CIPs). Besides, the pH values of the formation water are positively correlated with the concentration of H_2S in the condensate reservoirs, which may indicate that high pH might be another factor to promote and maintain TSR. Oil–source correlation of biomarkers in the sulfuretted condensates indicates the Cambrian source rocks could be the origin of condensates. The formation water in the condensate reservoirs of O_1y is similar to that in the Cambrian; therefore, the TSR of sulfate-CIPs likely occurred in the Cambrian. High H_2S-bearing condensates are mainly located near the No. 1 Fault and NE-SW strike-slip faults, which are the major migration pathway of deep fluids in the Tazhong Uplift. The redox between sulfate-CIPs and hydrocarbons is the generation mechanism of H_2S in the deep dolomite condensate reservoirs of the Tazhong Uplift. This finding should be helpful to predict the fluid properties of deep dolomite reservoirs.

Keywords Formation water · Sulfate-CIPs · TSR · Condensates · Dolomite reservoir · Tarim basin

1 Introduction

Formation water has been conventionally taken as a kind of undesirable by-product in petroleum exploration and development, while it would be one of the most crucial factors in hydrocarbon generation, migration and accumulation as well as the interaction between organic and inorganic geo-fluids. Detailed analysis of the geochemical properties of formation water could indicate the hydrodynamic conditions for hydrocarbon accumulation and the geochemical alteration of reservoirs (Jiang and Zhang 1999; Cai et al. 2001; Zha et al. 2003; Chen and Zha 2005, 2006a, b, 2008), which are of great geological significance for interpreting the reaction mechanisms between ions and hydrocarbons. And then, the origin of multi-phase reservoirs could be deduced from the secondary geochemical alteration (Land 1995; Davisson and Criss 1996; Varsanyi and Kovacs 1997; Su et al. 2011). Palmer (1984) published research on the hydrocarbon composition variation in carbonate reservoirs as a result of the formation water flowing across the reservoirs. In particular, dibenzothiophene would be mostly altered in its composition

✉ Jin Su
susujinjin@126.com

1 Key Laboratory of Petroleum Geochemistry, CNPC, Beijing 100083, China

2 State Key Laboratory for Enhancing Oil Recovery, Research Institute of Petroleum Exploration and Development, PetroChina, Beijing 100083, China

3 Research Institute of Tarim Oilfield Company, PetroChina, Korla 841000, Xinjiang, China

4 PetroChina Coalbed Methane Company Limited, Beijing 100028, China

Edited by Jie Hao

(Kuo 1994; Lafargue and Tluez 1996). In addition, high H$_2$S concentration from microbial activities around the oil–water contact would increase the acidity and density of crude oil, which negatively impacts on the commercial value of petroleum and as well as the safe exploitation and development of reservoirs.

With the extension of hydrocarbon exploration to deep dolomite rocks in recent decades in China, a large number of gas reservoirs with high H$_2$S have been discovered. In domestic and overseas investigations, high concentrations of H$_2$S have generally been attributed to the thermochemical sulfate reduction (TSR) between anhydrite and hydrocarbons (Worden et al. 2000; Jiang et al. 2015), during which liquid hydrocarbons are enriched with sulfur and the drying coefficient of gas reservoirs may exceed 99%. In actual geological conditions, the formation water could act as the reaction agent of TSR. In the Tazhong Uplift of the Tarim Basin, H$_2$S-bearing condensates have been discovered on a large scale in deep dolomite reservoirs, which differs greatly from the geochemical properties of normal condensate reservoirs around the world and also differs from gas reservoirs with high H$_2$S. Previous study of sulfur isotopes and concentration of individual sulfur compounds in the condensate reservoirs has shown that the condensate reservoirs with high H$_2$S may have experienced TSR (Jiang et al. 2008; Zhang et al. 2015). Various soluble sulfate species act as the initial oxidant of TSR, mainly including free sulfate ions (SO$_4^{2-}$) and bisulfate ions (HSO$_4^-$), solvent-shared ion-pairs (SIPs) and contact ion-pairs (CIPs) (Eigen and Tamm 1962; Atkinson and Petrucci 1966). Therefore, the salts of formation water may be critical to activate and affect TSR. Recently, a series of gold-tube hydrous pyrolysis experiments have been conducted to address the reaction mechanism and the factors affecting TSR, which include compositions of hydrocarbons, labile organosulfur compounds (LSC) and the early generated H$_2$S (Tang et al. 2005; Ellis et al. 2007; Amrani et al. 2008). Because the relative concentration of H$_2$S is not a reliable reaction parameter to access the reaction intensity of TSR (Ma et al. 2008; Jendu et al. 2015), a systematic study on the genetic relationship between water and mineral composition of the Ordovician formation and high sulfur condensates would avail to clarify the formation mechanism of high sulfur condensate in the actual geological conditions. These new insights could indicate the effect of ions in the formation water on the alteration of hydrocarbons in deep dolomite reservoirs, which would be useful to predict the fluid properties of deep stratum and promote the exploration of dolomite reservoirs.

2 Geological background

The Tarim Basin experienced three important stages of hydrocarbon accumulation, i.e., the later Eopaleozoic, the Neopaleozoic to the early Mesozoic, and the late Cenozoic (Zhang et al. 2011), each of which had great impacts on the accumulation and distribution of hydrocarbons in the marine reservoirs (Zhang and Huang 2005). The Tarim Basin has become a petroliferous basin which has undergone multistage secondary alterations (Sun and Püttmann 1996; Su et al. 2010; Sun et al. 2009a, b). The Tazhong Uplift, as an inherited paleo-uplift, is located in the central of the Tarim Basin and neighbors the Bachu Uplift in the west, the Tadong Low Uplift in the east, the Tangguzibasi depression in the south and the North depression in the north (Fig. 1). After the successful exploration in the reef and beach reservoirs of the upper Ordovician Lianglitage Formation (O$_3$l), a large-scale petroleum resource has been discovered in dolomite reservoirs of the Yingshan Formation (O$_1$y), in the lower Ordovician in the North Slope of the Tazhong Uplift. It has proved that the lower Ordovician dolomite reservoirs are mainly distributed and developed below an unconformity surface, about 120 m deep. The distribution of dolomite reservoirs is not controlled by structural highs/lows.

The dolomite reservoirs of the Yingshan Formation (O$_1$y) are characterized by various hydrocarbon and fluid phases, as well as large discrepancies in gas/oil ratio (GOR). Most high-yield wells produce large volumes of gas and condensate oil; in contrast, low-yield wells mainly produce crude oil with low GOR. Based on the PVT curve, the types of reservoirs in the Tazhong Uplift include unsaturated and saturated condensates, and unsaturated and saturated oil (Fig. 1), which are heterogeneously distributed across the Tazhong Uplift and are not controlled by structure and burial depth. The difference in reservoir phases indicates that the distribution of condensates in the whole Tazhong Uplift roughly correlates with the fluid properties and burial depth and the dolomite reservoirs are correlated with NE-SW faults.

3 Samples and methods

3.1 Collection of samples

In order to investigate the origin of H$_2$S, hydrocarbon samples of 13 wells in the Tazhong Uplift were collected (Table 1). The saturated and aromatic hydrocarbons were quantified by GC–MS, including steranes and hopanes, as well as thiophenes. Diamantanes were detected by

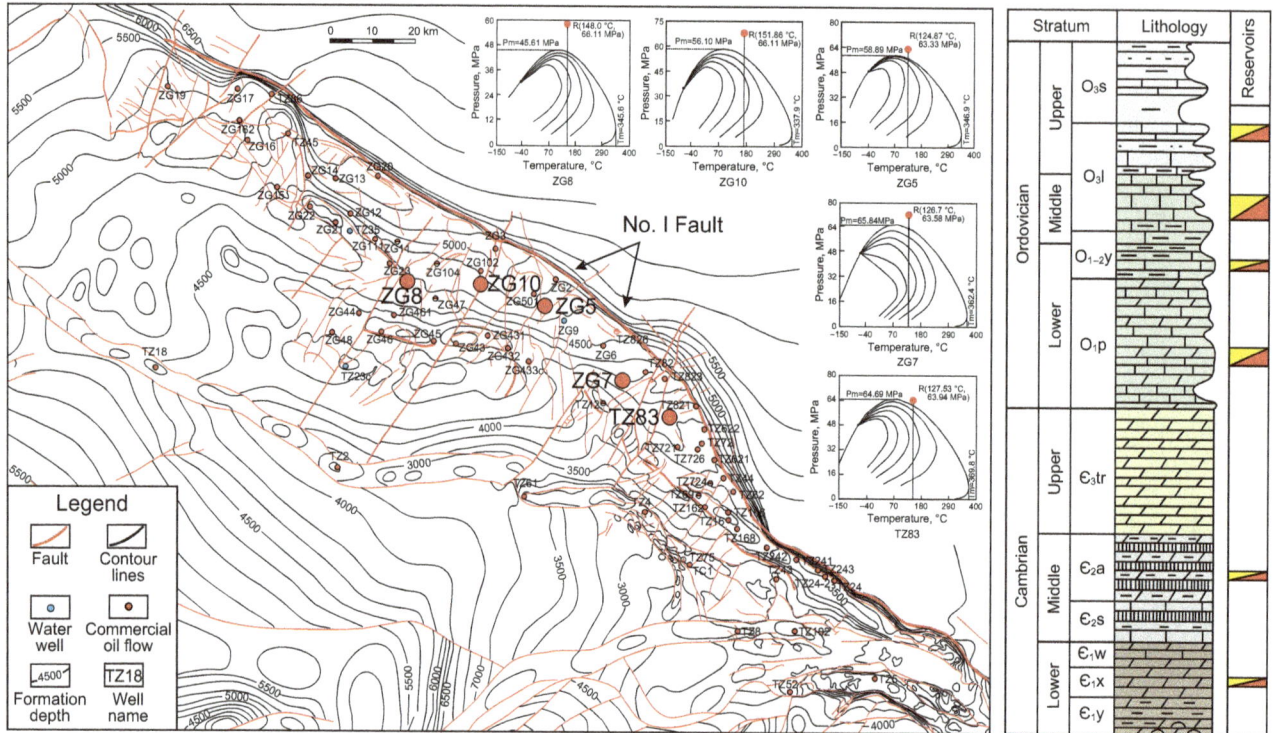

Fig. 1 Tectonic map of the Yingshan Formation and distribution of condensate reservoirs in the Tazhong Uplift

Table 1 Features of high sulfur condensate reservoir of Yingshan Formation in the Tazhong area

Well	Depth, m	H$_2$S content, mg/m^3	Bottom-hole		Critical condensate		Condensate oil content, g/m^3	Gas output, m^3/d
			Temperature, °C	Pressure, MPa	Pressure, MPa	Temperature, °C		
ZG14	6300	5103	141.80	72.51	62.40	361.1	305.6	130,921
ZG22	5736	2310	129.75	67.31	71.46	366.6	319.2	70,200
ZG12	6279	640	140.60	72.93	67.42	348.9	66.0	169,537
ZG11	6475	5300	141.39	73.88	60.73	332.6	376.1	95,540
ZG111	6250	3400	137.00	71.81	61.88	325.1	354.8	114,362
ZG8	6145	46,267	148.00	66.11	45.61	345.6	748.1	144,844
ZG441	5522	10,100	126.76	63.24	69.19	369.8	176.3	134,935
TZ201c	5779	20,700	126.68	63.58	65.84	362.4	195.0	74,270
ZG43	5334	36,537	124.87	63.33	50.89	346.9	451.6	83,679
ZG10	6309	36,540	151.86	68.11	56.10	337.9	386.9	218,703
ZG102	6287	77800	145.78	63.53	40.48	314.1	764.2	57,658
ZG103	6233	3867	140.30	71.58	50.29	327.0	378.5	74,366
ZG5	6460	49,480	150.60	75.09	51.02	297.3	312.2	130,518

H$_2$S content (mg/m^3) = the H$_2$S weight per m^3 condensate gas; condensate oil content (g/m^3) = the condensate oil weight per m^3 condensate gas

GC × GC to access the oil-cracking extent of hydrocarbons.

3.2 GC–MS analysis

GC–MS analysis was performed using a Trace GC Ultra gas chromatograph interfaced to a Thermo DSQII mass selective detector (MSD) for both saturated and aromatic hydrocarbon fractions. The HP-5 column is 30 m long × 0.25 mm ID × 0.25 μm thickness. Helium maintained at a constant flow rate of 1.0 cm^3/min was used as the carrier gas. The GC oven was programmed from 50 to 220 °C at 4 °C/min with an initial hold time of 5 min and then from 220 to 320 °C at 2 °C/min with a final hold time of 25 min. Samples were injected in the splitless mode at a constant temperature of 300 °C. For quantitative determination of saturated and aromatic hydrocarbons, known concentrations of standard compounds (d4-cholestane and d10-anthracene) were added to the condensate and oil prior to the fractionation.

3.3 Diamondoid hydrocarbon analysis

For GC × GC-TOF–MS analysis, a whole condensate sample was dissolved in a solution of 5% dichloromethane (DCM) in *n*-hexane. Two Leco Pegasus 4D GC × GC systems were used in this study coupled with a TOF–MS and a FID, respectively. They were equipped with an Agilent 6890 N GC (TOF–MS) and an Agilent 7890A GC (FID system) and configured with a split/splitless auto-injector and a dual-stage cryogenic modulator. Two capillary GC columns were fitted in the GC. The first dimension chromatographic separation was performed by a nonpolar Petro-column (50 m × 0.2 mm I.D., 0.5 μm film thickness). The second column was connected to the TOF–MS instrument via a DB-17HT column (3 m × 0.1 mm I.D., 0.1 μm film thickness). Temperature was programmed for 35 °C (10 min) in the primary GC oven. Then, the temperature was increased to 60 °C at 0.5 °C/min and held for 0.2 min, from 60 to 220 °C at 2 °C/min and held for 0.2 min, followed by an increase at 4 °C/min to the final temperature of 300 °C and kept constant for 5.0 min. The secondary oven was programmed 20 °C above the primary GC oven gradient. The sample injection temperature was 300 °C with an injection volume of 0.5 μL. The carrier gas

was helium with flow rate of 1.8 mL/min. The modulation period was 10 s with a 3.0-s hot pulse duration. The TOF–MS instrument was operated in the electron impact mode (70 eV) with a range of 40–520 Da. The ion source temperature was 240 °C, the detector voltage was set at 1475 V, and the acquisition rate was 100 spectra/s.

Instrument control and data processing were done using ChromaToF (Leco) software and Microsoft Excel. The deconvoluted spectra were compared with the National Institute of Standards and Technology (NIST) software library for compound identification. The quantification of adamantanes was calculated by comparison with an internal standard of d16-adamantane, while the quantification of diamantanes required a linear correlation among the ratios of various diamantane standard concentrations to the constant d16-adamantane concentration.

3.4 Constituent analysis of formation water

We have reviewed more than 200 compositions of formation water in Tazhong Uplift, which were analyzed in the chemical laboratory of the Tarim Oilfield Company, in order to determine the origin of formation water in the condensate reservoirs. The analysis method for oil and gas field water is conducted to industry standard SY/T 5523-2000. As well, 14 formation water samples were collected to measure the concentration of ions at ambient temperature. Since the temperature of the reservoirs is mostly from 125 to 150 °C, the measured ionic concentrations had been converted to in situ conditions using thermodynamic modeling (Table 2).

Table 2 Concentrations of main ions in the condensate reservoirs converted to in situ geological condition by thermodynamic modeling

Wells	Formation	Depth, m	pH	Cl^-, mg/L	SO_4^{2-}, mg/L	Ca^{2+}, mg/L	Mg^{2+}, mg/L	Salinity, mg/L	H_2S, mg/m^3
ZG9	O	6049.1	7.24	120,200	78	12,900	1420	197,700	616,000
ZG7	O	5865	7.81	102300	53	8890	532	168,100	47,500
ZG48	O_3	5498.1	6.85	88,030	400	7694	714	150,000	1000
ZG461	O	5479.64	6.24	56,990	891	19,800	810	6800	1300
ZG45	O_3	5637.2	7.33	42,100	894	4327	416	70,930	15,867
ZG44	O	5603.96	7.17	100,200	450	9635	770	169,500	10,600
ZG43	O	5380	8.21	6097	550	1576	1050	11,330	36,537
ZG26	O	6085.5	6.16	70,180	381	29,690	529	116,600	100
ZG164	O_3	6122.13	7.42	41,900	221	2014	399	72,020	7200
ZG163	O	6140	7.11	51,750	114	3364	461	85,210	6900
ZG162	O_1	6123	6.17	4624	1047	155	74	9607	21
ZG15-2	O_2	5918.5	7.11	63,510	209	4256	418	107,200	4900
ZG11	O_3	6165	6.83	64,570	97	3791	283	109,400	5300
ZG10	O_3	6198	6.91	107,000	31	11,900	772	179,200	36,540

4 Results and discussion

4.1 Geochemical features of high sulfur condensate reservoirs

For condensates in the O_1y reservoir of Tazhong Uplift, the burial depth is more than 5500 m. The reservoir temperature may rise beyond 130 °C with a gradient of 2.03 °C/100 m, and the pressure coefficient is 1.18. The reservoirs are weakly over-pressured (Table 1). The concentration of condensate oil in the reservoir is generally higher than 100 g/m^3 with a peak of 764.2 g/m^3 in well ZG102. According to the difference between formation pressure and dew point pressure, the condensates were classified as unsaturated (pressure difference > 0) and saturated reservoirs (pressure difference $= 0$). Most of the unsaturated condensates are distributed near strike-slip faults, while the saturated condensates are distributed far away from faults.

Another distinct feature of O_1y condensates is their high concentration of H_2S (Fig. 2), which is different from the condensate in the O_3l reservoirs around the Tazhong No. 1 structure. For the reservoirs around the Tazhong No. 1 fault, the concentration of H_2S correlates well with gas production. According to the previous literature (Zhang et al. 2015), the H_2S concentration in the condensate reservoirs of the Lianglitage Formation (O_3l) is presumed to originate from the charge of high-maturity gases in the late Himalayan (Zhao et al. 2009). In contrast, the H_2S in the condensates of the Yingshan Formation (O_1y) may have not dissolved and migrated in the gas phase. The production volume of formation water and the concentration of H_2S in O_1y condensate reservoirs show that the high concentration of H_2S is correlated well with active formation water (Fig. 2). So, it is proposed that the origin of H_2S may be related to the geochemical properties of O_1y formation water. Therefore, a knowledge of the geochemical properties of formation water and the accumulation process of condensates will be essential to clarify the origin of H_2S in the condensates of deep dolomite reservoirs.

Previous studies confirmed that two sets of source rocks mainly developed in the Tarim Basin, which, respectively, developed in the Ordovician platform margin slope facies and Cambrian basin facies. The planktonic algae organic phase of Cambrian source rock is distributed in the eastern Manjia'er sag. A series of biomarkers (triaromatic dinosterane and 4-methyl-24-ethyl cholestane) indicating complex acritarchs, planktonic algae, such as diatoms and dinoflagellates, have been found in a large number of Cambrian source rocks. The mud mound organic facies of Ordovician source rock are mainly distributed in the platform edges and transition slopes. Some biomarkers of sponge and gloeocapsa, such as 24-norcholestane and 4-methyl-24-propyl cholestane have been detected in the lime-mud mounds (Zhang and Huang 2005). These molecular fossils have become an important marker of oil–source correlation in the Tarim Basin. Through the distribution of these special biomarkers in the condensates and normal oils in the Ordovician Yingshan Formation of Tazhong Uplift, the concentration of 4, 23, 24-trimethyl triaromatic dinosterane in the condensates was higher than that in the normal oils, while the concentration of

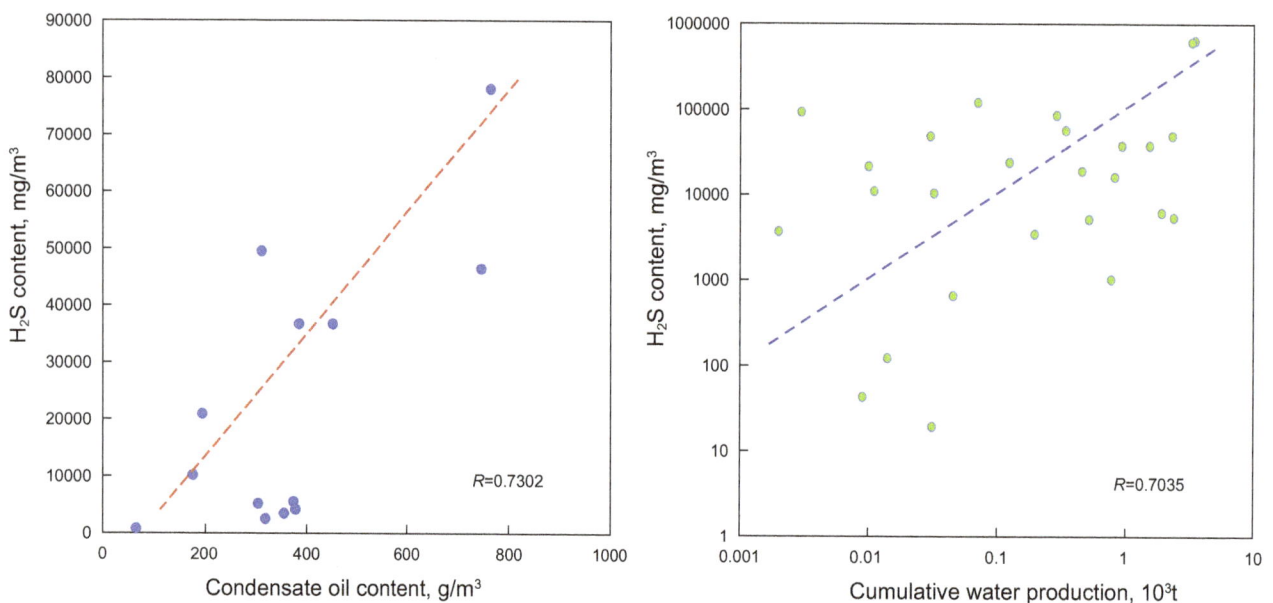

Fig. 2 Correlation between H_2S content and condensate oil content as well as cumulative water output of the Yingshan condensate gas reservoir in the Tazhong area

3-methyl-24-methyl triaromatic dinosterane in normal oils increases significantly (Fig. 3). This indicates that the condensates in the lower Ordovician Yingshan Formation (O_1y) of Tazhong Uplift mainly migrated from the deep Cambrian source rocks.

The exploration area around well ZG43 is one of the most hydrocarbon-rich tectonic zones in the Yingshan Formation (O_1y) and also one of the reservoirs with the highest H_2S concentration in the Tazhong Uplift. Well ZG43, for example, produced natural gas of 83,679 m^3 per day with a condensate concentration of 451.6 g/m^3 and an H_2S concentration of 36,537 mg/m^3. The condensate reservoirs distributed along strike-slip faults have high oil/gas yields and high GOR, while oil reservoirs located

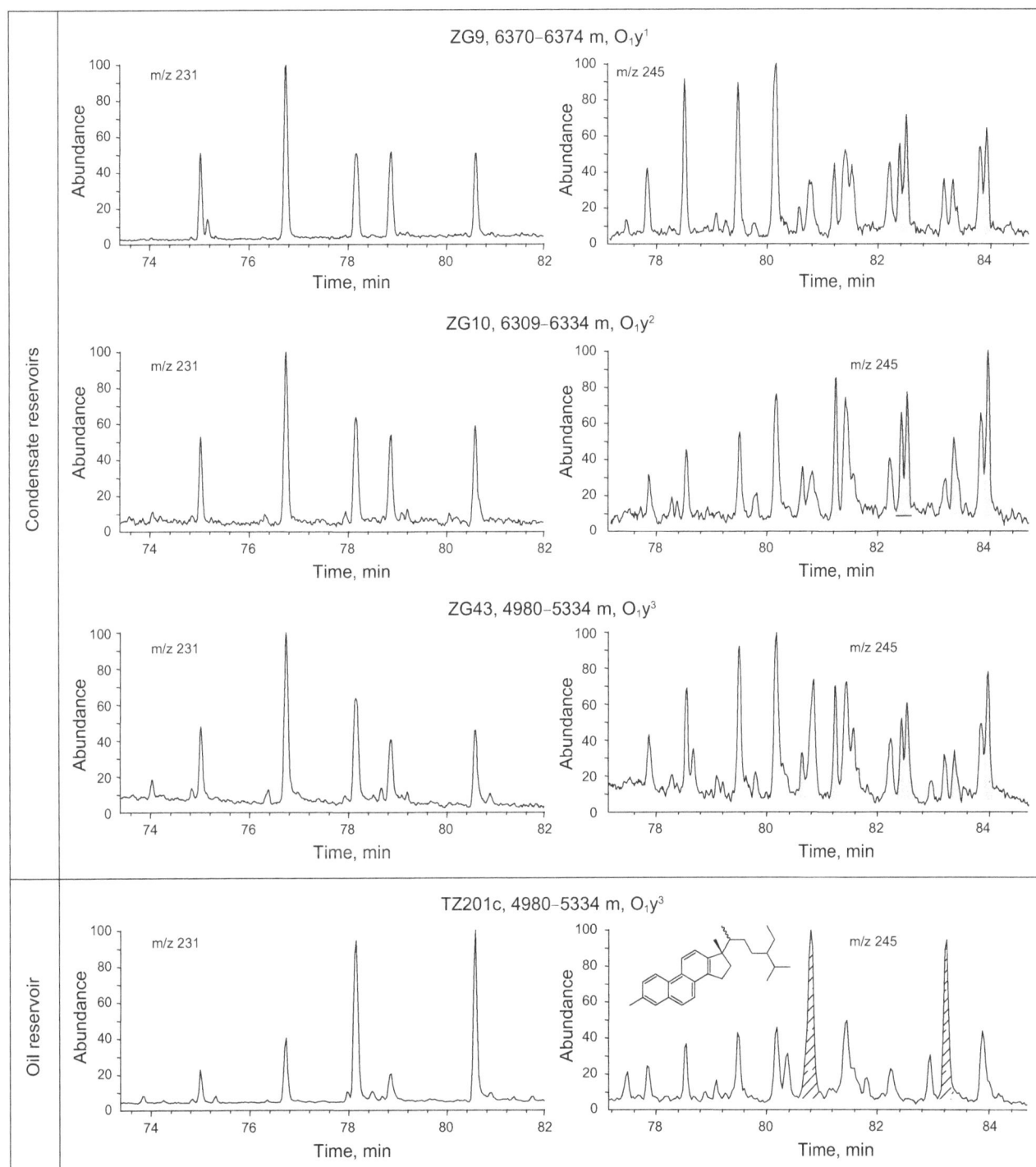

Fig. 3 Biomarker characteristics of multi-phase reservoirs in the lower Ordovician Yingshan Formation

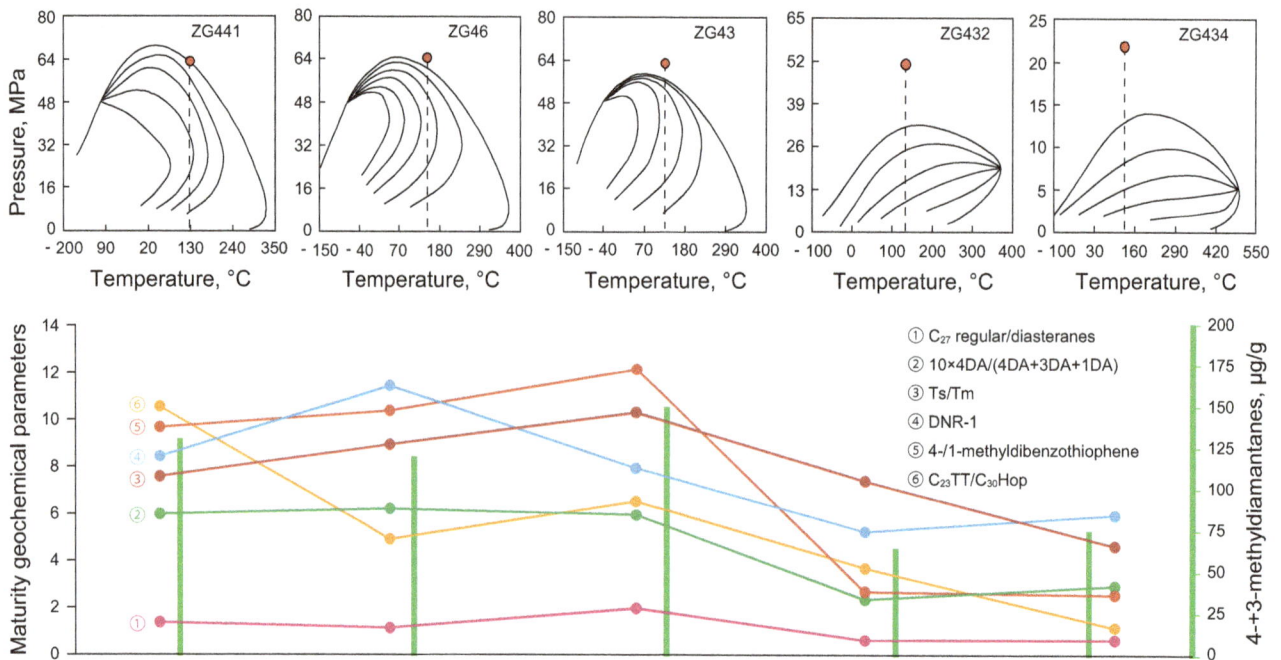

Fig. 4 Comparison of phase and maturity between condensate and normal oils in the ZG-43 block of the Yingshan Formation in the Tazhong Uplift

far away from faults have low oil/gas yields and low GOR. This indicates that the strike-slip faults might act as pathways for gas charging. The intensity of gas charging has been proved to be main mechanism by which the early oil reservoir transferred to the coexisting condensate and oil reservoirs (Zhang et al. 2011). The analysis on the relative concentrations of individual compound in condensates shows that the maturity of condensates is evidently higher than that of the oils (Fig. 4). In particular, the concentration of diamantanes reflects the extent of oil cracking (Dahl et al. 1999; Zhang et al. 2011). The concentration of diamantanes is 150 μg/g in the condensates of well ZG43 with an H_2S concentration of 36.537 mg/m^3, which is much higher than that in conventional oil. So, it is indicated that high concentrations of H_2S may originate from oil cracking or thermal chemical alteration, which have also led to the higher maturity of condensates compared with conventional oil.

The oil cracking commonly includes two main processes, i.e., thermal cracking of oil and thermochemical sulfate reduction (TSR). It is revealed that the well bottom temperature of O_1y condensate reservoirs does not exceed 160°C. Previous research considered that thermal cracking would not occur unless the temperature exceeds at least 190°C (Price 1980; Zhang et al. 2008). So, it could be concluded that the high maturity of condensates and high concentration of H_2S in the O_1y reservoirs were not due to liquid hydrocarbon thermal cracking. So, it can be inferred that the high concentration of diamantanes and H_2S in the condensate reservoirs is likely to derive from the TSR.

Further detailed analyses on the composition of the Ordovician formation water would be needed to diagnose whether the geochemical conditions in the Tazhong Uplift were suitable to trigger the TSR and accumulate H_2S of TSR-origin.

4.2 Relationships between H_2S-origin and formation water in the condensate reservoirs

According to the statistics, the formation water of the lower Ordovician—Cambrian differs from that of the upper Ordovician in the concentration of principal ions (Fig. 5). In particular, the concentrations of SO_4^{2-} in the O_3l formation water are higher than those in the lower Ordovician—Cambrian formation water. On the contrary, the Mg^{2+} concentration is higher in the formation water of O_1y-Cambrian. Both of SO_4^{2-} and Mg^{2+} are significant for thermochemical sulfate reduction (Zhang et al. 2011). It has been noted that the SO_4^{2-} commonly exists in the form of contact ion-pairs, which are the actual oxidant of TSR at geological temperatures. The decrease in SO_4^{2-} concentration in the lower Ordovician to Cambrian formation water may be due to the consumption of SO_4^{2-} in the TSR (Su et al. 2016). The concentration of Mg^{2+} and Ca^{2+} is lower than 1.1 and 20 g/L in the upper Ordovician formation water and much higher in the lower Ordovician to Cambrian formation water. In particular, the concentration of Mg^{2+} goes up beyond 1.5 g/L, which is proposed to activate TSR in actual geochemical conditions. Therefore,

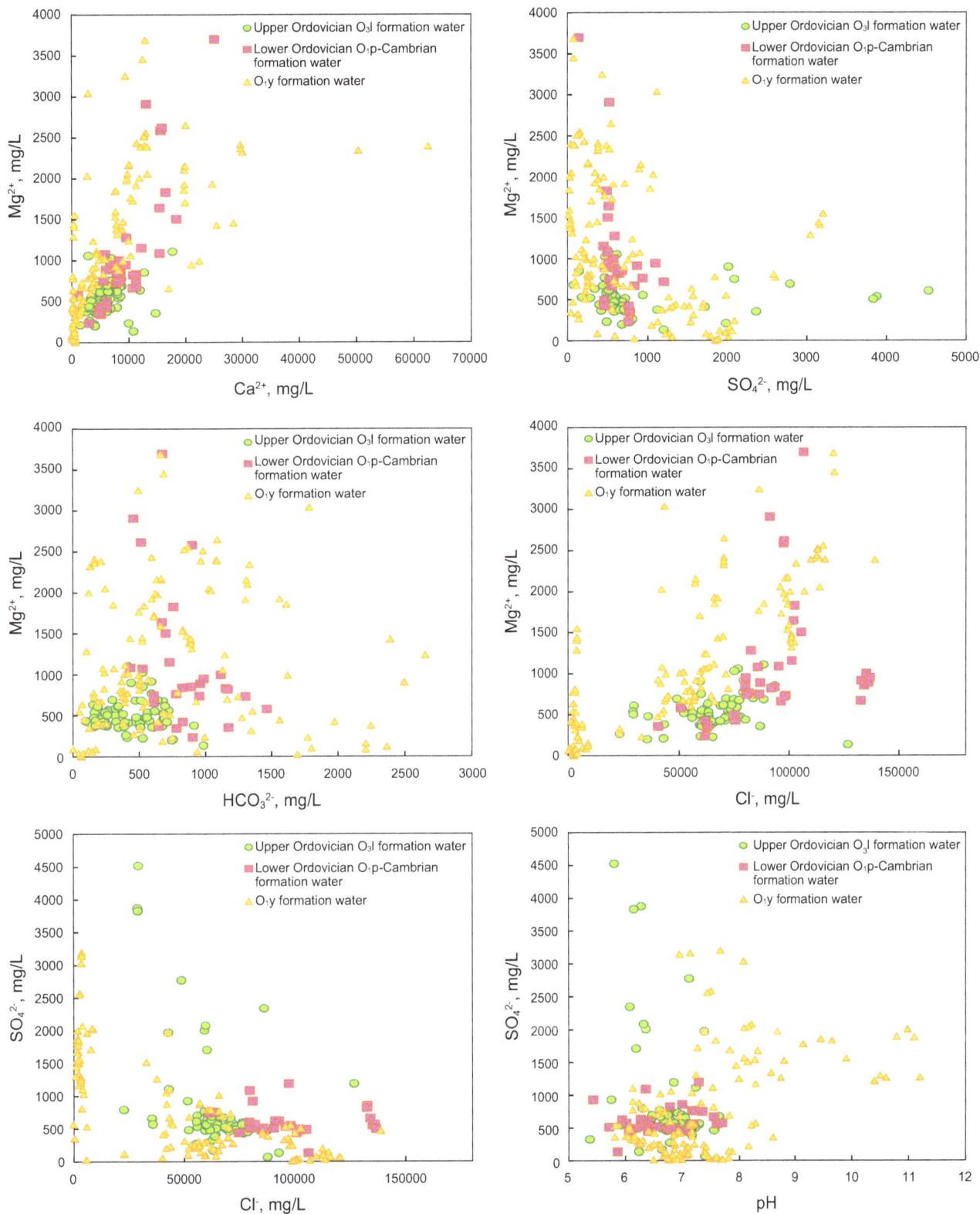

Fig. 5 Comparison of formation water ion concentration between the upper Ordovician Lianglitage Formation and the lower Ordovician Penglaiba Formation—Cambrian in the Tazhong Uplift

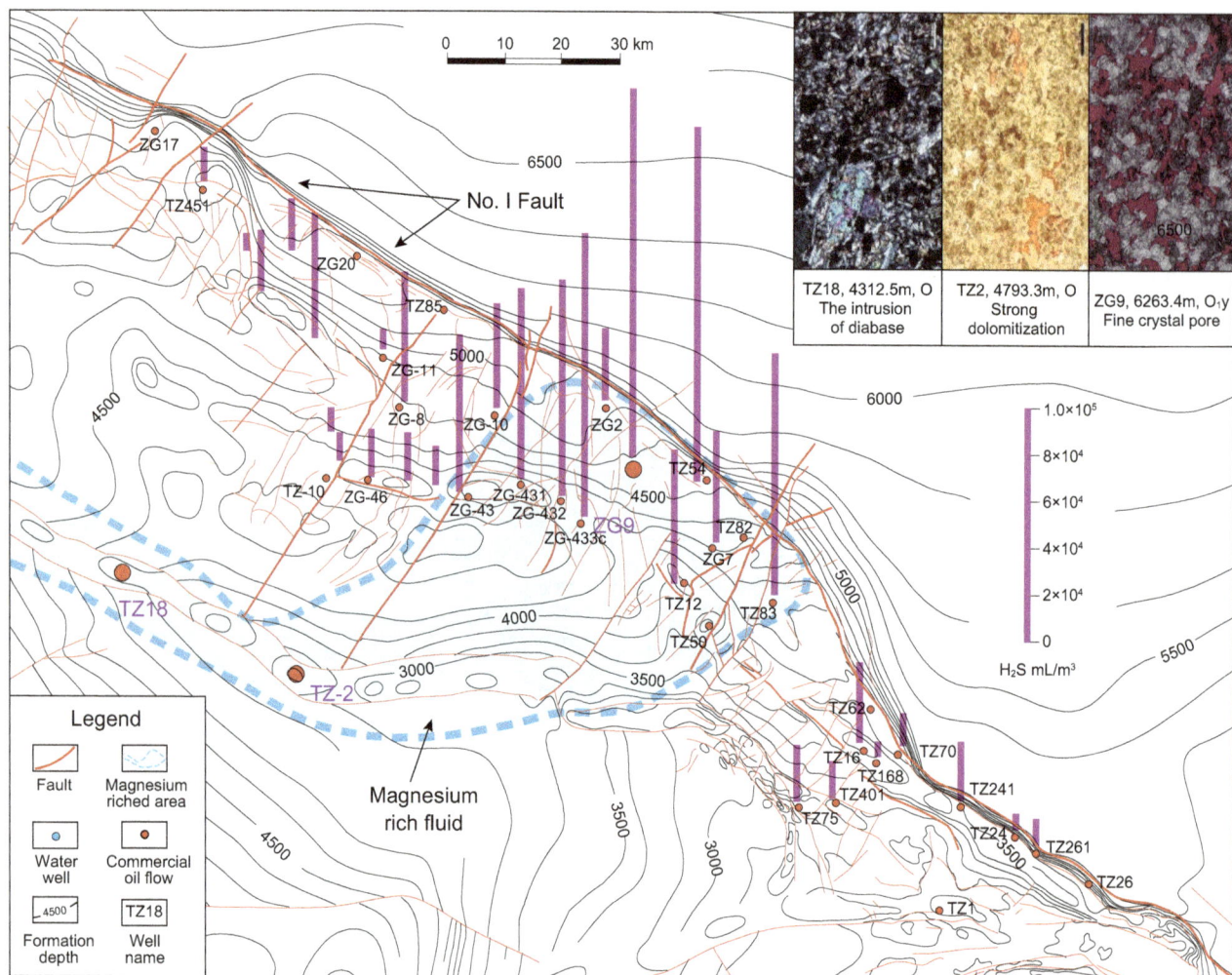

Fig. 6 Relationship between the content of H$_2$S in the condensate reservoirs and Mg^{2+}-enriching fluid in the Ordovician of Tazhong Uplift

the deep dolomite reservoirs of lower Ordovician could provide a favorable setting for TSR.

The Tazhong Uplift has experienced multi-period hydrothermal invasion and dolomitization. So that Mg^{2+} is rich in the formation water of the tectonic area experiencing strong dolomitization and hydrothermal activity, which provides the geochemical conditions for the TSR of sulfate-CIPs in the Tazhong Uplift. Intrusive diabases are seen in well TZ-18, strong dolomitization happened in well Tazhong-2, and a number of fine crystalline porous dolomites have been found in the lower Ordovician Yingshan Formation of well ZG-9 (Fig. 6). The most active region of magnesium-rich fluid was mainly located in the Tazhong Uplift and the middle section of the No. 1 fault zone (Fig. 6). On the other hand, the highest concentration of H$_2$S in the Tazhong area is also mainly distributed in the same zone as the magnesium-rich fluid. The concentration of H$_2$S was more than 5% (m^3/m^3), coinciding with the region of reservoirs with strong dolomitization (Fig. 6).

This indicates a very close relationship between the formation of H$_2$S and the activity of Mg^{2+}-enriching fluid in the lower Ordovician Yingshan Formation.

4.3 The formation mechanism of high H$_2$S in the condensate reservoirs

The production data have shown that the concentration of H$_2$S is not clearly related to the gas volume of condensate reservoirs in O$_1$y reservoirs, but correlated to the volume of formation water and the content of condensates. In the condensate reservoirs of Well-ZG7 and ZG9 blocks with the highest production volume of formation water, the H$_2$S concentration is up to 600 g/m^3 (Table 3). Through comparing the properties and distribution of formation water in the Tazhong Uplift, it is found that the formation water in the O$_1$y dolomite reservoirs connects with the fluids in the Cambrian formations. The Tazhong No. 1 fault and strike-slip faults of Tazhong Uplift may act as the major

Table 3 Statistics of fluid production and H_2S content in the Yingshan condensate reservoir of the Tazhong Uplift

Well	Formation	GOR, m^3/m^3	Output			H_2S content, mg/m^3
			Oil, 10^3t	Water, 10^3t	Gas, 10^3t	
ZG9	O_1y^1	0	0.000	3.497	0.256	616,000
ZG501	O_1y^1	237	8.392	0.000	0.283	32,200
ZG51	O_1y^2	1500	1.791	0.337	0.269	55,100
ZG6	O_1y^1	245	1.521	3.310	0.037	593,000
ZG7	O_1y^2	2024	3.503	2.332	0.594	47,500
ZG10	O_1y^2	2372	28.984	0.931	6.616	36,500
ZG102	O_1y^1	1688	1.403	0.000	0.195	77,800
ZG103	O_1y^1	2049	8.448	0.000	1.723	3870
ZG11	O_1y^2	3300	8.771	2.398	2.586	5300
ZG111	O_1y^2	2815	18.089	0.193	5.126	3400
ZG12	O_1y^2	13,500	0.029	0.045	0.040	640
ZG13	O_1y^1	185	10.596	1.921	0.594	6080
ZG14	O_1y^2	1800	8.536	0.512	1.526	5100
ZG14-1	O_1y^2	3064	4.849	0.002	1.955	3600
ZG21	O_1y^2	886	0.057	0.031	0.005	19
ZG22	O_1y^2	2300	3.385	0.123	0.671	23,100
ZG23CH	O_1y^2	0	1.282	0.009	0.047	42
ZG8	O_1y^2	1168	18.97	0.030	2.409	46,300
TZ201C	O_1y^3	3900	2.881	0.010	1.144	20,700
ZG43	O_1y^3	1205	26.556	1.545	3.689	36,500
ZG431	O_1y^3	589	7.627	0.286	0.458	82,100
ZG432	O_1y^3	176	6.054	0.003	0.102	89,400
ZG433C	O_1y^3	307	16.197	0.07	0.368	116,000
ZG441	O_1y^3	3008	2.939	0.032	0.962	10,100
ZG44C	O_1y^3	3260	0.073	0.011	0.024	10,600
ZG45	O_1y^3	778	7.079	0.817	0.583	15,900
ZG46	O_1y^3	10,572	0.609	0.451	0.641	18,400
ZG462	O_1y^3	1924	7.274	0.014	1.329	120
ZG48	O_1y^4	4294	0.860	0.767	0.373	1000

migration pathway for the formation water in gas condensate reservoirs, and contribute high H_2S in the condensate reservoirs around the faults.

Much previous research has proved TSR between hydrocarbons and sulfate could induce oil-cracking processes in actual geological conditions (Worden and Smalley 1996; Wei et al. 2012). It is generally agreed that the sulfate contact ion-pair would be the dominant mechanism to trigger thermochemical sulfate reduction (Rudolph et al. 2003; Amrani et al. 2008). So, samples of formation water were collected to test the concentration of ions at ambient temperature. Then, the measured ionic concentrations had been converted to in situ conditions by thermodynamic modeling, in order to discuss the TSR reaction mechanism with the oxidant of sulfate-CIPs in the highly sulfuretted condensate reservoirs (Table 3). Mg^{2+} with high ionic strength is encircled completely by water molecules in aqueous systems, so SO_4^{2-} could not be bound with Mg^{2+} directly and would form contact ion-pairs (CIPs) to accelerate TSR reactions (Azimi et al. 2007; Leusbrock et al. 2008). Based on the theory of CIPs, good correlation between the concentration of H_2S and Mg^{2+} in formation water (Fig. 7) has indicated the TSR-origin of H_2S in the condensate reservoirs. In addition, the concentration of principal negative ions, i.e., Cl^- and HCO_3^-, is separately lower than 100 and 1.0 g/L in the upper Ordovician formation water and higher than 100 g/L and 1.0 g/L in the lower Ordovician—Cambrian formation water. It has shown that the concentration of H_2S in the O_1y gas condensates has increased with the alkalinity of formation water (Fig. 7), which demonstrated that the pH of formation water may drop down in the process of forming CIP from magnesium sulfate (He et al. 2014). Alkaline formation water would guarantee the rightward reaction to form H_2S in large amounts. Therefore, it could be inferred that TSR with sulfate-CIPs might also occur in the deep Cambrian dolomite reservoirs.

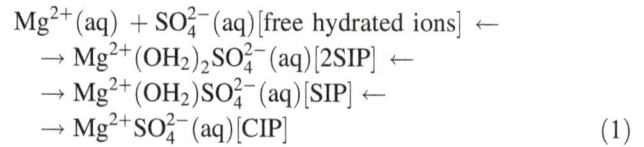

$$
\begin{aligned}
Mg^{2+}(aq) + SO_4^{2-}(aq)\,[\text{free hydrated ions}] &\leftarrow \\
\rightarrow Mg^{2+}(OH_2)_2SO_4^{2-}(aq)\,[2SIP] &\leftarrow \\
\rightarrow Mg^{2+}(OH_2)SO_4^{2-}(aq)\,[SIP] &\leftarrow \\
\rightarrow Mg^{2+}SO_4^{2-}(aq)\,[CIP] &
\end{aligned}
\tag{1}
$$

With the initiation of TSR, the concentration of H_2S increases and concentration of SO_4^{2-} decreases. Therefore, it could be demonstrated that H_2S in the condensates of lower Ordovician dolomite reservoirs does originate from redox reaction between hydrocarbons and sulfate-CIPs.

5 Conclusions

The condensates in the dolomite reservoirs of the lower Ordovician in the Tazhong Uplift are generally characterized by various properties and phases of hydrocarbons. The concentration of H_2S in the condensate reservoirs increases with the production volume of formation water. Both the Mg^{2+} concentration and pH in the O_1y formation water are all higher than those in the upper Ordovician reservoirs and correlate well with the H_2S concentration in the gas condensate reservoirs. The decrease in SO_4^{2-} concentration in the O_1y condensates is due to the consumption of SO_4^{2-} during TSR, which is the formation mechanism of H_2S in the O_1y condensates. The pH values of the formation water are positively correlated with the H_2S concentration in the condensate of the lower Ordovician dolomite reservoirs, which shows that high alkalinity of the formation water is another important factor to initiate and promote the TSR of sulfate-CIPs. It is thus inferred the deep dolomite reservoirs have favorable geological conditions for TSR. The properties of the formation water in the Cambrian are similar to those of the high H_2S-bearing condensate in the O_1y reservoirs. This indicated that the sulfur condensates of the O_1y reservoirs originated from the Cambrian source rocks based on triaromatic dinosterane and 4-methyl-24-ethyl cholestane. Therefore, it can be inferred that the H_2S concentration of Cambrian dolomite reservoirs might be higher than that of the O_1y reservoirs.

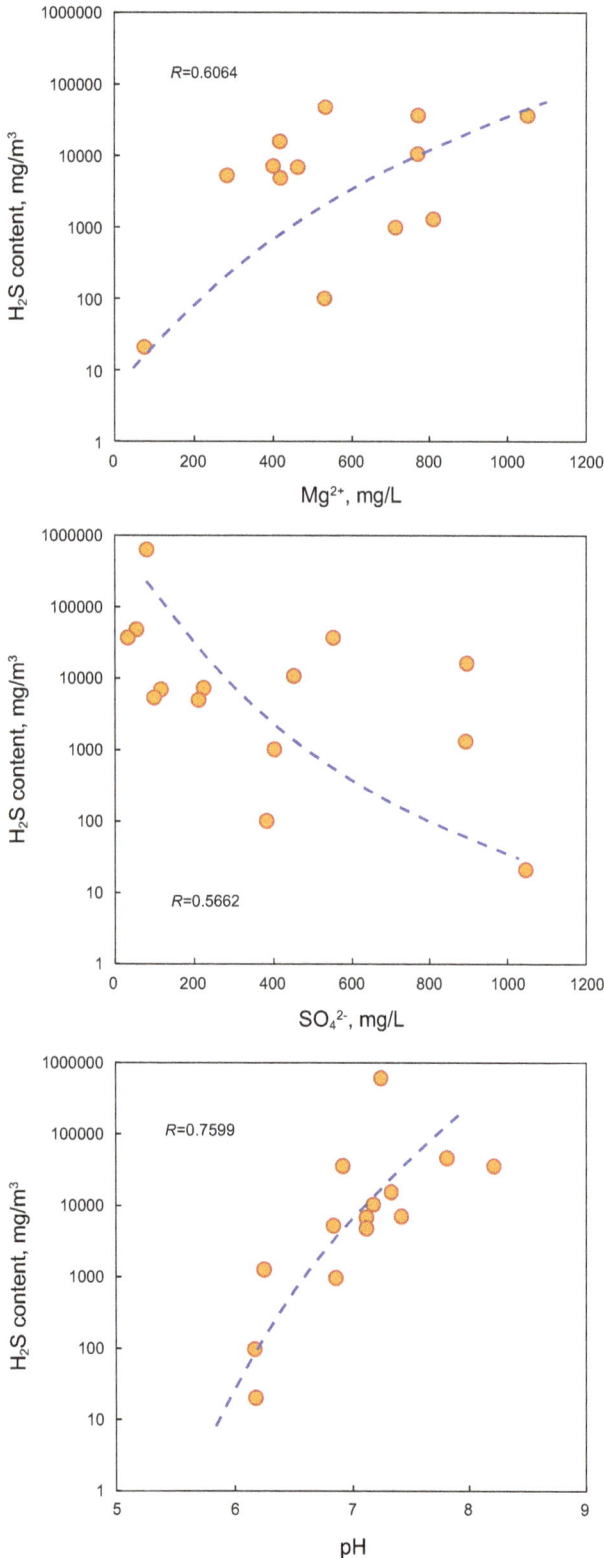

Fig. 7 Scatter point correlation between H_2S content in condensate reservoirs and ion concentration of underground water

expressed in Eq. (1) to generate sufficient ion-pairs and activate reaction in Eq. (1) as well to produce sulfate with various structures which would promote the TSR reaction

Acknowledgements The study is funded by the Natural Science Foundation of China (NSFC, Project No. 41473020) and the CNPC International Cooperation Project (Grant No. 2011A-0203-01). The extraction, separation, GC and GC–MS analyses were performed in the Key Laboratory of Petroleum Geology (KLPG), PetroChina. The Tarim Oilfield Company is thanked for providing the background geological information and data on formation water. The anonymous reviewers are gratefully acknowledged for their constructive comments that substantially improved the quality of this manuscript.

References

Amrani A, Zhang TW, Ma QS, Ellis GS, Tang YC. The role of labile sulfur compounds in thermochemical sulfate reduction. Geochim Cosmochim Acta. 2008;72:2960–72.

Atkinson G, Petrucci S. Ion association of magnesium sulfate in water at 25°C. J Phys Chem. 1966;70:3122–8.

Azimi G, Papangelakis VG, Dutrizac JE. Modelling of calcium sulphate solubility in concentrated multi-component sulphate solutions. Fluid Phase Equilib. 2007;260(2):300–15.

Cai CF, Franks SG, Aagaard P. Origin and migration of brines from Paleozoic strata in central Tarim, China: constraints from $^{87}Sr/^{86}Sr$, δD, $\delta^{18}O$ and water chemistry. Appl Geochem. 2001;16(9–10):1269–84.

Chen ZH. Mechanism of overpressured fluid compartment and its controlling on hydrocarbon migration and accumulation in faulted lacustrine basin: a case study from the Dongying Sag, Bohai Bay Basin. Chin J Geol. 2008;43(1):50–64 (**in Chinese**).

Chen ZH, Zha M. Sedimentary characteristics of source rocks in fluctuation from facies. J Lake Sci. 2006a;18(1):19–25.

Chen ZH, Zha M. Over-pressured fluid compartment and hydrocarbon migration and accumulation in Dongying depression. Acta Sedimentol Sin. 2006b;24(4):607–13 (**in Chinese**).

Chen ZH, Zha M. Geological and geochemical characteristics of hydrocarbon—expulsion from source rocks in Dongying depression. Geochimica. 2005;34(1):79–86.

Dahl JE, Moldowan JM, Peters KE, et al. Diamondoid hydrocarbons as indicators of natural oil cracking. Nature. 1999;399:54–7.

Davisson ML, Criss RE. Na–Ca–Cl relations in basinal fluids. Geochimica et Cosmochim Acta. 1996;60(15):2743–52.

Eigen M, Tamm K. Sound absorption in electrolytes as a consequence of chemical relaxation. I. Relaxation theory of stepwise dissociation. Z Elektrochem. 1962;66:107–21.

Ellis GS, Zhang T, Ma QS, et al. Kinetics and mechanism of hydrocarbon oxidation by thermochemical sulfate reduction. In: 23th international meeting on Organic Geochemistry, Torquay, United Kingdom. 2007.

He K, Zhang SC, Mi JK, Hu GY. The speciation of aqueous sulfate and its implication on the initiation mechanisms of TSR at different temperatures. Appl Geochem. 2014;43:121–31.

Jendu PD, Titley PA, Worden RH. Enrichment of nitrogen and ^{13}C of methane in natural gases from the Khuff Formation, Saudi Arabia, caused by thermochemical sulfate reduction. Org Geochem. 2015;82:54–68.

Jiang L, Worden RH, Cai CF. Generation of isotopically and compositionally distinct water during thermochemical sulfate reduction (TSR) in carbonate reservoirs: Triassic Feixianguan Formation, Sichuan Basin, China. Geochimica et Cosmochim Acta. 2015;165:249–62.

Jiang NH, Zhu GY, Zhang SC. Detection of 2-thiaadamantanes in the oil from Well TZ-83 in Tarim Basin and its geological implication. Chin Sci Bull. 2008;53(3):396–401.

Jiang YL, Zhang Y. Main geological factors controlling the accumulation and distribution of oil and gas in a complicated fault—block region. Pet Explor Dev. 1999;26(5):39–42.

Kuo LH. An experimental study of crude oil alteration in reservoir rocks by water-washing. Org Geochem. 1994;21:465–79.

Lafargue E, Tluez PL. Effect of waterwashing on light and compositional heterogeneity. Org Geochem. 1996;24:1141–50.

Land LS. Na-Ca-Cl saline formation waters, Frio Formation (Oligocene), south Texas, USA: products of diagenesis. Geochim et Cosmochim Acta. 1995;59(11):2163–74.

Leusbrock I, Metz S, Rexwinkel G, et al. Quantitative approaches for the description of solubilities of inorganic compounds in near-critical and supercritical water. J Supercrit Fluids. 2008;47(2):117–27.

Ma QS, Ellis GS, Amrani A, et al. Theoretical study on the reactivity of sulfate species with hydrocarbons. Geochim Cosmochim Acta. 2008;72:4565–76.

Palmer SE. Effect of water washing on C_{15}-hydrocarbon fraction of crude oils from northwest Palawan, Philippines. AAPG Bull. 1984;68:137–149.

Price LC. Crude oil degradation as an explanation of the depth rule. Chem Geol. 1980;28:1–30.

Rudolph WW, Irmer G, Hefter GT. Raman spectroscopic investigation of speciation in $MgSO_4(aq)$. Phys Hem Hem Phys. 2003;5:5253–61.

Sun YZ, Püttmann W. Relationship between metal enrichment and organic composition in Kupferschiefer of structure-controlled mineralization from Oberkatz Schwelle. Appl Geochem. 1996;11:567–81.

Sun YZ, Jiao WW, Zhang SC, et al. Gold enrichment mechanism in crude oils and source rocks in Jiyang Depression. Energy Explor Exploit. 2009a;27(2):133–42.

Sun YZ, Liu CY, Lin MY, et al. Geochemical evidences of natural gas migration and releasing in the Ordos Basin, China. Energy Explor Exploit. 2009b;27(1):1–13.

Su J, Zhang SC, Yang HJ, et al. Evidence of the organic geochemistry and petrology for the adjustment process of primary reservoir: insight into the adjustment mechanism of Hadexun oil field visa Xiang-3 well. Acta Pet Sinica. 2011;27(6):1886–98 (**in Chinese**).

Su J, Zhang SC, Yang HJ, et al. Control of fault system to formation of effective carbonate reservoir and the rules of petroleum accumulation. Acta Pet Sinica. 2010;31(2):196–203 (**in Chinese**).

Su J, Zhang SC, Huang HP, et al. New insights into the formation mechanism of high hydrogensulfide–bearing gas condensates: case study of Lower Ordovician dolomite reservoirs in the Tazhong uplift, Tarim Basin. AAPG Bull. 2016;100(6):893–916.

Tang YC, Ellis GS, Zhang TW, et al. Effect of aqueous chemistry on the thermal stability of hydrocarbons in petroleum reservoirs. Geochim Cosmochim Acta. 2005;69:A559.

Varsanyi I, Kovacs LO. Chemical evolution of groundwater in the River Danube deposits in the southern part of the Pannonian Basin (Hungary). Appl Geochem. 1997;12(5):625–36.

Wei ZB, Walters CC, Moldowan JM, et al. Thiadiamondoids as proxies for the extent of thermochemical sulfate reduction. Org Geochem. 2012;44:53–70.

Worden RH, Smalley PC. H_2S-producing reactions in deep carbonate gas reservoirs: Khuff Formation, Abu Dhabi. Chem Geol. 1996;133:157–71.

Worden RH, Smalley PC, Cross MM. The influences of rock fabric and mineralogy upon thermochemical sulfate reduction: Khuff. Formation, Abu Dhabi. J Sediment Res. 2000;70:1210–21.

Zha M, Chen ZH, Zhang NF, et al. Hydrochemical conditions and hydrocarbon accumulation in the Luliang uplift in Junggar Basin. Chin J Geol. 2003;38(3):315–22.

Zhang SC, Huang HP. Geochemistry of Palaeozoic marine petroleum from the Tarim Basin, NW China: part 1. Oil family classification. Org Geochem. 2005;36(8):1204–14.

Zhang SC, Huang HP, Su J, et al. Geochemistry of Paleozoic marine petroleum from the Tarim Basin, NW China: part 5. Effect of maturation, TSR and mixing on the occurrence and distribution of alkyldibenzothiophenes. Org Geochem. 2015;86:5–18.

Zhang SC, Su J, Wang XW, et al. Geochemistry of Palaeozoic marine petroleum from the Tarim Basin, NW China: part 3. Thermal cracking of liquid hydrocarbons and gas washing as the major mechanisms for deep gas condensate accumulations. Org Geochem. 2011;42(10):1394–410.

Zhang TW, Amrani A, Ellis GS, et al. Experimental investigation on thermochemical sulfate reduction by H_2S initiation. Geochimica et Cosmochim Acta. 2008;72:3518–30.

Zhao WZ, Zhu GY, Zhang SC, et al. Relationship between the later strong gas-charging and the improvement of the reservoir capacity in deep Ordovician carbonate reservoir in Tazhong area, Tarim Basin. Chin Sci Bull. 2009;54(17):3076–89.

Fracture prediction in the tight-oil reservoirs of the Triassic Yanchang Formation in the Ordos Basin, northern China

Wen-Tao Zhao[1,2,3] · **Gui-Ting Hou**[1]

Abstract It is important to predict the fracture distribution in the tight reservoirs of the Ordos Basin because fracturing is very crucial for the reconstruction of the low-permeability reservoirs. Three-dimensional finite element models are used to predict the fracture orientation and distribution of the Triassic Yanchang Formation in the Longdong area, southern Ordos Basin. The numerical modeling is based on the distribution of sand bodies in the Chang 7_1 and 7_2 members, and the different forces that have been exerted along each boundary of the basin in the Late Mesozoic and the Cenozoic. The calculated results demonstrate that the fracture orientations in the Late Mesozoic and the Cenozoic are NW–EW and NNE–ENE, respectively. In this paper, the two-factor method is applied to analyze the distribution of fracture density. The distribution maps of predicted fracture density in the Chang 7_1 and 7_2 members are obtained, indicating that the tectonic movement in the Late Mesozoic has a greater influence on the fracture development than that in the Cenozoic. The average fracture densities in the Chang 7_1 and 7_2 members are similar, but there are differences in their distributions. Compared with other geological elements, the lithology and the layer thickness are the primary factors that control the stress distribution in the study area, which further determine the fracture distribution in the stable Ordos Basin. The predicted fracture density and the two-factor method can be utilized to guide future exploration in the tight-sand reservoirs.

Keywords Ordos Basin · Yanchang Formation · Fracture prediction · Finite element modeling · Two-factor method · Tight-sand reservoirs

1 Introduction

Unconventional oil and gas resources, such as tight gas, tight oil and shale oil, have been successfully developed commercially in the USA, Canada, Australia and some other countries. The production of tight oil soared from 30 million tons in 2011 to 96.9 million tons in 2012 in the USA by using new unconventional technologies (Du et al. 2014). In China, the Ordos Basin, the Junggar Basin, the Songliao Basin, the Sichuan Basin, the Qaidam Basin, etc., have abundant tight-oil resources with an output of 97 million tons, accounting for 22% of the nationwide total oil output (Jia et al. 2014). In the Ordos Basin, the tight-oil reservoirs in the Triassic Yanchang Formation have become a major target of petroleum exploration and development in recent years (e.g., Guo et al. 2012; Yao et al. 2013).

Since tight-oil reservoirs in the Ordos Basin are of low-permeability ($<2 \times 10^{-3}$ μm^2) and low-porosity ($<10\%$) overall, fracturing is crucial for the reservoir reconstruction, even though the reservoirs are formed with complicated mechanisms (e.g., Yao et al. 2013; Ezulike and Dehghanpour 2014). Therefore, it is important to predict the natural fracture distribution in reservoirs, including

✉ Gui-Ting Hou
gthou@pku.edu.cn

[1] The Key Laboratory of Orogenic Belts and Crustal Evolution of Ministry of Education, School of Earth and Space Sciences, Peking University, Beijing 100871, China

[2] China Huaneng Clean Energy Research Institute, Beijing 102209, China

[3] PetroChina Research Institute of Petroleum Exploration and Development, Beijing 100083, China

Edited by Jie Hao

their orientation and density, for future exploration and development (e.g., Smart et al. 2009). Previous studies have focused on the geometrical or kinematic models, such as analyses of seismic techniques or of the layer curvature (e.g., Zahm et al. 2010; Pearce et al. 2011; Tong and Yin 2011), and fracture prediction in the Ordos Basin has also been involved in some papers (e.g., Ju et al. 2014a). However, since earlier fracture prediction was mainly carried out through layer curvature or two-dimensional (2D) models, which cannot meet the demands for the tight-oil study and exploration, it is necessary to build three-dimensional (3D) mechanical models in order to achieve the accuracy needed for further research on the unconventional petroleum.

Various factors, such as the proximity of faults, the curvature of folds, the layer thickness and the lithology, are deemed to control the fracture development in tight reservoirs (e.g., Ju et al. 2013), and the anisotropy or heterogeneity should also be considered in the modeling (e.g., Glukhmanchuk and Vasilevskiy 2013). However, it is difficult for 2D geomechanical models to fully consider all these factors, and the modeling results cannot be used successfully for exploration and production. Therefore, 3D models will be utilized in this paper, which take the lithology, the thickness and the stress fields into consideration.

The study area in this paper, namely the Longdong area, is located in the southern Ordos Basin, where research on structural fractures in the tight reservoirs is still deficient (e.g., Ren et al. 2014; Li et al. 2015). The structural fractures in the Longdong area were mainly formed after the Late Triassic, as a result of multiple-stage tectonic events in the Late Mesozoic and the Cenozoic. These extensively developed fractures are mostly unfilled and effective, which noticeably improve the permeability of tight reservoirs in the Ordos Basin.

2 Geological background

The Ordos Basin, covering an area of 2.6×10^5 km^2, is a large N–S trending basin in the western North China Craton, which is located between the Siberian Craton and the South China Craton (Hou et al. 2010) (Fig. 1). Three orogenic belts have been developed along different boundaries of the stable basin, including the Yinshan Mountain in the north, the Qinling Orogen in the south and the Liupanshan Mountain in the southwest (e.g., Nutman et al. 2011) (Fig. 2). The basement of the basin is composed of Archean rocks with Proterozoic sedimentary cover. Although the margin underwent multiple tectonic activities, the central part is still stable and is covered by shallow Paleozoic marine carbonate sediment (Kusky and

Li 2009). Some small-scale paleo-faults exist, but no large faults have been found within the basin (Wan and Zeng 2002; Yang et al. 2013).

Contrary to the evolution of the eastern North China Craton including thickening, thinning and destruction, the Ordos Basin has evolved from three Mesoproterozoic aulacogens to a Paleozoic–Mesozoic cratonic basin since the Middle Proterozoic (Menzies et al. 2007; Yang et al. 2013; Wang et al. 2014b). The interior part of the basin is characterized by horizontal or gently dipping strata (<3°), especially for the Mesozoic and Cenozoic strata, whereas the strata along the margins have been subjected to significant folding and faulting since the Late Triassic.

Two distinct tectonic events took place from the Late Mesozoic to the Cenozoic, resulting in two different stress fields in these periods. In the Late Mesozoic, namely from the Early Jurassic to the Late Cretaceous, the long-distance effect of subduction of the Izanagi Plate turned from north-northwestward to northwestward when the force arrived at the Ordos Basin, resulting in the WNW-trending stress fields and the structural fractures in NW–EW trends (e.g., Wan 1994; Hou et al. 2010; Sun et al. 2014; Zhao et al. 2016); while in the Cenozoic, the predominant tectonic event became the northeastward collision between the Indian and the Eurasian Plate, which led to the NE-trending stress fields and the structural fractures in NNE–ENE trends (e.g., Yuan et al. 2007; Wang et al. 2014b). Two episodes of fractures are developed under distinct tectonic events, so the stress fields of different periods should be taken into consideration during the fracture prediction.

3 Fracture measurement

The parameters of fracture characteristics are important in the exploration and development of fractured tight reservoirs. The fracture density is one of the significant indicators to reflect the failure degree of rocks, which can be divided into three types, including the linear density, the surface density and the bulk density of fractures. In this paper, the surface density is utilized to describe the fracture distribution in the Ordos Basin. The surface density is defined as the ratio between the cumulative fracture length and the cross-sectional area of the matrix, which can better reflect the degrees of fracture development and be measured more effectively than others (Golf-Racht 1982). The fracture density from core observations can be calculated as:

$$f = \frac{\sum l_i}{S} = \frac{\sum l_i}{2\pi r^2 + 2\pi r \times L} \tag{1}$$

where f is the fracture surface density, l_i is the length of each structural fracture, S is the surface area of the observed core, r is the radius and L is the length of the core.

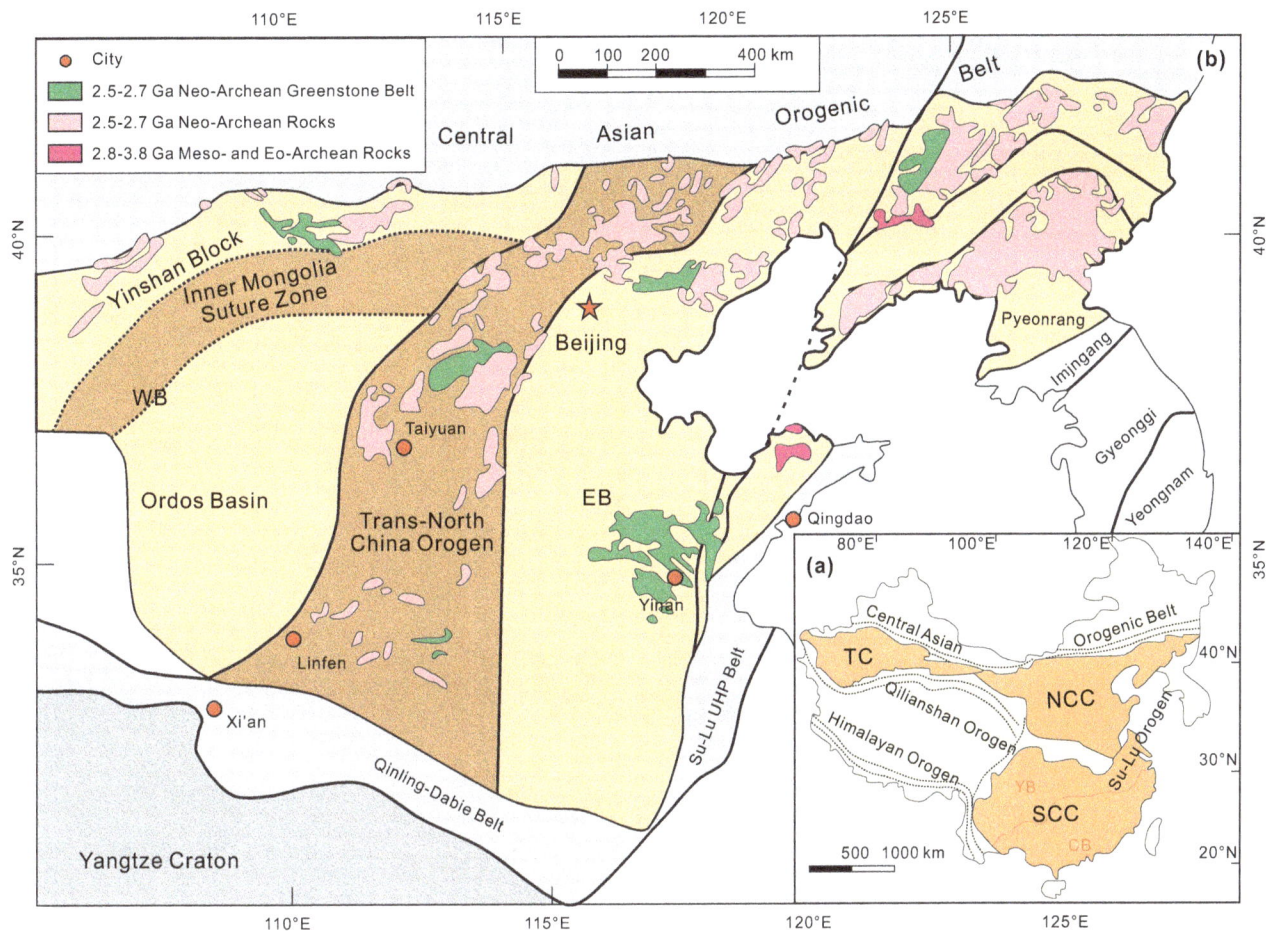

Fig. 1 A geological summary of the Ordos Basin. **a** Tectonic framework of the major cratonic blocks in China (after Kusky 2011 and Santosh et al. 2012); **b** generalized geological and tectonic map of the North China Craton (after Zhao et al. 2005). *TC* Tarim Craton, *NCC* North China Craton, *SCC* South China Craton, *YB* Yangtze Block, *CB* Cathaysia Block, *WB* Western Block, *EB* Eastern Block

In this paper, the Longdong area was selected as the study area to carry out fracture measurements (Fig. 2). Sixty-six wells were chosen to study the distribution of structural fractures in the Chang 7_1 and 7_2 members (Fig. 3). As shown in Fig. 4, the fracture density in the tight-sandstone cores is relatively low (smaller than $0.5 \ m^{-1}$), representing the general condition in the Longdong area. The fracture density of the Chang 7_2 member is more concentrative than that of Chang 7_1 member, even though their average densities are similar in general ($0.071 \ m^{-1}$ for the Chang 7_1 member and $0.081 \ m^{-1}$ for the Chang 7_2 member). The difference of fracture distribution between the Chang 7_1 and 7_2 members is obvious: The highest fracture density in the Chang 7_1 member lies in the Laocheng and Qingyang areas, while that in the Chang 7_2 member lies in the Laocheng, the Qingyang and the Zhengning areas (Fig. 4).

4 Modeling approach

Methods such as geological analysis, physical modeling and numerical simulation including the finite element method (FEM) can be applied in the study of stress fields, which is the foundation of fracture prediction. In this study, the finite element software ANSYS is used to calculate the stress field and predict the fracture distribution (Velázquez et al. 2009; Jarosinski et al. 2011). The basic concept of FEM is that a geological body can be discretized into finite continuous elements connected by nodes. The geometrical and mechanical parameters allocated on each element are consistent with the properties of real rocks. The continuous field function of the geological area is first transformed into linear functions at every node that contain displacement, stress and strain variables resulting from the applied forces (Jiu et al. 2013), and then all these elements are used to obtain the stress distribution over the entire area.

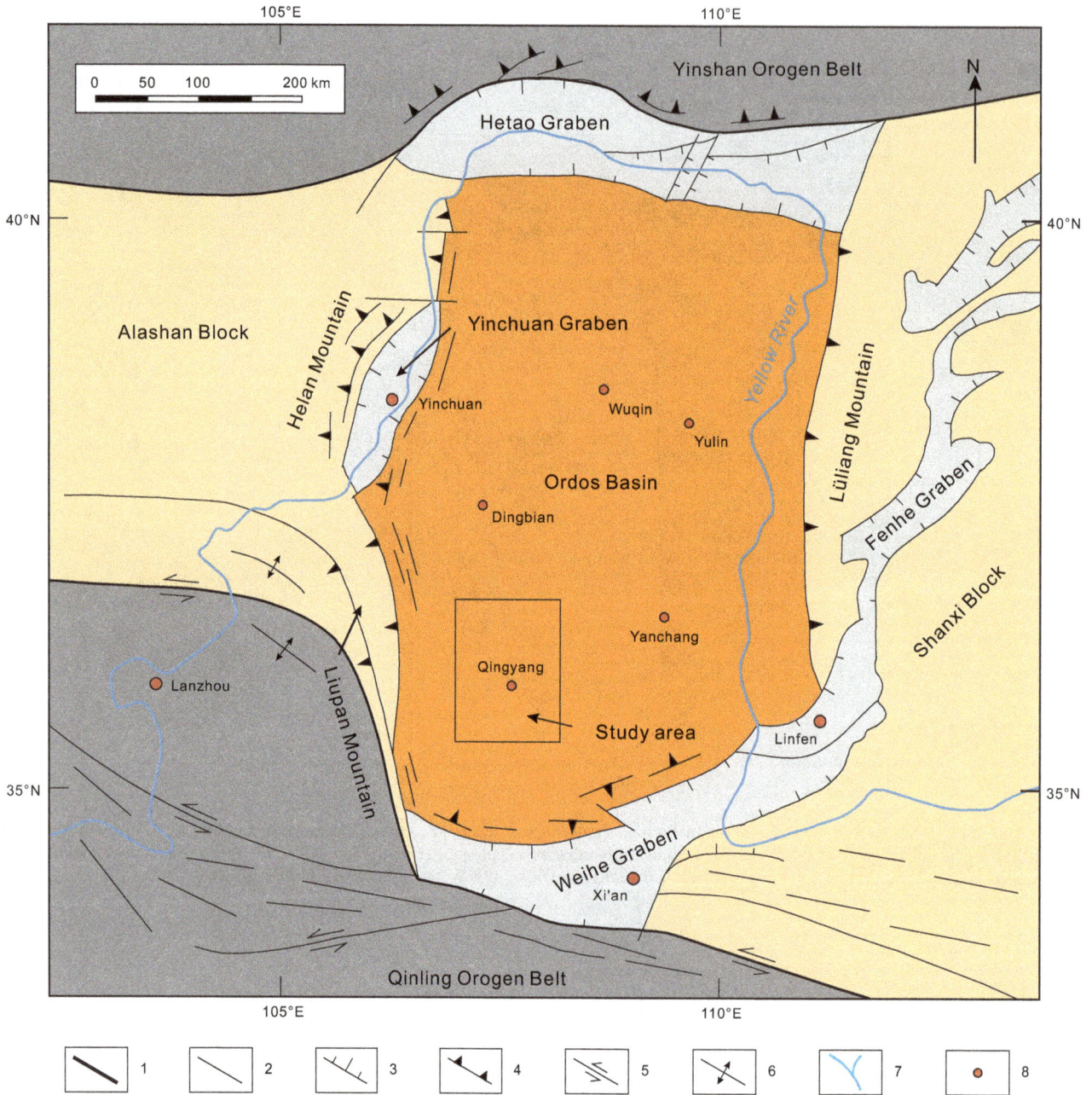

Fig. 2 Tectonic framework of the Ordos Basin. The *orange area* denotes the Ordos Basin, and the *light yellow* ones denote the adjacent blocks. The *light gray* areas represent the graben system along the margins, and the *dark gray* ones represent the orogenic belts around the basin. The *black frame* denotes the study area (the Longdong area) (after Darby and Ritts 2002) (*1* Boundary of orogenic belt, *2* Fault, *3* Normal fault, *4* Reversed fault, *5* Strike-slip fault, *6* Fold, *7* River, *8* City)

4.1 Geometrical model

The Ordos Basin is a near-rectangular basin in the western part of the North China Craton (Li and Li 2008; Tang et al. 2012) (Fig. 2). Although the Ordos Basin underwent multi-stage tectonic movements in the Late Mesozoic–Cenozoic eras, the deformation was confined to the western margin and no significant tectonic events occurred in the central part. Therefore, the outline of the basin remained unchanged in these periods (Sun et al. 2014) (Fig. 5a). Since no large faults or folds have been recorded inside the Ordos Basin, the sedimentary facies, the lithology and the distribution of sand bodies are the key factors to determine the fracture development.

In this paper, the Chang 7_1 and 7_2 members in the Triassic Yanchang Formation, the major tight-oil members in

Epoch	Formation	Subsection	Lithology column	Sedimentary facies	Thickness, m
Middle Jurassic	Anding			Swamp	80–150
	Zhiluo			Fluvial	20–40
Lower Jurassic	Yan'an	Yan 1– Yan 10		Fluvial-lacustrine-swamp	250–300
Upper Triassic	Yanchang	Chang 1			0–245
		Chang 2		Fluvial-lacustrine	120–160
		Chang 3			100–170
		Chang 4+5			90–130
		Chang 6			180–200
		Chang 7		Deep lacustrine	
		Chang 8		Lacustrine	100–190
		Chang 9			
		Chang 10		Fluvial	200–320

Conglomerate   Sandstone   Sandy mudstone

Mudstone   Oil shale   Coal

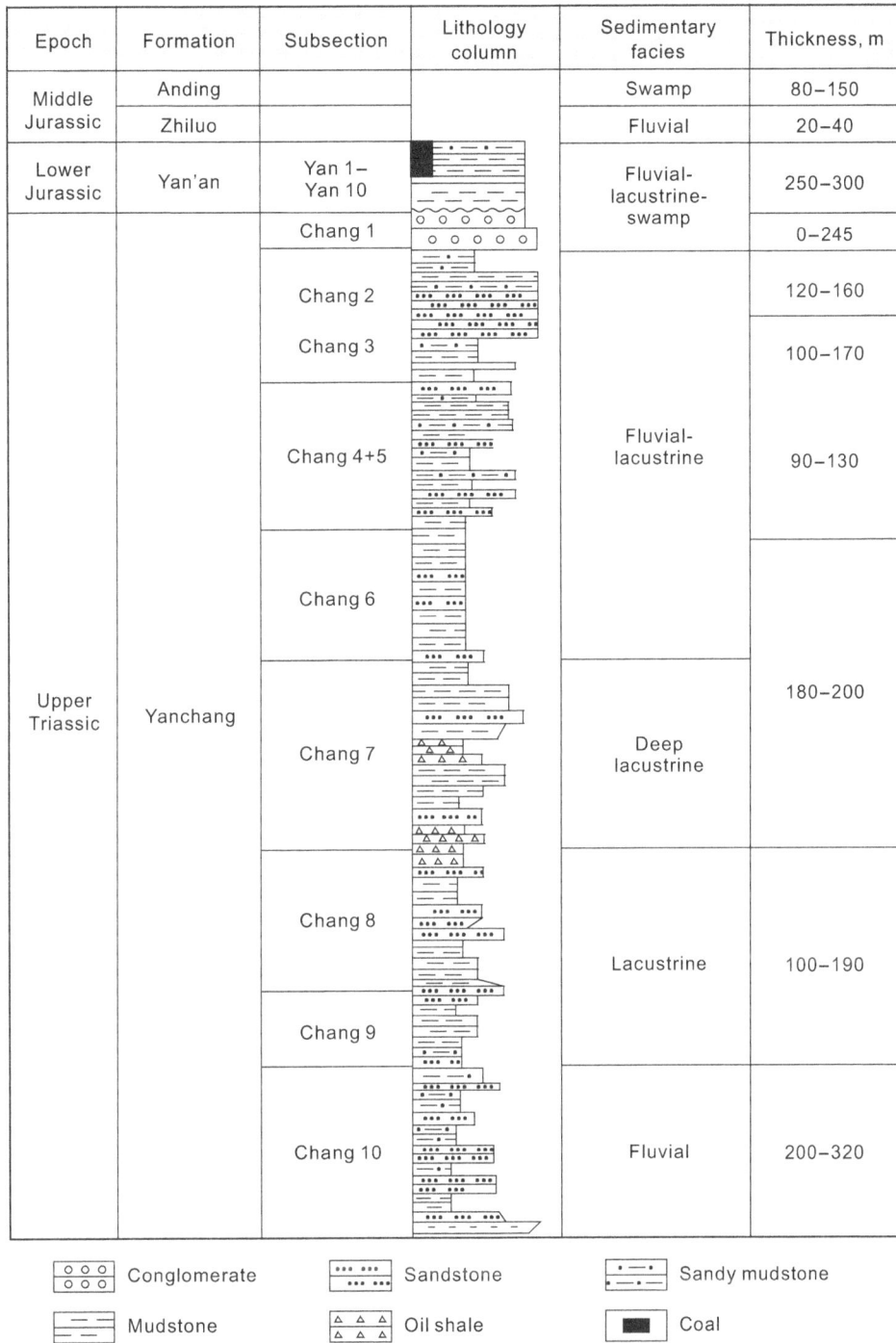

Fig. 3 Synthetical stratigraphic column and depositional environment of the Upper Triassic—Middle Jurassic in the Ordos Basin (after Duan et al. 2008)

the Ordos Basin, are selected as the study strata, and the Longdong area is chosen to discuss the stress fields and the fracture distribution (Figs. 2, 3, 5). Since the Yanchang Formation is characterized by strong heterogeneity with facies change, the simplified model with only one rock mechanical property is no longer suitable for the complex interior of the Ordos Basin (Yang and Deng 2013). Based

on the sandstone-mudstone ratio, it is assumed that the ratios between the sandstone and mudstone layers are 0.43–4.26 (average 1.27) in the Chang 7_1 member and 0.54–9.00 (average 1.70) in the Chang 7_2 member (e.g., Guo et al. 2012; Li et al. 2015), and multiple-layer constructions (four sandstone layers in Chang 7_1 member and three sandstone layers in Chang 7_2 member) are applied in

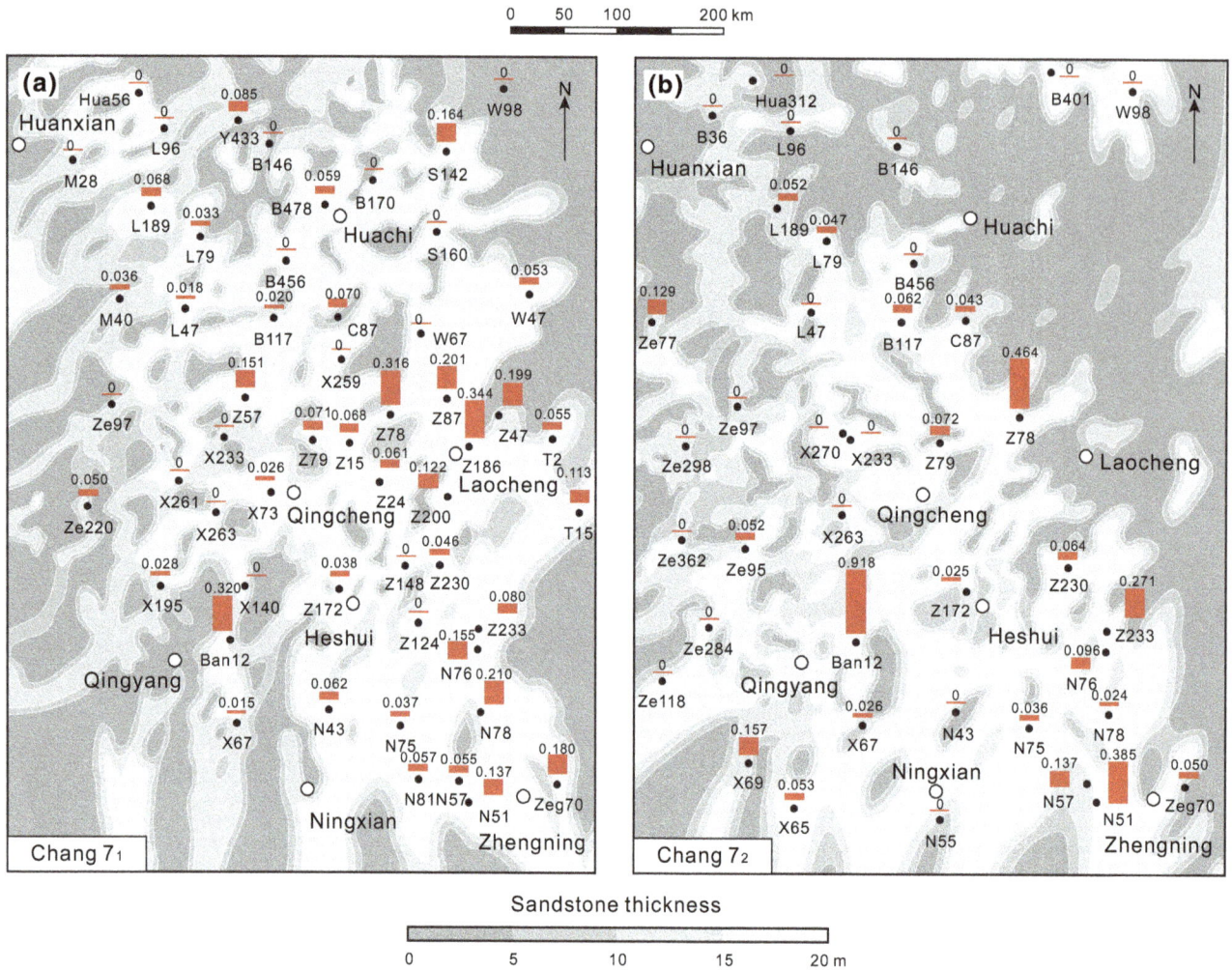

Fig. 4 Distribution of the measured fracture density and the sandstone thickness in **a** the Chang 7_1 and **b** Chang 7_2 members in the Longdong area. The *black solid dots* denote the observed wells, while the *hollow dots* denote the cities. Measured fracture densities in the Chang 7_1 and 7_2 members are marked with *red cylinders*, and the *contours* represent the sandstone thickness in the study area

the geometrical model to simulate the sandstone-mudstone interlayers (Fig. 5). To avoid the boundary effect, forces in the models are set on the boundaries of the Ordos Basin, and the study area is nested inside the basin (Fig. 5). The sandstone layers, which are also the main layers in fracture development, in the middle of each member, are selected to display the modeling results in the following text, representing the general situation of fracture development in the Chang 7_1 and 7_2 members.

4.2 Boundary conditions and modeling

In order to predict the fracture distribution of the Ordos Basin, it is assumed that the upper crustal thickness of the basin in the Late Mesozoic–Cenozoic era is 25 km (e.g., Liu et al. 2006). The top of the model is set as a free surface, and the entire model is subjected to gravity load. The average density of the upper crust, which mainly

consists of sedimentary cover, greenschist and granite, is 2750 kg/m^3 (Hou et al. 2010; Wang et al. 2014b). Based on the velocities of P and S waves, the calculated average Poisson's ratio is 0.20 and the average Young's modulus is 80 GPa for the whole basin (Liu et al. 2006).

To subtly depict the distribution of structural fractures in the Longdong area, four more kinds of material elements are involved in the 3D geometrical model, including the sandstones/mudstones of the Chang 7_1 member and the sandstones/mudstones of the Chang 7_2 member. Tri-axial rock mechanical experiments were carried out by the Institute of Acoustics, Chinese Academy of Sciences, on 62 core samples collected from observed wells in the Longdong area (Fig. 4). In order to simulate the real conditions underground, in these experiments, confining pressures corresponding to the original depth of the Yanchang Formation are applied in the radial directions, and vertical pressures are applied in the axial directions of all samples.

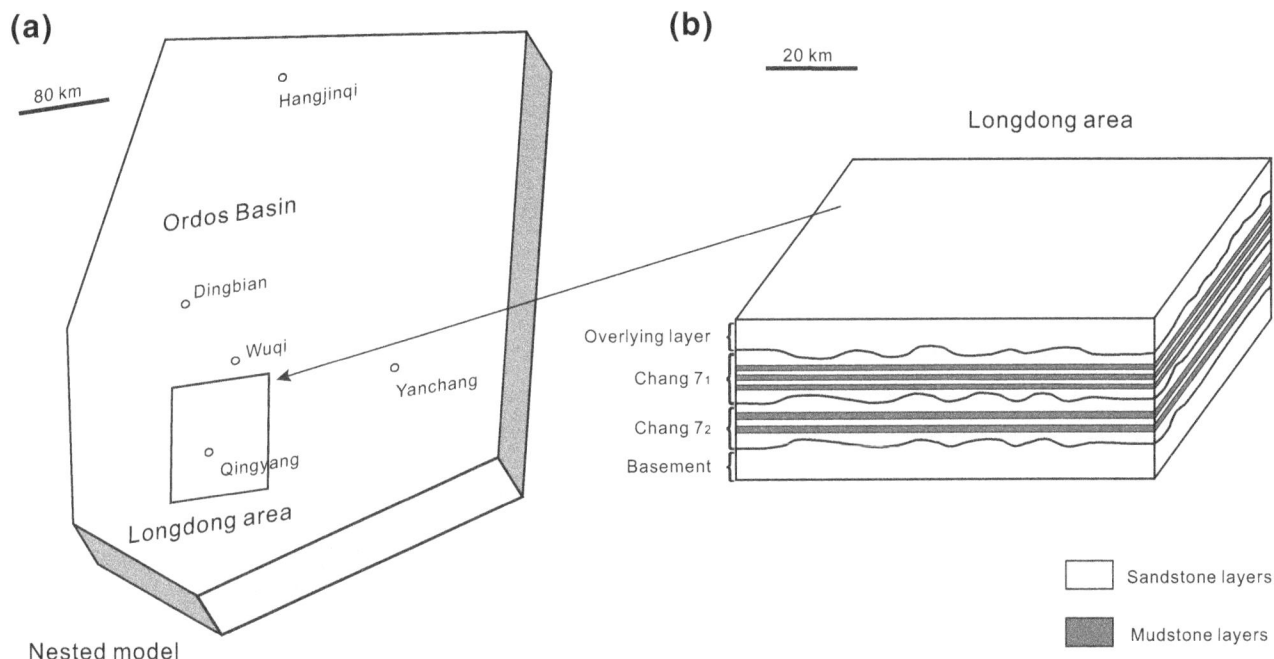

Fig. 5 Simplified geometrical model of the Longdong area (**b**) and its nested model (**a**) in the Ordos Basin

Through statistical analysis and geological classification, five sets of rock mechanical properties, the average density, Young's modulus, Poisson's ratio, internal friction angle and cohesion, are listed in Table 1 by layer and lithology.

Because the stress fields in the Late Mesozoic and the Cenozoic are strikingly different and both of them had a significant effect on the fracture development in the Ordos Basin, the boundary conditions during these two episodes along the basin need to be defined (Zhao et al. 2016).

As a result of intense compression from the Early Jurassic to the Middle Cretaceous, the Yinshan Orogen Belt was developed as thrust faults with dextral shearing features in the Late Mesozoic (Darby and Ritts 2002; Zhang et al. 2007; Faure et al. 2012). A uniform direction and a constant magnitude of a 40 MPa normal component with a 10 MPa dextral shearing component are applied along the northern side of the Ordos Basin (L1) (Fig. 6a). The east-dipping thrusts, the NWW-dipping back-thrusts and the associated folds developed along the Lüliang Mountain in the Jurassic show that the stress regime in the eastern margin was related to the long-distance effect of the

push from the northwestward subduction of the Izanagi Plate in the Late Mesozoic (Zhang et al. 2007; Hou and Hari 2014). Hence, it is a compressive boundary with a sinistral shearing component along the eastern edge of the basin. A deviatoric stress of an approximately 150 MPa normal component with a 45 MPa shearing component is set along the eastern boundary (L2) (Fig. 6a). Based on the paleo-magnetic constraints, geological evidence and $^{40}Ar/^{39}Ar$ and U–Pb dating, it can be assumed that in the southern part of the Ordos Basin, the Qinling Ocean finally closed during the Late Jurassic-Early Cretaceous period. This indicates that the collision between the North China Craton and the South China Craton continued up to the Cretaceous period (Huang et al. 2005; Liu et al. 2015). And due to this collision, thrust faults with sinistral strike-slip features were developed along the northern margin of the Qinling Orogen Belt (Malaspina et al. 2006; Yuan et al. 2007). Therefore, a constant magnitude of normal stress (60 MPa) with sinistral shearing stress (30 MPa) is applied along the southern margin of the basin (L3) (Fig. 6a). In the western and southern margins, the long-distance effect of collision from the Qiangtang Massif affected the Ordos

Table 1 Rock mechanical parameters for the numerical modeling of the Chang 7_1 and 7_2 members in the Longdong area

Model position	Lithology	Density, g/cm^3	Young's modulus, GPa	Poisson's ratio	Internal friction angle, °	Cohesion, MPa
Chang 7_1	Sandstone	2.562	26.971	0.229	36.00	37.62
Chang 7_1	Mudstone	2.457	24.296	0.269	24.02	40.34
Chang 7_2	Sandstone	2.639	25.629	0.229	37.80	31.13
Chang 7_2	Mudstone	2.485	24.943	0.269	24.08	40.12

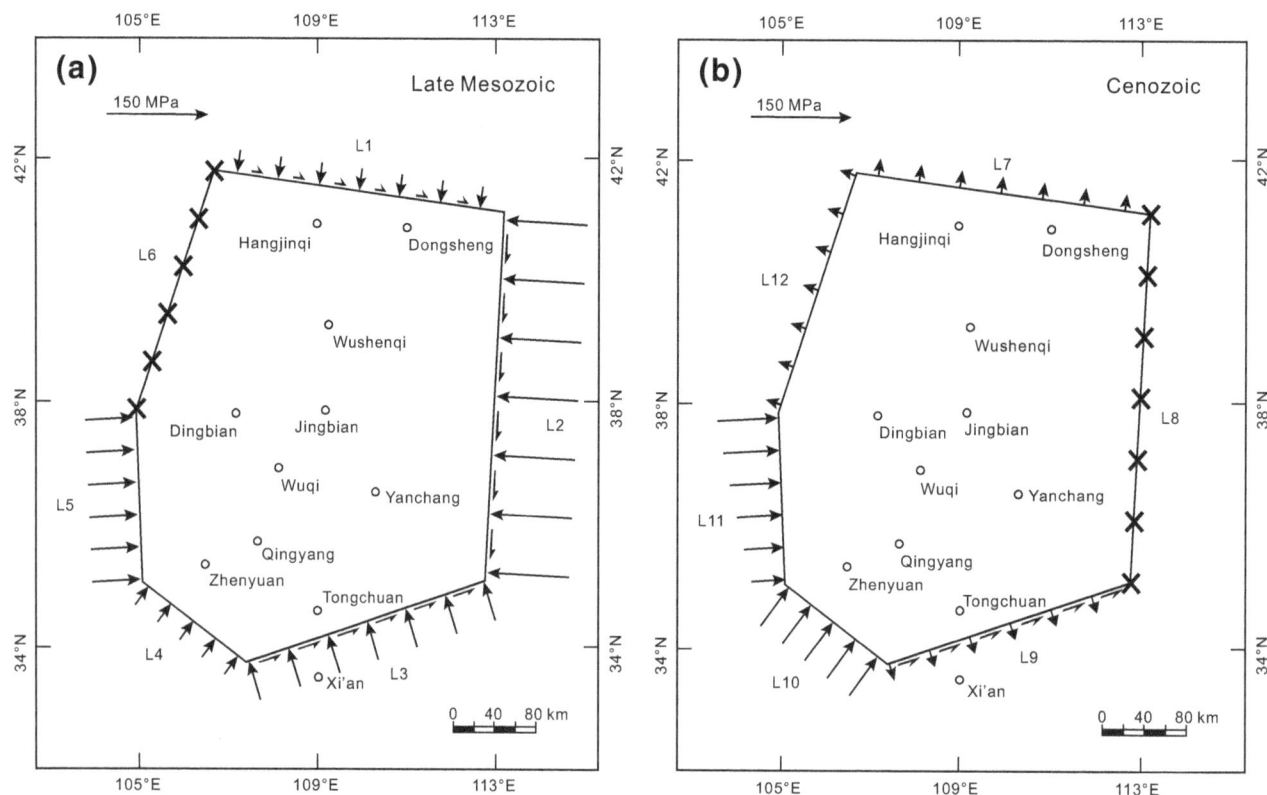

Fig. 6 Boundary conditions of the Ordos Basin in **a** the Late Mesozoic and **b** the Cenozoic eras. *Arrows* indicate the tectonic forces and the length represents the stress magnitude. The *crosses* represent the fixed boundaries in different models

Basin (Zhang et al. 2007; Li and Li 2008), so a compressive traction with a uniform direction and a constant magnitude of deviatoric stress of 30 MPa on the southwestern boundary (L4) and 75 MPa on the western boundary (L5) are applied along the basin (Fig. 6a). On the basis of SHRIMP zircon U–Pb ages and other geochronological data, it can be presumed that the closure of the Paleo-Asian Ocean finally took place after the Early Permian. Due to this episode of closure, the northward movement of the Alashan Block (Fig. 2) was arrested by the Siberian Craton in the Late Mesozoic (e.g., Zheng et al. 2014). The final closure of the Qilian Ocean took place at the end of the Ordovician, and after that, the Qaidam Block, which was adjacent to the Alashan Block, restricted the southward movement of the Alashan Block (Song et al. 2013). The nonidentical apparent polar wandering paths of the Tarim Block and the Alashan Block up to the Jurassic period clearly indicates that the amalgamation of these two blocks might have occurred during the Jurassic (Gilder et al. 2008). As a result of amalgamation in the Jurassic, the wedge-shaped Alashan Block was trapped between the Siberian Plate, the Qaidam Block, the Tarim Block and the Ordos Block (Zhang et al. 2007). Therefore, the northwestern boundary (L6) is kept fixed as the Alashan Block

was locked by the adjacent blocks in the Late Mesozoic (Fig. 6a).

The stress field in the Cenozoic era, which is regarded as a consequence of the Indo-Asian collision, is strikingly different from that in the Late Mesozoic era (e.g., Darby and Ritts 2002; Bao et al. 2013) (Fig. 6b). During the Cenozoic, the extension along the margins of the Ordos Basin triggered the formation of the Hetao, the Weibei and the Yinchuan Grabens, which in turn transposed reverse faults to normal faults in the Helan Mountain and the Qinling Mountain (Rao et al. 2014). Therefore, a tensile traction with a uniform direction and a constant magnitude of 5 MPa is applied on the northern, the southern and the northwestern margins, respectively (L7, L9 and L12) (Fig. 6b). The subduction of northwestern Pacific Plate restricted the further eastward movement of the Ordos Basin (Fournier et al. 2004; Schellart and Lister 2005). The current GPS horizontal velocity field map shows that the eastward velocity of the Shanxi Block (Fig. 2) is relatively smaller than that of the Ordos Basin (e.g., Zhu and Shi 2011; Wang et al. 2014c). The velocity differences between the Shanxi Block and the Ordos Basin suggest that the northeastward motion of the Ordos Basin, which was pushed by the Tibet Plateau, was restricted by the Shanxi

Table 2 Shortening rates of different profiles in the mid-south section of the western margin (L11) in the Ordos Basin (*Source*: Feng et al. 2013)

Profiles	Shortening rate, %	Average shortening rate, %
Tianshuibao (A–A′)	30.4–50.6	42.4
Shibangou (B–B′)	32.8	32.8
Shajingzi (C–C′)	16.5–38.6	29.3
Pengyang (D–D′)	12.9–17.9	15.4

Block due to the westward subduction of the Pacific Plate in the Cenozoic (Hou et al. 2010). Accordingly, the eastern edge of the basin is kept fixed for the Cenozoic era (L8) (Fig. 6b). On the basis of massive fault-striation data, it can be interpreted that the southern margin, namely the Weihe Graben, turned into a sinistral shearing tensile boundary (e.g., Mercier et al. 2013; Rao et al. 2014), and hence, a constant left-lateral shearing stress of 30 MPa is set on the southeastern border of the basin (L9) (Fig. 6b).

Due to the impact of collision between the Indian Plate and the Eurasian Plate, the Liupanshan Thrust-Fold Belt (namely the Liupan Mountain in Fig. 2) was developed along the southwestern margin of the Ordos Basin, which resulted in the transformation of the west-southwestern margin into a strongly compressive boundary during the Cenozoic era (Yuan et al. 2007; Li and Li 2008). When the western boundary of the basin is taken into consideration, as the shortening rate of the northern section (Tianshuibao Profile: 30.4%–50.6%) is greater than that of the southern one (Pengyang Profile: 12.9%–17.9%) (Feng et al. 2013) (Table 2; Fig. 7), a compressive traction with a uniform direction and a gradient magnitude from 80 to 55 MPa is applied on the western boundary (L11), whereas a compressive traction with a constant magnitude of 80 MPa is applied on the southwestern margin (L10) (Fig. 6b).

4.3 Theory of fracture prediction

Lagrangian formulations are used in ANSYS to simulate the three-dimensional, plane strain deformation, applying

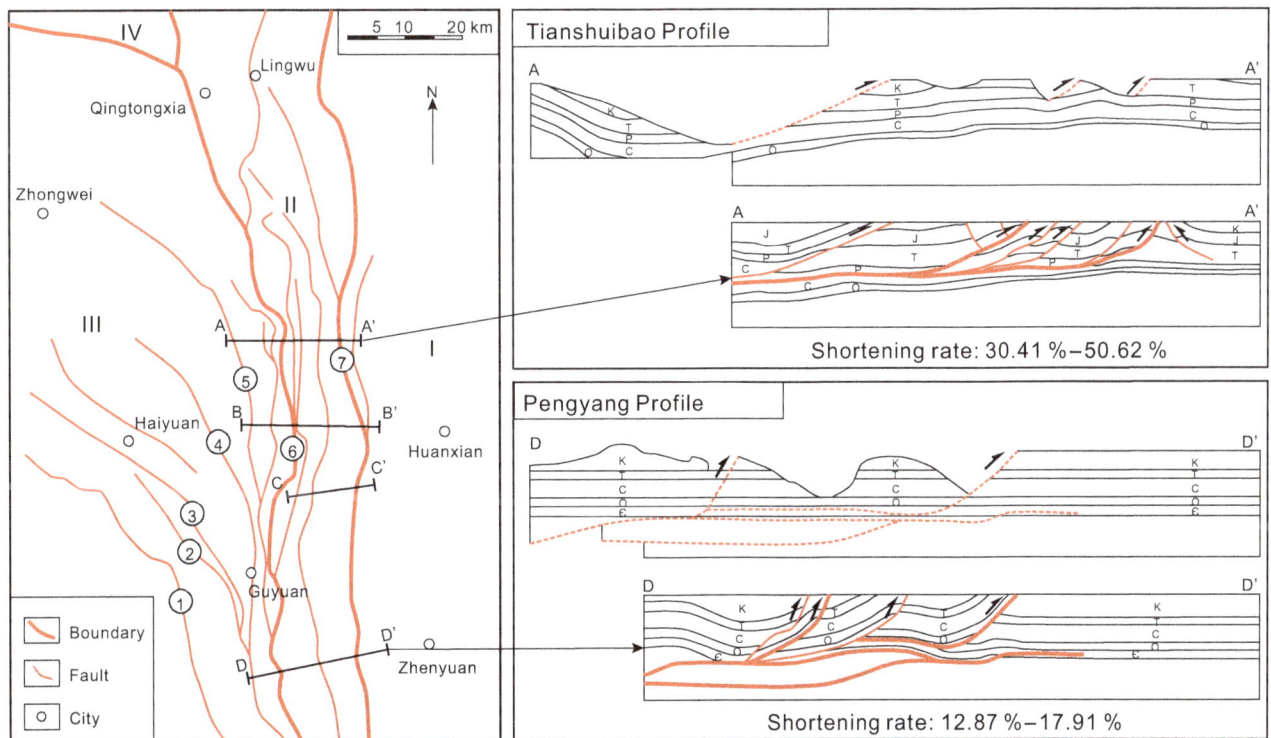

Fig. 7 Maps of tectonic units, two relevant profiles and their corresponding balanced sections in the mid-south section along the western margin (L11) of the Ordos Basin (*1* Western Liupanshan Fault, *2* Eastern Liupanshan Fault, *3* Haiyuan Fault, *4* Qingshuihe Fault, *5* Yantongshan-Yaoshan Fault, *6* Qingtongxia-Guyuan Fault, *7* Hui'anbao-Shajingzi Fault) (modified after Feng et al. 2013). Area: *I* Tianhuan Depression, *II* Thrust Belt of Western Margin, *III* Qilianshan Orogen, *IV* Alashan Block. Age: *O* Ordovician, *C* Carboniferous, *P* Permian, *T* Triassic, *J* Jurassic, *K* Cretaceous

8-node isotropic elements to represent each lithological layer. The mechanical behavior in the elastic domain is dominated by the generalized Hook's law. As the Yanchang Formation is generally less than 3000 m in depth where the plastic deformation is not obvious and the structural fractures in the Chang 7_1 and 7_2 members are chiefly shearing fractures based on field measurements and core observations (Fig. 8), the mechanical behavior follows the elastic model, which is described by the generalized Hook's law.

Various methods for fracture prediction have been proposed in previous literature, such as the conventional logging method, the stress field method, the principle curvature method, the geostatistical method, etc. (e.g., Savage et al. 2010; Zahm et al. 2010; Jiu et al. 2013). The two-factor method, involving the rupture value and the strain energy density, is used in this paper to predict the distribution of structural fractures in the Ordos Basin (Ding et al. 1998).

4.3.1 Rupture value

Tensile fractures and shearing fractures conform to different criteria. Griffith's criterion, which is derived from the micro-mechanism, is an effective criterion to predict the development and the distribution of tensile fractures; however, this criterion, which in nature is equivalent with the theory of maximum tensional stress, is only suitable for the tensile fractures (Griffith 1920). Although tensile fractures are found in some areas of the Ordos Basin, they are limited to the contact surfaces of sandstone and mudstone layers, and more than 95% of structural fractures in the Longdong area are shearing fractures, whose rupture is controlled by the Mohr–Coulomb failure criterion (Xie et al. 2008). Therefore, only Mohr–Coulomb failure criterion is taken into consideration in this study, which follows the equation (Coulomb 1776):

$$[\tau] = C + \sigma_n \times \tan \varphi \qquad (2)$$

where $[\tau]$ represents the critical shearing stress, C represents the cohesion, σ_n represents the stress normal to the shearing fractures and φ represents the internal friction angle (Table 1). Shearing fracture is triggered once the shearing stress exceeds the critical shearing stress ($[\tau]$) in Eq. (2). σ_n can be obtained via the maximum principal stress (σ_1) and the minimum principal stress (σ_3) according to Wang et al. (2004):

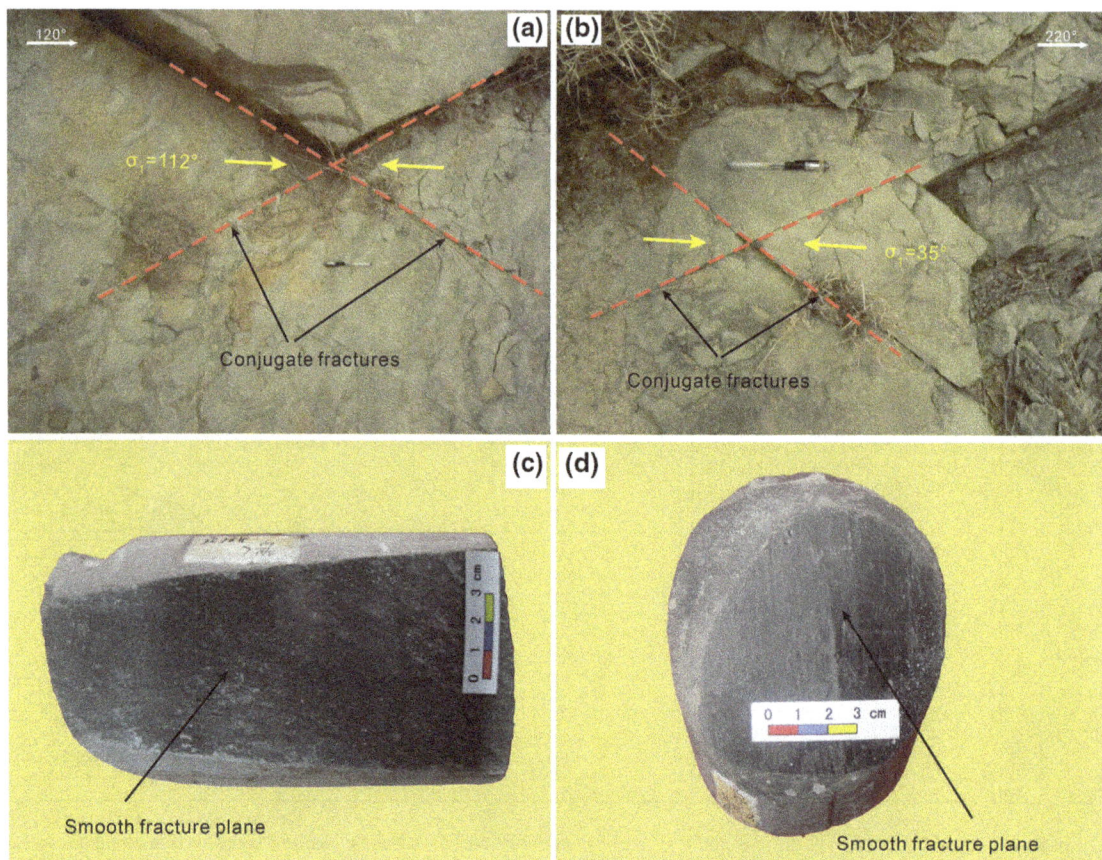

Fig. 8 Photographs of structural fractures in outcrops and cores of the Ordos Basin. **a** *Conjugate fractures* indicate the maximum principal compressive stress of WNW orientation in the Late Mesozoic; **b** *conjugate fractures* indicate the maximum principal compressive stress of NE orientation in the Cenozoic; **c** near-vertical fracture plane in a core from the Longdong area; and **d** moderate-dipping fracture plane in a core from the Longdong area

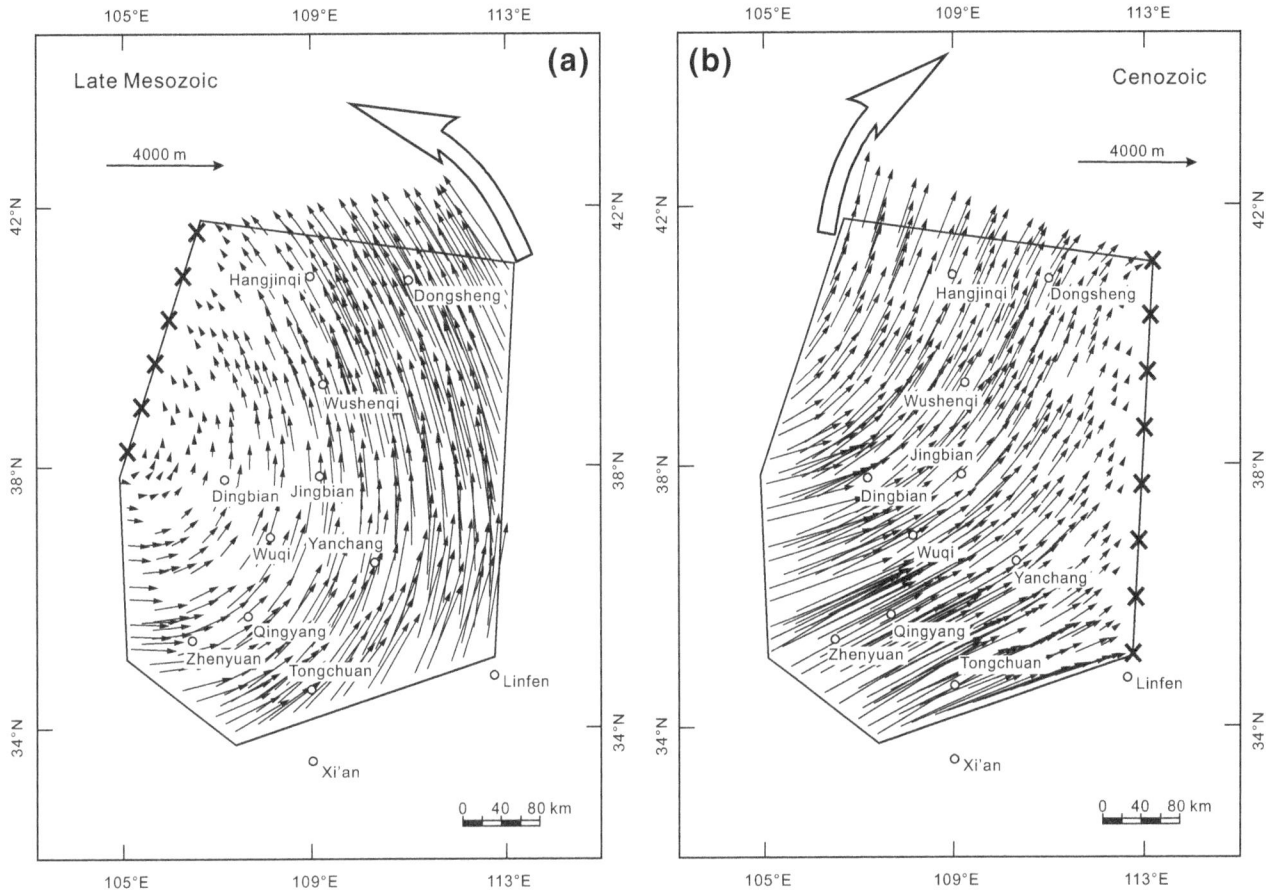

Fig. 9 Displacement fields of the two models for the Ordos Basin in the Late Mesozoic and the Cenozoic eras. *Thin black arrows* indicate the calculated displacement directions and their lengths indicate the magnitude of displacement. *Thick black arrows* outside the basin denote the rotation of the basin during these periods. *Black crosses* represent the fixed boundaries in different models

$$\sigma_n = (\sigma_1 + \sigma_3)/2 - (\sigma_1 - \sigma_3) \times \sin\varphi/2 \qquad (3)$$

The shearing stress (τ_n) can also be obtained via the two principal stresses according to Wang et al. (2004):

$$\tau_n = (\sigma_1 - \sigma_3) \times \cos\varphi/2 \qquad (4)$$

Following the Mohr–Coulomb failure criterion, the rock will break when the shearing stress is equal or greater than the critical shearing stress in Eq. (2), so the rupture value (I) is introduced in order to measure the probability of rock's rupture according to Ding et al. (1998):

$$I = \tau_n/[\tau] \qquad (5)$$

The possibility of rock's failure is very small when the rupture value (I) is far smaller than 1, whereas the possibility is relatively larger when the rupture value (I) exceeds 1. The fracture density (f) and the rupture value (I) may have a positive correlation, so the rupture value (I) is an effective index for fracture prediction through empirical formulas established between them.

4.3.2 Strain energy density

It is generally accepted that the rocks with relatively high strain energy density are more likely to develop structural fractures than those with a lower one. The strain energy density, namely the strain energy per unit volume, is described as follows (Prince and Rhodes 1966):

$$U = [\sigma_X^2 + \sigma_Y^2 + \sigma_Z^2 - 2v(\sigma_X\sigma_Y + \sigma_Y\sigma_Z + \sigma_Z\sigma_X) \\ + 2(1+v)(\tau_{XY}^2 + \tau_{YZ}^2 + \tau_{ZX}^2)]/2E \qquad (6)$$

where U is the strain energy density, v is Poisson's ratio, σ_X, σ_Y and σ_Z are the normal stress components in x, y and z directions, respectively, and τ_{XY}, τ_{YZ} and τ_{ZX} are the shearing stress components in the corresponding directions. Strain energy density (U) could be utilized to indicate the fracture distribution.

Rupture value (I) stands for the possibility of rock failure, whereas the strain energy density (U) stands for the developing ability of structural fractures. In this study,

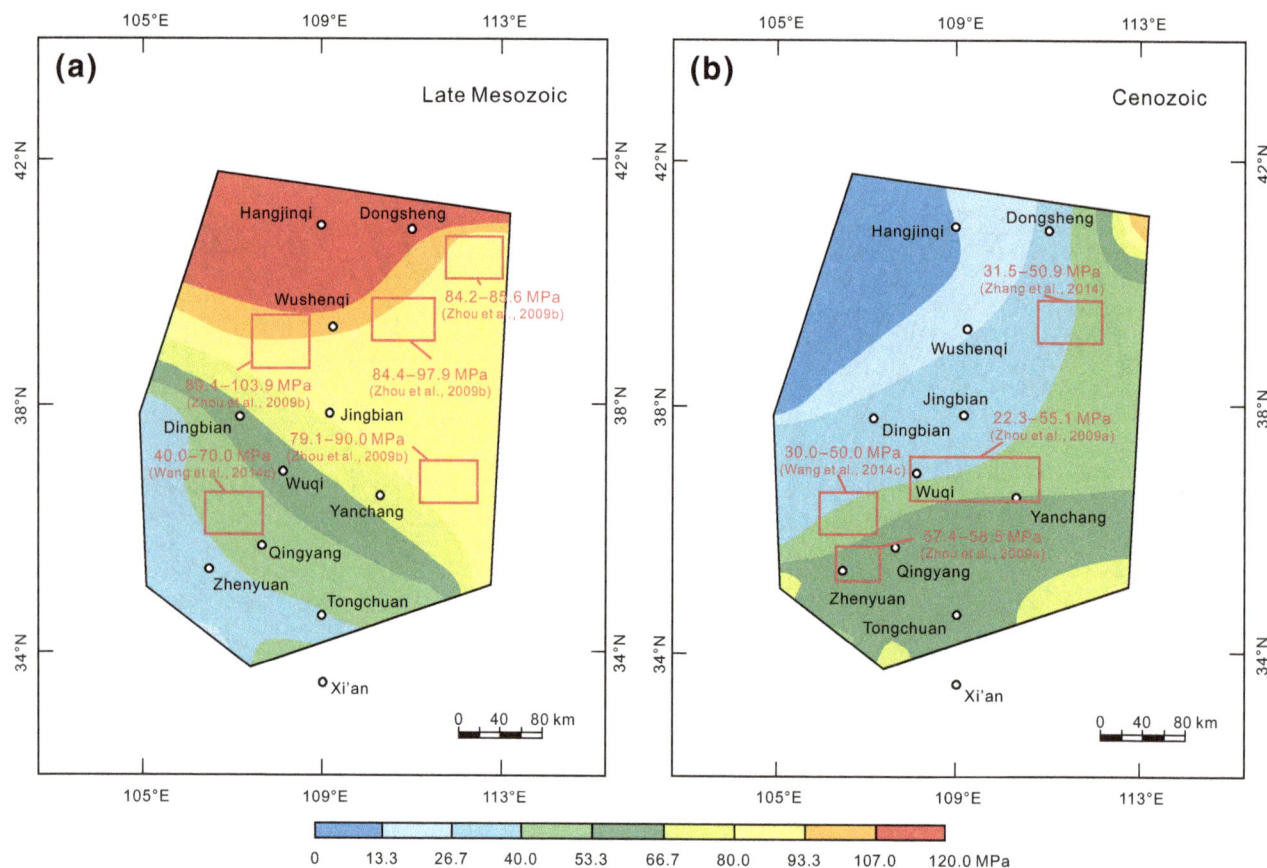

Fig. 10 Maximum principal stress distribution of the two models in **a** the Late Mesozoic and **b** the Cenozoic eras. *Red frames* denote the areas where the calculated stress magnitudes match well with the Acoustic Emission paleo-stress magnitudes in earlier literature (Zhou et al. 2009a, b; Wang et al. 2014a; Zhang et al. 2014)

syntheses of the rupture value and the strain energy density, namely the two-factor method, are applied, in order to build finite element models for fracture prediction in the Ordos Basin (Ding et al. 1998).

5 Results and analyses

Because the orientation and the distribution of structural fractures are the key elements in fracture prediction, the fracture orientation and the estimated density have been calculated with the finite element modeling and will be compared with the observed data in outcrops and cores. With the two-factor method, modeling results, including the principal compressive stress orientations, the rupture values, the strain energy density and the fracture density, are presented as maps, which can imply the relative degrees of fracture development in the Longdong area.

5.1 Validity of models

Since reliable numerical models are the basis of further study on the fracture prediction in the Longdong area, it is

necessary to verify the correctness of the two models proposed in this paper, including the Late Mesozoic and the Cenozoic ones, by comparing the results of finite element modeling with earlier published data.

The calculated displacement directions reveal that the relative rotation directions in these periods are (1) anti-clockwise from the Early Jurassic to the Cretaceous and (2) clockwise in the Cenozoic era (Fig. 9). These results are in good agreement with earlier findings (e.g., Pei et al. 2011; Li et al. 2014; Yang et al. 2014).

Acoustic Emission (AE) is an important technique in rock mechanics and experimental seismology, which can offer rock mechanical parameters, such as the maximum principal stress magnitudes generated in the geological history. The maximum principal stress magnitudes of the Late Mesozoic era after pore-pressure correction range from 40.0 to 103.9 MPa in the Yanhewan, the Dingbian, the Dongsheng areas, etc. (Fig. 10a). The Cenozoic stress magnitudes remain in a limited range of 22.3–58.5 MPa within the Wuqi-Yanhewan, the Zhenyuan, the Wushengqi areas, etc. (Zhou et al. 2009a, b; Wang et al. 2014a; Zhang et al. 2014) (Fig. 10b). The calculated maximum principal stress magnitudes in the Late Mesozoic and the Cenozoic

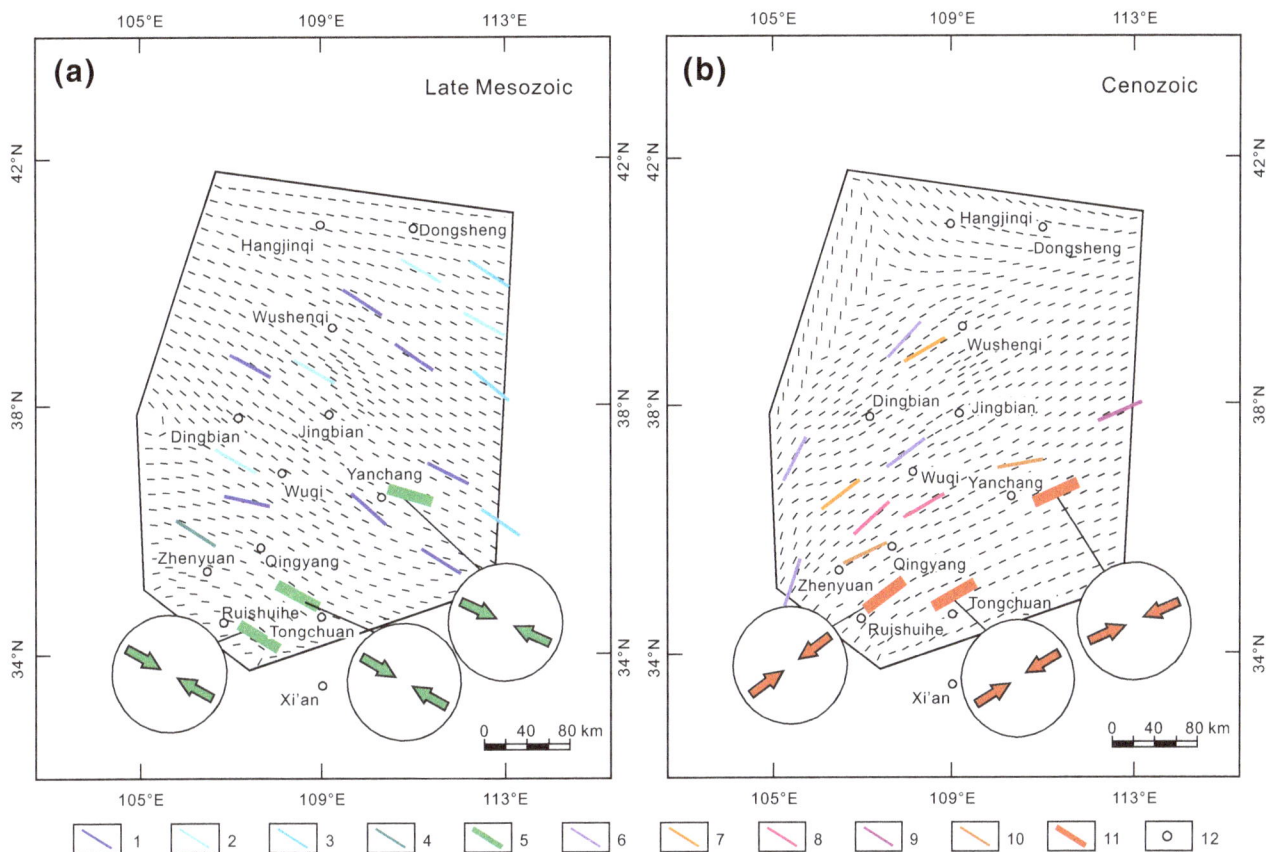

Fig. 11 Maximum principal compressive stress trajectory maps of the two models for the Ordos Basin in the **a** Late Mesozoic and **b** Cenozoic eras. *Green and red arrows* represent the two major orientations of horizontal maximum principal stress (S_{Hmax}) through conjugate joint measurements. *Short black bars* indicate the calculated S_{Hmax}, and *bars in other colors* represent the observed S_{Hmax} in previous literature. *1* Late Mesozoic S_{Hmax} from Wan (1994), *2* Late Mesozoic S_{Hmax} from Hou et al. (2010), *3* Late Mesozoic S_{Hmax} from Sun et al. (2014), *4* Late Mesozoic S_{Hmax} from Zhou et al. (2009b), *5* Late Mesozoic S_{Hmax} deduced from conjugate joints in our field measurements, *6* Cenozoic S_{Hmax} from Wang et al. (2008), *7* Cenozoic S_{Hmax} from Xie et al. (2011), *8* Cenozoic S_{Hmax} from Sun et al. (2014), *9* Cenozoic S_{Hmax} from Yang et al. (2014), *10* Cenozoic S_{Hmax} from Zhou et al. (2009a), *11* Cenozoic S_{Hmax} deduced from conjugate joints in our field measurements, *12* City

are in agreement with the range of stress magnitudes measured by AE technology (Fig. 10). The above-mentioned evidence strengthens the validity of our calculated results in the models.

In addition, earlier published stress orientation data (Wan 1994; Hou et al. 2010; Sun et al. 2014) are also used as evidence to substantiate our models (Fig. 11a). These stress orientation data suggest that the dominant orientation of maximum principal compressive stress in the Late Mesozoic is WNW. Current stress field data can also be utilized to interpret the Cenozoic stress fields because the basin has been stable during this period (Wang et al. 2008; Xie et al. 2011; Sun et al. 2014; Yang et al. 2014). Based on the borehole collapse and multiple strain analyses in the Yanhewan area, it can be inferred that the dominant orientation of maximum principle compressive stress in the Cenozoic is NE (Zhou et al. 2009a). All these orientations are presented in the stereonets (Fig. 11). The differences between the calculated orientations of maximum

compressive stress and the measured ones, including the stress orientations in previous literature (e.g., Wan 1994; Hou et al. 2010; Sun et al. 2014) and the measured data in the present study, are in general less than 5°, proving the reliability of the Late Mesozoic and Cenozoic models (Fig. 11).

Evidence including the rotation directions, the measured maximum principal stress magnitudes and the previous stress data is gathered to prove the authenticity of the two stress fields in the Late Mesozoic–Cenozoic models, and it is found that the calculated results are reliable. Despite slight differences between the calculated and observed maximum principal compressive stress, the modeling results of the Late Mesozoic stress fields indicate that the orientation of the maximum principal compressive stress in the Ordos Basin is WNW, whereas in the Cenozoic model, the orientation is NE. Based on the above-mentioned proofs, the validity of the two models in the Late Mesozoic and the Cenozoic can be corroborated.

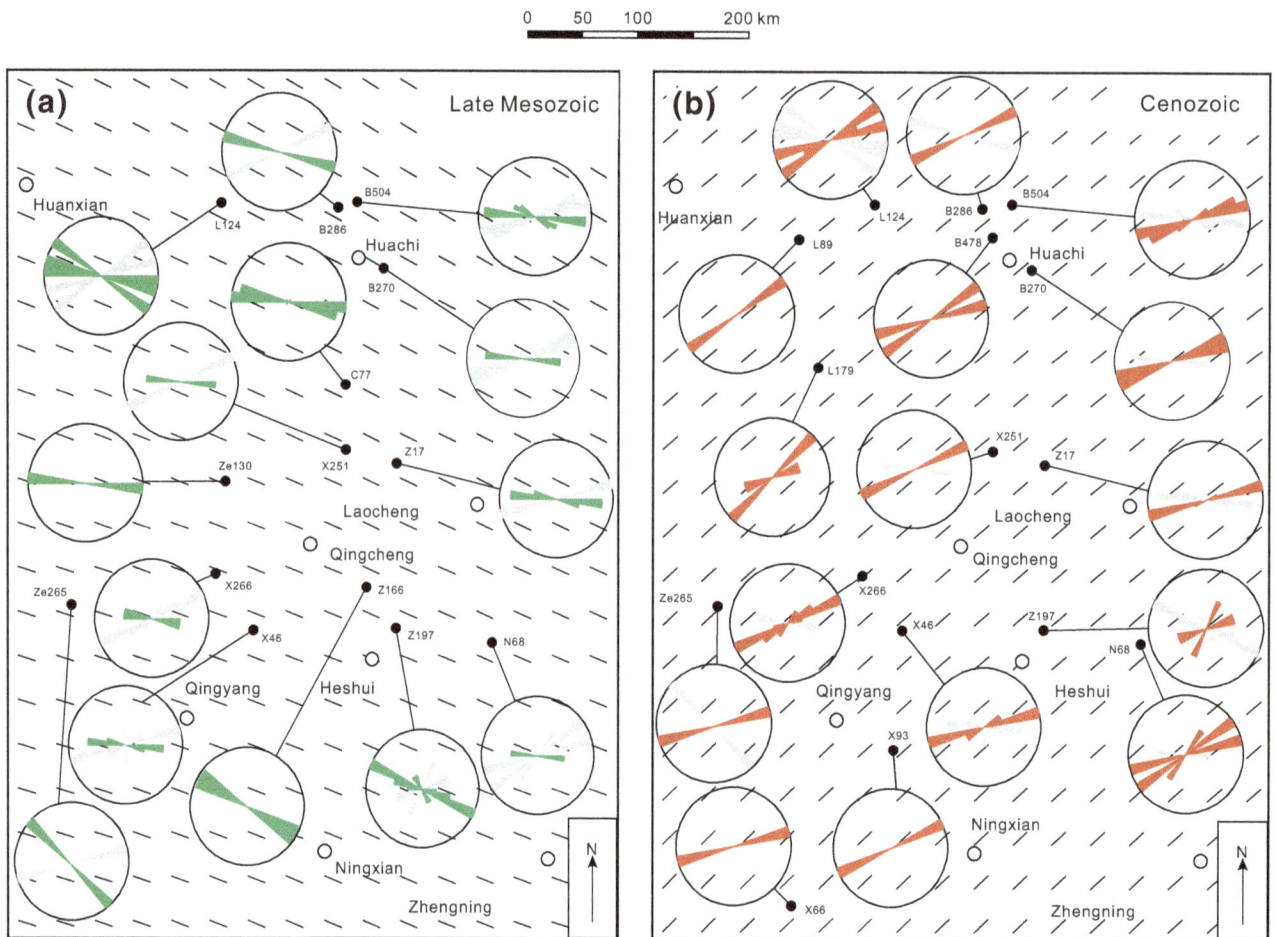

Fig. 12 Calculated maximum principal compressive stress orientations in **a** the Late Mesozoic and **b** the Cenozoic within the Longdong area are compared with the observed fracture orientations in 19 wells. The observed orientations are obtained from imaging logging (FMI technology) data, which are shown as *rose diagrams* in the figures. The *green rose diagrams* denote the structural fractures in NW–EW trends, whereas the *red ones* denote those in NNE–ENE trends

5.2 Maximum principal stress orientations

Tectonic events of different episodes have distinct effects on the principal stress orientations in the Ordos Basin. Since there is little difference between the Chang 7_1 and 7_2 members except for lithology and layer thickness, the pattern of principal stress orientation during the same period is similar in each layer of the Longdong area. Thus, the Chang 7_1 member is taken as an example to demonstrate the distribution of maximum principal compressive stress in the study area (Fig. 12).

On the basis of paleo-magnetic evidence in earlier studies, although the Ordos Basin experienced rotation in different directions from the Late Mesozoic to the Cenozoic, the rotation angle of the basin is less than 5° in the Late Mesozoic–Cenozoic eras (e.g., Huang et al. 2005). Therefore, the present stress data, including fracture trends and Formation Microscanner Image (FMI) data, can also be utilized to indicate the stress orientations in the Late Mesozoic–Cenozoic. From numerical modeling, the orientations of calculated maximum compressive stress in the Late Mesozoic are mainly WNW, while those in the Cenozoic are mainly NE (Zhao et al. 2013, 2016) (Fig. 12). In outcrops and cores, the observed fractures developed in the Late Mesozoic are chiefly in NW–EW trends and those in the Cenozoic are chiefly in NNE–ENE trends (Fig. 8). Our field measurements also corroborate that the ENE-trending structural fractures developed later than the NW-trending ones. Therefore, it can be concluded that the NW to EW fractures were developed in a Late Mesozoic stress field, whereas the NNE to ENE ones were developed in a Cenozoic stress field. Despite tiny differences between the calculated and the observed data, in general, modeling results fit well with the dominant orientations of observed fractures which are obtained from the FMI technology (Fig. 12).

Structural fractures in the Ordos Basin were developed in multiple orientations under different stress fields, primarily in the Late Mesozoic and the Cenozoic episodes, and this intersection pattern will contribute to wider opening and better connectivity of the fractures. The formed fracture networks provide a path for fluid transmission and enhance the permeability, which will have notably improved the fractured tight-oil reservoirs in the Ordos Basin (e.g., Izadi and Elsworth 2014).

5.3 Rupture values

Since the rupture value is an important parameter to indicate the fracture development in the study area, comparison between the calculated rupture values and the observed core fracture density is informative to help analyze the reliability of the models (Figs. 13, 14).

In the maps of rupture values in the Chang 7_1 member during the Late Mesozoic–Cenozoic era, the highest rupture values are situated in the east and center of the study area, mainly concentrated in the Qingyang, the Laocheng and the Zhengning areas (Fig. 13a, b), while the highest rupture values in the Chang 7_2 member are chiefly situated in the mid-southern area, particularly in the Qingyang-Heshui and the Ningxian areas (Fig. 13c,

d). The distribution of sand bodies and the thickness of sandstone layers have a distinct impact on the distribution of rupture values within the Longdong area. Both in the Chang 7_1 and 7_2 members, the rupture values are relatively higher where sand bodies are developed and the thickness of sandstone layers is relatively larger, due to the brittleness of sandstones (Fig. 4). The regional stress fields during different periods also influence the rupture values, resulting in the Cenozoic rupture values being smaller than the Late Mesozoic ones. However, the influence of regional stress fields is not as remarkable as that of lithology, because regional stress fields determine only the magnitudes, not the distribution of rupture values in the Chang 7_1 and 7_2 members within the study area (Fig. 13).

5.4 Strain energy density

Because rocks with higher strain energy density are more likely to form structural fractures than those with a lower one, the strain energy density can be used as another parameter to predict the fracture density.

Similar to the rupture value, there is obvious positive correlation between the strain energy density and the thickness of sandstone layers. The strain energy density is

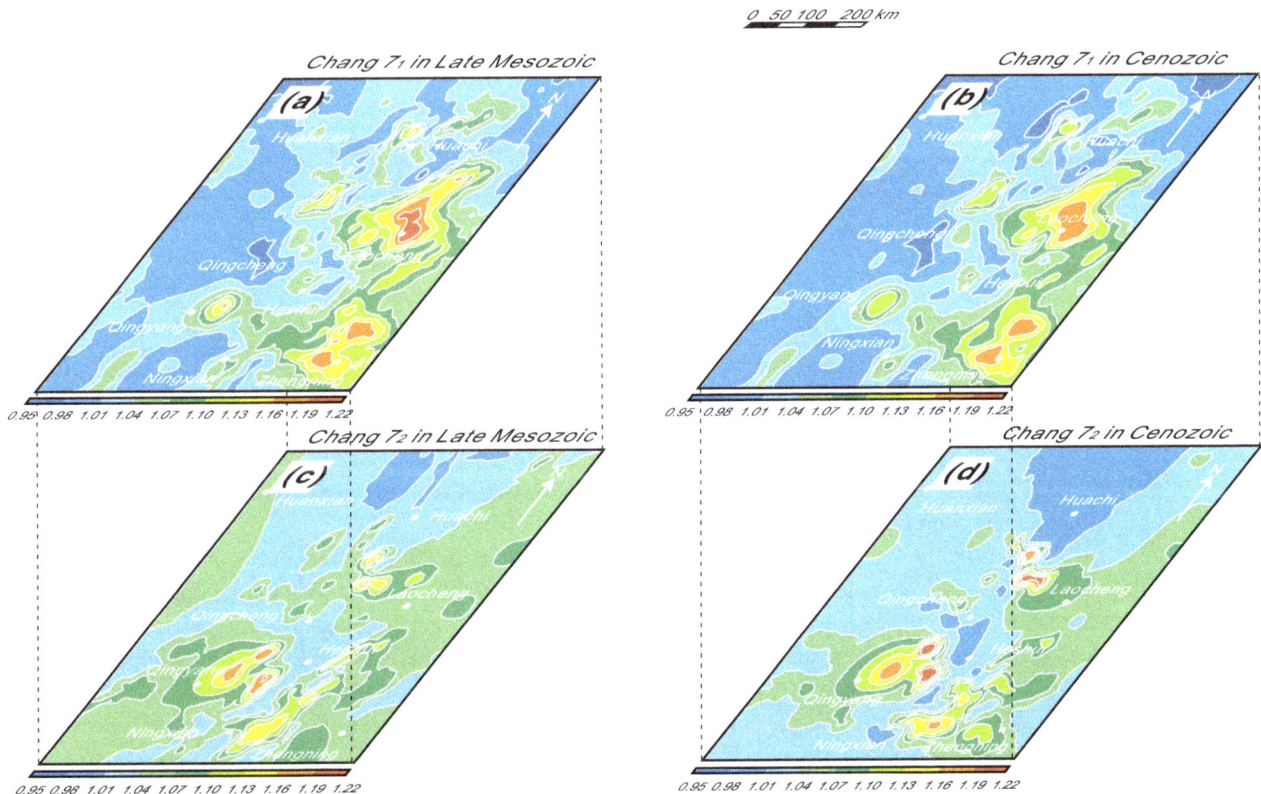

Fig. 13 Distribution of rupture value of the Late Mesozoic and the Cenozoic in the Chang 7_1 and 7_2 members within the Longdong area. **a** Rupture values of the Late Mesozoic in the Chang 7_1 member, **b** rupture values of the Cenozoic in the Chang 7_1 member, **c** rupture values of the Late Mesozoic in the Chang 7_2 member and **d** rupture values of the Cenozoic in the Chang 7_2 member

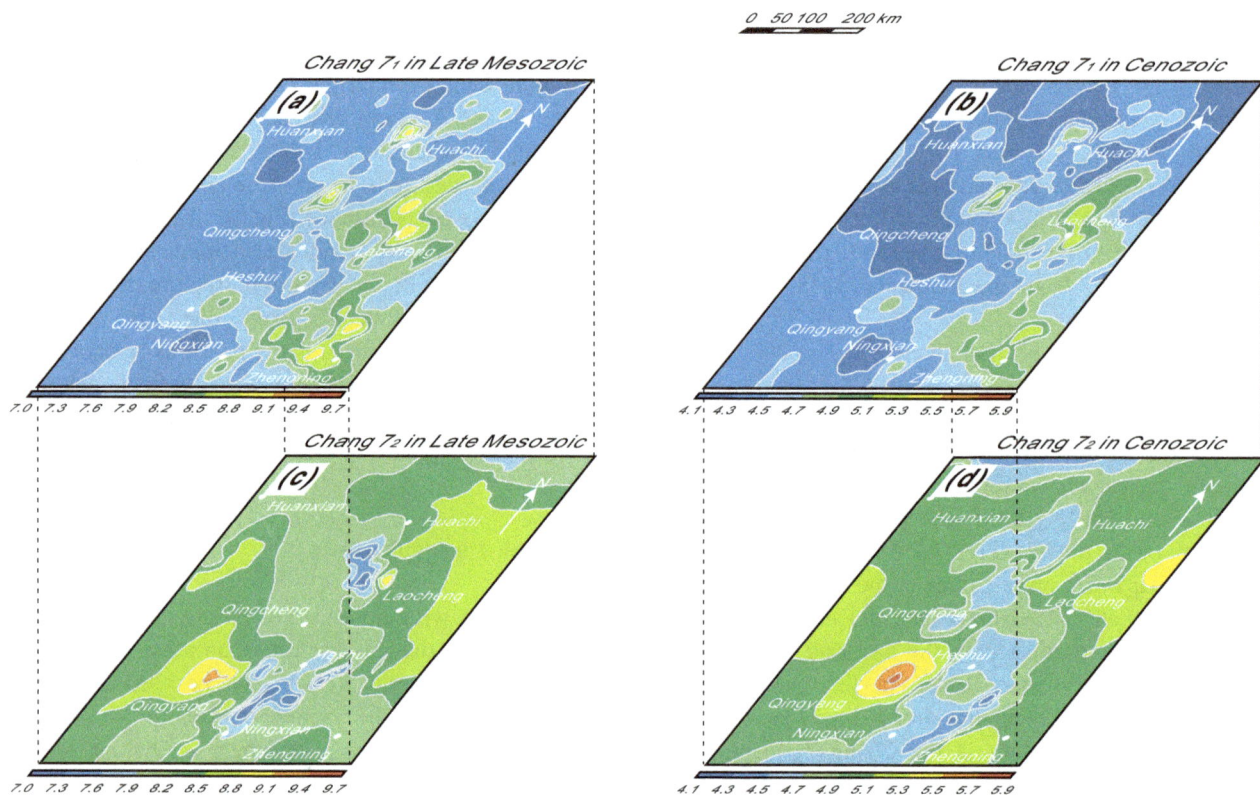

Fig. 14 Distribution of strain energy density (10^4 J/m^3) of the Late Mesozoic and the Cenozoic in Chang 7_1 and 7_2 members within the Longdong area. **a** Strain energy density of the Late Mesozoic in the Chang 7_1 member, **b** strain energy density of the Cenozoic in the Chang 7_1 member, **c** strain energy density of the Late Mesozoic in the Chang 7_2 member and **d** strain energy density of the Cenozoic in the Chang 7_2 member

Table 3 Curve-fitting relationships of the measured fracture densities, the calculated rupture values and the strain energy densities of Chang 7_1 and 7_2 members in the Longdong area	Layers	Curve-fitting relations	Correlation coefficient
	Chang 7_1	$D_M = 3.493\ I^2 - 0.049\ U^2 - 6.241\ I + 0.695\ U + 0.270$	0.947
		$D_C = 3.581\ I^2 - 0.123\ U^2 - 7.105\ I + 3.240\ U + 1.054$	0.904
	Chang 7_2	$D_M = 48.429\ I^2 - 0.039\ U^2 - 100.308\ I + 0.734\ U + 48.585$	0.871
		$D_C = 22.944\ I^2 + 0.450\ U^2 - 46.876\ I - 4.094\ U + 33.241$	0.941

D_M (m^{-1}) and D_C (m^{-1}) represent the measured fracture densities in cores of the Late Mesozoic and the Cenozoic periods, respectively. I and U denote the calculated rupture values and the strain energy densities (10^4 J/m^3), respectively

higher where the sand bodies are developed as a whole (Figs. 4, 14). Although the Cenozoic stress field of the Ordos Basin is strikingly different from the Late Mesozoic one, the impact of regional stress is mainly limited to the magnitudes, not the distribution of strain energy density in the Longdong area. The distribution of strain energy density in the Late Mesozoic and the Cenozoic periods is similar, but the Late Mesozoic strain energy density is larger than the Cenozoic one both in the Chang 7_1 and 7_2 members, implying that the strain energy density is more influenced by the movement in the Late Mesozoic than that in the Cenozoic (Fig. 14).

5.5 Predicted fracture distribution

In order to predict the fracture distribution in the Yanchang Formation within the Longdong area, connection between the calculated and the measured fracture density in cores must be established to study their relationship. In this paper, the two-factor method is utilized to compare the calculated data (including the rupture value and the strain energy density) and the measured fracture density (Ding et al. 1998). Since structural fractures in the Ordos Basin were chiefly developed during two stages of stress fields, namely the Late Mesozoic and the Cenozoic ones, two

Table 4 Overview of predicted and measured fracture densities in the Chang 7_1 member in the Longdong area

Well name	Measured density, m^{-1}	Predicted density, m^{-1}	Absolute error, m^{-1}	Relative error, %	Well name	Measured density, m^{-1}	Predicted density, m^{-1}	Absolute error, m^{-1}	Relative error, %
B117	0.020	0.045	0.025	**122**	W98	0.000	0.000	0.000	–
B146	0.000	0.005	0.005	–	X140	0.000	0.000	0.000	–
B170	0.000	0.008	0.008	–	X195	0.028	0.015	0.013	46
B456	0.000	0.012	0.012	–	X233	0.000	0.000	0.000	–
B478	0.059	0.085	0.026	44	X259	0.000	0.027	0.027	–
Ban12	0.320	0.305	0.015	5	X261	0.000	0.005	0.005	–
C87	0.070	0.058	0.012	18	X263	0.000	0.000	0.000	–
Hua56	0.000	0.007	0.007	–	X67	0.015	0.008	0.007	47
L189	0.068	0.069	0.001	2	X73	0.026	0.057	0.031	**118**
L47	0.018	0.042	0.023	**128**	Y433	0.085	0.052	0.033	39
L79	0.033	0.023	0.010	29	Z124	0.000	0.013	0.013	–
L96	0.000	0.008	0.008	–	Z148	0.000	0.020	0.020	–
M28	0.000	0.005	0.005	–	Z15	0.068	0.099	0.031	46
M40	0.036	0.022	0.014	39	Z172	0.038	0.048	0.010	26
N43	0.062	0.090	0.029	46	Z186	0.344	0.372	0.028	8
N51	0.137	0.110	0.027	20	Z200	0.122	0.112	0.009	8
N57	0.055	0.080	0.025	45	Z230	0.046	0.054	0.009	19
N75	0.037	0.047	0.010	27	Z233	0.080	0.084	0.004	5
N76	0.155	0.155	0.000	0	Z24	0.061	0.086	0.025	41
N78	0.210	0.176	0.033	16	Z47	0.199	0.132	**0.067**	34
N81	0.057	0.084	0.027	47	Z57	0.151	0.121	0.030	20
S142	0.164	0.136	0.028	17	Z78	0.316	0.231	**0.085**	27
S160	0.000	0.000	0.000	–	Z79	0.071	0.036	0.035	49
T15	0.113	0.097	0.016	14	Z87	0.201	0.234	0.033	16
T2	0.055	0.063	0.008	14	Ze220	0.050	0.026	0.024	48
W47	0.053	0.079	0.026	49	Ze97	0.000	0.000	0.000	–
W67	0.000	0.024	0.024	–	Zeg70	0.180	0.132	0.048	27

"–" means that the relative errors do not exist in these wells because the corresponding measured fracture densities are 0 m^{-1}, and large errors, including absolute and relative errors, are denoted in bold type in this table

episodes of fractures should be fitted separately and then be added up by weight.

By multiple regression analyses, bi-quadratic relationships between the rupture values, the strain energy density and the measured fracture density in the Chang 7_1 and 7_2 members of different episodes have been built and the empirical formulas are shown in Table 3. Correlation coefficients in all curve-fitting relationships are larger than 0.87, which means that there is a significant correlation between the calculated and the measured data.

To further illustrate the reliability of our models, error analyses are carried out as follows. Both the absolute error and the relative error were applied to reflect the accuracy of fracture prediction. Absolute error is calculated by:

$$\Delta = \left| D_p - D_m \right| \tag{7}$$

And relative error can be described as follows:

$$\varepsilon = \frac{\left| D_p - D_m \right|}{D_m} \times 100\% \tag{8}$$

Δ denotes the absolute error and ε denotes the relative error. D_P and D_M represent the predicted and the measured fracture densities, respectively. Generally, when ε is less than 50%, we can consider that the predicted data match the measured ones and the modeling results are reliable to a certain extent.

The differences between the measured and the predicted fracture densities are shown in Tables 4 and 5. For most of the wells in the Chang 7_1 member, the predicted and the measured data match quite well. In the 54 measured wells, only 2 wells exceed 0.05 m^{-1} in absolute errors and 3 wells exceed 50% in relative errors (Table 4). The differences between them may be caused by the stress concentration in some areas, such as Well Z47 and Z78, where numerous fractures are found. As for the Chang 7_2

Table 5 Overview of predicted and measured fracture densities in the Chang 7_2 member in the Longdong area

Well name	Measured density, m^{-1}	Predicted density, m^{-1}	Absolute error, m^{-1}	Relative error, %	Well name	Measured density, m^{-1}	Predicted density, m^{-1}	Absolute error, m^{-1}	Relative error, %
B117	0.062	0.083	0.021	34	X233	0.000	0.049	0.049	–
B146	0.000	0.033	0.033	–	X263	0.000	0.004	0.004	–
B36	0.000	0.000	0.000	–	X270	0.000	0.041	0.041	–
B401	0.000	0.026	0.026	–	X65	0.053	0.040	0.013	25
B456	0.000	0.000	0.000	–	X67	0.026	0.031	0.005	20
Ban12	0.918	0.903	0.016	2	X69	0.157	0.080	**0.077**	49
C87	0.043	0.036	0.007	15.	Z172	0.025	0.022	0.003	12
Hua312	0.000	0.000	0.000	–	Z230	0.064	0.037	0.027	42
L189	0.052	0.037	0.015	29	Z233	0.271	0.109	**0.162**	**60**
L47	0.000	0.009	0.009	–	Z78	0.464	0.464	0.000	0
L79	0.047	0.039	0.008	16	Z79	0.072	0.000	**0.072**	**100**
L96	0.000	0.000	0.000	–	Ze118	0.000	0.045	0.045	–
N43	0.000	0.000	0.000	–	Ze284	0.000	0.023	0.023	–
N51	0.385	0.337	0.048	12	Ze298	0.000	0.097	**0.097**	–
N55	0.000	0.009	0.009	–	Ze362	0.000	0.131	**0.131**	–
N57	0.137	0.100	0.036	27	Ze77	0.129	0.072	**0.057**	44
N75	0.036	0.149	**0.113**	**317**	Ze95	0.052	0.036	0.016	30
N76	0.093	0.038	**0.055**	**59**	Ze97	0.000	0.029	0.029	–
N78	0.024	0.056	0.032	**132**	Zeg70	0.050	0.036	0.014	27
W98	0.000	0.044	0.044	–					

"–" means that the relative errors do not exist in these wells because the corresponding measured fracture densities are 0 m^{-1}, and large errors, including absolute and relative errors, are denoted in bold type in this table

member, predicted data of only 8 wells in the 39 measured wells are more than 0.05 m^{-1} in absolute errors, and data of only 5 wells are more than 50% in relative errors (Table 5). Most of these wells are with extraordinarily high fracture density, which results in large errors between the predicted and the measured fracture densities. Some large errors may be caused by non-structural factors, such as various sedimentary phenomena. Cross bedding and lenticular bedding appeared widely in Well Ze77, etc., which may lead to the difference between the predicted and measured data. Despite these differences, the tendency of predicted fracture distribution is still in accordance with the measured one. In short, the errors between the predicted and the measured fracture densities are within acceptable limits, implying that the modeling results are suitable for the fracture prediction in the Yanchang Formation of the Ordos Basin.

6 Discussion

As is shown in the maps of maximum principal compressive stress orientations in the Chang 7_1 and 7_2 members in the Longdong area, the dominant orientations of the Late Mesozoic fractures are NW–EW (Fig. 12a) and those of

the Cenozoic ones are NNE–ENE (Fig. 12b) which are consistent with the regional stress fields of the Ordos Basin in the corresponding periods (e.g., Zhang et al. 2003). In the maps of predicted fracture density in different periods, the average density of the Cenozoic fractures is larger than that of the Late Mesozoic ones (Fig. 15). By comparison between the distribution maps of predicted total fracture densities in the Chang 7_1 and 7_2 members within the study area (Fig. 16), the predicted fracture density in each member is alike as a whole; however, their fracture distributions are significantly distinct. In the Chang 7_1 member, the maximum fracture density is located in the center and the east of the Longdong area (Fig. 16a), while in the Chang 7_2 member, the maximum density is situated in the southern-central section of the study area (Fig. 16b).

In addition, by comparing the predicted fracture density with the distribution of sand bodies, their similarity reveals that the lithology is a key factor in controlling the fracture distribution in the Ordos Basin. Structural fractures are more likely to be developed in the sandstones rather than in the mudstones. Where thicker sandstone layers are developed, the fracture density is relatively higher than other areas (Figs. 4, 16). However, there is still a difference between the predicted fracture distribution and the outline of sand bodies, indicating that the regional stress field also

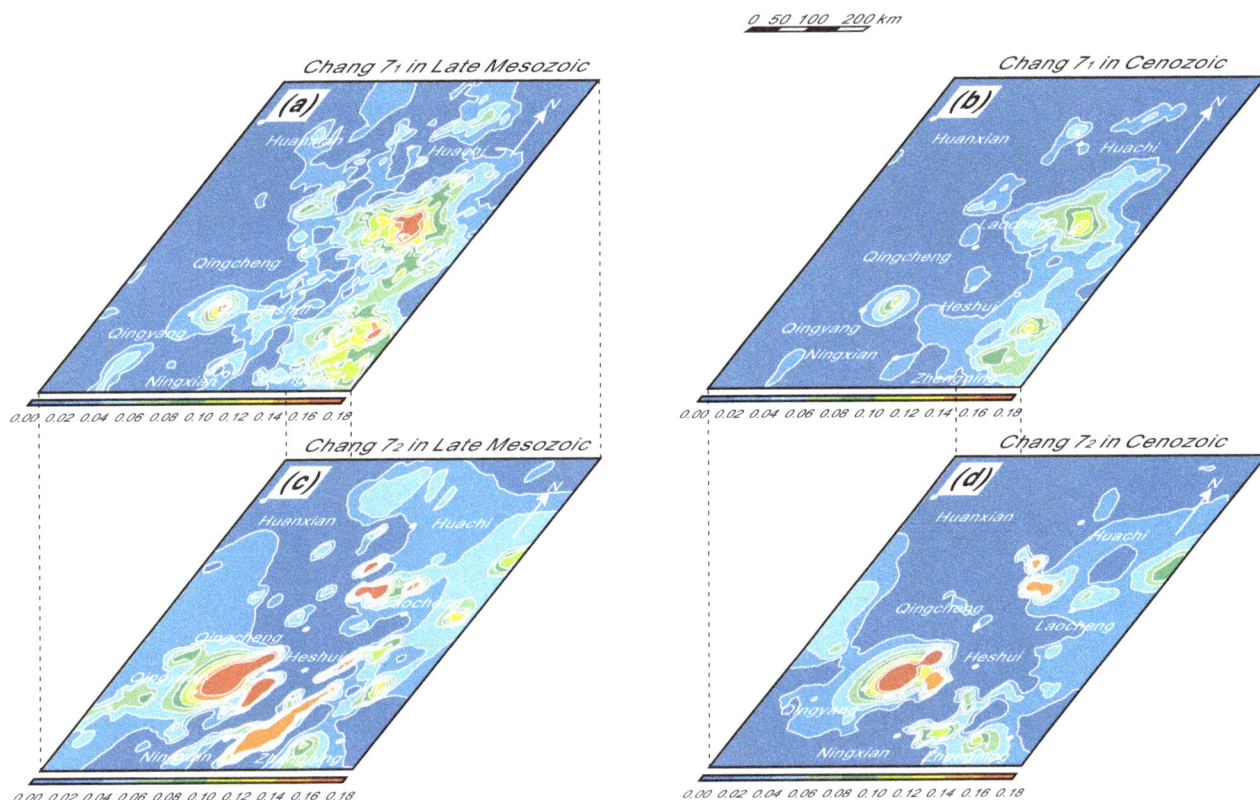

Fig. 15 Distribution of predicted fracture density (m^{-1}) of the Late Mesozoic and the Cenozoic in the Chang 7_1 and 7_2 members within the Longdong area. **a** Predicted fracture density of the Late Mesozoic in the Chang 7_1 member, **b** predicted fracture density of the Cenozoic in the Chang 7_1 member, **c** predicted fracture density of the Late Mesozoic in the Chang 7_2 member and **d** predicted fracture density of the Cenozoic in the Chang 7_2 member

plays a role in the fracture development, even though its influence is limited compared with the lithology and the layer thickness.

In brief, the stress fields determine the overall fracture orientations, and the lithology distribution and the thickness of sandstone layers in the study area play a predominant role in the distribution of predicted fracture density. Some potential factors which are not covered in these numerical models may restrict the accuracy of predicted results, including:

1. The complicated heterogeneity of each layer;
2. The extreme stress in some areas;
3. The interaction between the two episodes of structural fractures; and
4. The influence of deep paleo-faults.

Since the modeling is on a relatively large-scale while the outline of sand bodies is depicted in considerable detail, the modeling results, including the rupture values and the strain energy density, can still be used to guide further exploration in spite of the four above-mentioned restrictions. Meanwhile, the qualitative fracture prediction obtained from the numerical modeling may also be applicable. These results

act as a reference for future regional-scale petroleum exploration, while the method of fracture prediction, including the two-factor method and the empirical formulas can be used at well scale. Structural fractures play an important role in reconstructing the tight clastic reservoirs, especially in their permeability (Réda 2013).

The controlling factors of fracture development are complex owing to the complicated geological background. Fault systems can be a vital factor in developing fractures where tectonic movements are strong such as the Kuqa Depression of the northern Tarim Basin in the northwestern China (Ju et al. 2014b) and the Upper Rhine Graben in France and Germany (Johanna et al. 2015); flow may notably promote fracture development where fluid flow or lava flow appears (e.g., Agosta et al. 2010). However, in the Ordos Basin, where the tectonic events are rather weak and the dips of the Mesozoic–Cenozoic strata are less than 3°, the lithology and the layer thickness are the dominant factors in governing the distribution of fracture density. The relationship between the lithology and the fracture density is still obscure, but it may be related to the difference of rock physical parameters (Table 1) according to previous study (e.g., Zeng et al. 2008). The different grain

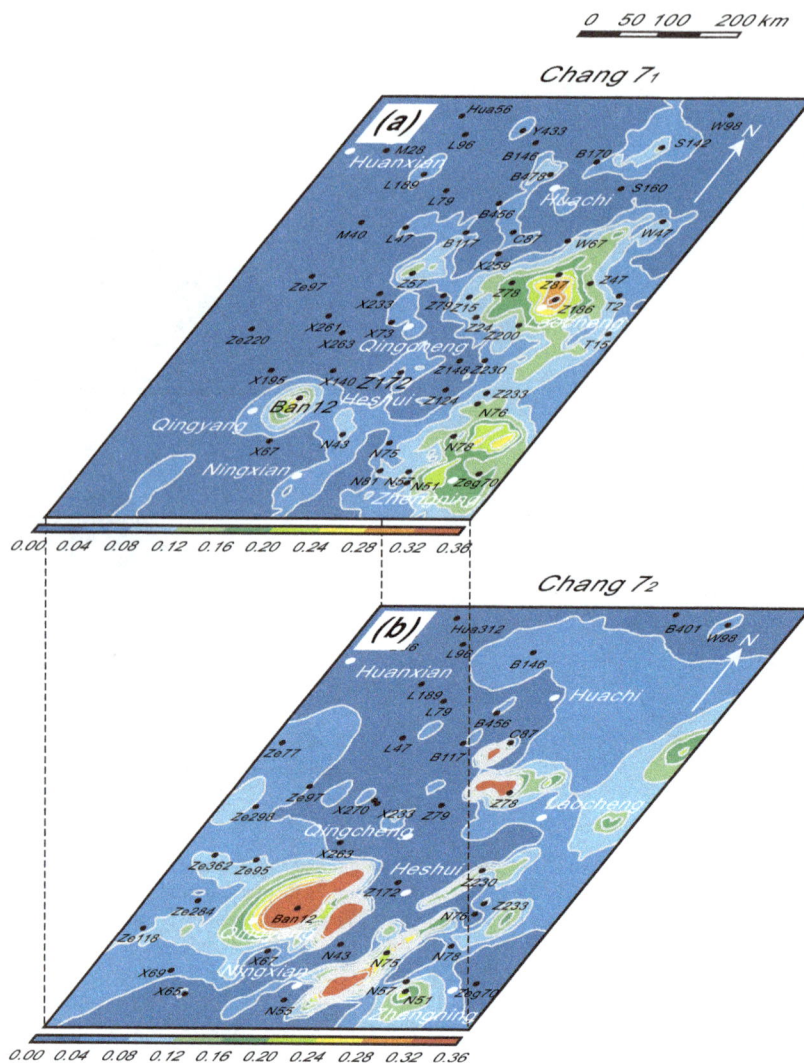

Fig. 16 Distribution of predicted total fracture density (m^{-1}) in **a** the Chang 7_1 and **b** Chang 7_2 members within the Longdong area. *Black solid dots* represent the measured wells in the study area

sizes in various clastic rocks may be the micro-mechanism that causes the distribution of fracture density in the Ordos Basin (Zhao et al. 2013; Ju et al. 2015).

7 Conclusions

The predicted fracture distribution provides a clear view of the fracture concentration and fracture development. Several primary conclusions can be drawn from the modeling results:

1. A finite element modeling technique, applying the two-factor method, is suitable for the fracture prediction of the Ordos Basin, based on comparison between the calculated and the measured fracture densities of the Chang 7_1 and 7_2 members in the Longdong area.

2. Two episodes of structural fractures have been developed since the Late Triassic: The dominant orientations of the Late Mesozoic fractures in the Yanchang Formation are NW–EW, whereas those of the Cenozoic fractures are NNE–ENE, both of which are in agreement with the modeling results.

3. Structural fractures in the Ordos Basin are controlled by the regional stress fields, and the lithology and the layer thickness have a significant impact on the distribution of structural fractures, because the stress distribution will be affected by the inhomogeneity of lithology and layer thickness. This conclusion is shown in the similarity between the maps of predicted fracture density and observed sand bodies in the Yanchang Formation within the study area.

4. The average fracture density is close in the Chang 7_1 and 7_2 members, but there are obvious differences in

their fracture distributions. In the Chang 7_1 member, the maximum fracture density is concentrated in the center and the east of the Longdong area, particularly in the Qingyang, the Laocheng and the Zhengning areas (up to 1.5 m^{-1}), while in the Chang 7_2 member, the maximum value is located in the central and southern part of the area.

5. The modeling results and the predicted fracture density can be utilized to guide future regional exploration, and the method of fracture prediction, namely the two-factor method, can be referred for further study of the tight-sand reservoirs.

Acknowledgements The authors would like to thank Drs. Wei Ju, Peng Zhang, Yan Zhan and Xuan Yu for their help in the core observation and modeling. This research was funded by the National Natural Science Foundations of China (Grant Nos. 40772121 and 41530207), State Key Projects of Petroleum (Nos. 2008ZX05029-001, 2011ZX05029-001 and 2014A0213) and Research and Development Foundations of the Huaneng Clean Energy Research Institute (TY-15-CERI02).

References

Agosta F, Alessandroni M, Antonellini M, et al. From fractures to flow: a field-based quantitative analysis of an outcropping carbonate reservoir. Tectonophysics. 2010;490(3–4):197–213. doi:10.1016/j.tecto.2010.05.005.

Bao XW, Song XD, Xu MJ, et al. Crust and upper mantle structure of the North China Craton and the NE Tibetan Plateau and its tectonic implications. Earth Planet Sci Lett. 2013;369–370:129–37. doi:10.1016/j.epsl.2013.03.015.

Coulomb CA. Essai sur une application des regles des maximas et minmas a quelques problemes de statique relatifs a l'architecture, vol. 7. Divers Savanta: Mem. Acad. Roy. Pres. 1776.

Darby BJ, Ritts BD. Mesozoic contractional deformation in the middle of the Asian tectonic collage the intraplate Western Ordos fold thrust belt, China. Earth Planet Sci Lett. 2002;205(1–2):13–24. doi:10.1016/S0012-821X(02)01026-9.

Ding ZY, Qian XL, Huo H, et al. A new method for quantitative prediction of tectonic fractures—the two-factor method. Oil Gas Geol. 1998;19(1):1–8 (**in Chinese**).

Du JH, Liu H, Ma DS, et al. Discussion on effective development techniques for continental tight oil in China. Pet Explor Dev. 2014;41(2):217–24. doi:10.1016/S1876-3804(14)60025-2.

Duan Y, Wang CY, Zheng CY, et al. Geochemical study of crude oils from the Xifeng Oilfield of the Ordos Basin, China. J Asian Earth Sci. 2008;31:341–56. doi:10.1016/j.jseaes.2007.05.003.

Ezulike DO, Dehghanpour H. A model for simultaneous matrix depletion into natural and hydraulic fracture networks. J Nat Gas Sci Eng. 2014;16:57–69. doi:10.1016/j.jngse.2013.11.004.

Faure M, Lin W, Chen Y. Is the Jurassic (Mesozoic) intraplate tectonics of North China due to westward indentation of North China block? Terra Nova. 2012;24(6):456–66. doi:10.1111/ter.12002.

Feng JP, Ouyang ZY, Huang ZL. The application of balance geological section technology in the mid-south section of the western margin of Ordos Basin. Geotecton Metallog. 2013;37(3):393–7 (**in Chinese**).

Fournier M, Jolivet L, Davy P, et al. Back-arc extension and collision: an experimental approach to the tectonics of Asia. Geophys J Int. 2004;157:871–89. doi:10.1111/j.1365-246X.2004.02223.x.

Gilder SA, Gomez J, Chen Y, et al. A new paleogeographic configuration of the Eurasian landmass resolves a paleomagnetic paradox of the Tarim Basin (China). Tectonics. 2008;27(1):1256. doi:10.1029/2007TC002155.

Glukhmanchuk ED, Vasilevskiy AN. Description of fracture zones based on the structural inhomogeneity of the reflector deformation field. Russ Geol Geophys. 2013;54(1):82–6. doi:10.1016/j.rgg.2012.12.007.

Golf-Racht TDV. Fundamentals of fractured reservoir engineering. Amsterdam: Elsevier; 1982. p. 1–12.

Griffith AA. Phenomena of rupture and flow in solids. Fish Manage Ecol. 1920;16(2):130–8. doi:10.1098/rsta.1921.0006.

Guo YR, Liu JB, Yang H, et al. Hydrocarbon accumulation mechanism of low permeable tight lithologic oil fields in the Yanchang Formation, Ordos Basin, China. Pet Explor Dev. 2012;39(4):447–56. doi:10.1016/S1876-3804(12)60061-5.

Hou GT, Hari KR. Mesozoic–Cenozoic extension of the Bohai Sea: contribution to the destruction of North China Craton. Front Earth Sci. 2014;8(2):202–15. doi:10.1007/s11707-014-0413-3.

Hou GT, Wang YX, Hari KR. The Late Triassic and Late Jurassic stress fields and tectonic transmission of North China Craton. J Geodyn. 2010;50(3–4):318–24. doi:10.1016/j.jog.2009.11.007.

Huang BC, Shi RP, Wang YC, et al. Palaeomagnetic investigation on early-middle Triassic sediments of the North China block: a new early Triassic palaeopole and its tectonic implications. Geophys J Int. 2005;160:101–13. doi:10.1111/j.1365-246X.2005.02496.x.

Izadi G, Elsworth D. Reservoir stimulation and induced seismicity: roles of fluid pressure and thermal transients on reactivated fractured networks. Geothermics. 2014;51:368–79. doi:10.1016/j.geothermics.2014.01.014.

Jarosinski M, Beekman F, Matenco L, et al. Mechanics of basin inversion: finite element modeling of the Pannonian Basin system. Tectonophysics. 2011;502(1–2):121–45. doi:10.1016/j.tecto.2009.09.015.

Jia CZ, Zhang YF, Zhao X. Prospects of and challenges to natural gas industry development in China. Nat Gas Ind B. 2014;1(1):1–13. doi:10.1016/j.ngib.2014.10.001.

Jiu K, Ding WL, Huang WH, et al. Simulation of paleotectonic stress fields within Paleogene shale reservoirs and prediction of favorable zones for fracture development within the Zhanhua Depression, Bohai Bay Basin, east China. J Pet Sci Eng. 2013;110:119–31. doi:10.1016/j.petrol.2013.09.002.

Johanna FB, Silke M, Sonja LP. Architecture, fracture system, mechanical properties and permeability structure of a fault zone in Lower Triassic sandstone, Upper Rhine Graben. Tectonophysics. 2015;647–648:132–45. doi:10.1016/j.tecto.2015.02.014.

Ju W, Hou GT, Hari KR. Mechanics of mafic dyke swarms in the Deccan Large Igneous Province: palaeostress field modeling. J Geodyn. 2013;66:79–91. doi:10.1016/j.jog.2013.02.002.

Ju W, Hou GT, Feng SB, et al. Quantitative prediction of the Yanchang Formation Chang 6_3 reservoir tectonic fracture in the Qingcheng-Heshui area, Ordos Basin. Earth Sci Front. 2014a;21(6):310–20 (**in Chinese**).

Ju W, Hou GT, Zhang B. Insights into the damage zones in fault-bend folds from geomechanical models and field data. Tectonophysics. 2014b;610:182–94. doi:10.1016/j.tecto.2013.11.022.

Ju W, Sun WF, Hou GT. Insights into the tectonic fractures in the Yanchang Formation interbedded sandstone-mudstone of the Ordos Basin based on core data and geomechanical models. Acta Geol Sin (Engl Ed). 2015;89(6):1986–97. doi:10.1111/1755-6724.12612.

Kusky TM, Li JH. Paleoproterozoic tectonic evolution of the North China Craton. J Asian Earth Sci. 2009;22(4):383–97. doi:10.1016/S1367-9120(03)00071-3.

Kusky TM. Geophysical and geological tests of tectonic models of the North China Craton. Gondwana Res. 2011;20(1):26–35. doi:10.1016/j.gr.2011.01.004.

Li HB, Guo HK, Yang ZM, et al. Tight oil occurrence space of Triassic Chang 7 member in Northern Shaanxi Area, Ordos Basin, NW China. Pet Explor Dev. 2015;42(3):434–8. doi:10.1016/S1876-3804(15)30036-7.

Li RX, Li YZ. Tectonic evolution of the western margin of the Ordos Basin (Central China). Russ Geol Geophys. 2008;49(1):23–7. doi:10.1016/j.rgg.2007.12.002.

Li YH, Wang QL, Cui DX, et al. One feature of the activated southern Ordos Block: the Ziwuling small earthquake cluster. Geod Geodyn. 2014;5(3):16–22. doi:10.3724/SP.J.1246.2014.03016.

Liu MJ, Mooney WD, Li SL, et al. Crustal structure of the northeastern margin of the Tibetan plateau from the Songpan-Ganzi terrane to the Ordos basin. Tectonophysics. 2006;420(1–2):253–66. doi:10.1016/j.tecto.2006.01.025.

Liu SF, Li WP, Wang K, et al. Late Mesozoic development of the southern Qinling-Dabieshan foreland fold-thrust belt, Central China, and its role in continent-continent collision. Tectonophysics. 2015;644–645(3):220–34. doi:10.1016/j.tecto.2015.01.015.

Malaspina N, Hermann J, Scambelluri M, et al. Multistage metasomatism in ultrahigh-pressure mafic rocks from the North Dabie Complex (China). Lithos. 2006;90(1–2):19–42. doi:10.1016/j.lithos.2006.01.002.

Menzies M, Xu YG, Zhang HF, et al. Integration of geology, geophysics and geochemistry: a key to understanding the North China Craton. Lithos. 2007;96(1–2):1–21. doi:10.1016/j.lithos.2006.09.008.

Mercier JL, Vergely P, Zhang YQ, et al. Structural records of the Late Cretaceous–Cenozoic extension in Eastern China and the kinematics of the Southern Tan-Lu and Qinling Fault Zone (Anhui and Shaanxi provinces, PR China). Tectonophysics. 2013;582:50–75. doi:10.1016/j.tecto.2012.09.015.

Nutman AP, Wan YS, Du LL, et al. Multistage late Neoarchaean crustal evolution of the North China Craton, eastern Hebei. Precambr Res. 2011;189(1–2):43–65. doi:10.1016/j.precamres.2011.04.005.

Pearce MA, Jones RR, Smith SAF, et al. Quantification of fold curvature and fracturing using terrestrial laser scanning. AAPG Bull. 2011;57:2367–85. doi:10.1306/11051010026.

Pei JL, Sun ZM, Liu J, et al. A paleomagnetic study from the Late Jurassic volcanics (155 Ma), North China: implications for the width of Mongol-Okhotsk Ocean. Tectonophysics. 2011;510(3–4):370–80. doi:10.1016/j.tecto.2011.08.008.

Prince NJ, Rhodes FH. Fault and joint development in brittle and semi-brittle rock. London: Pergamon Press; 1966. p. 110–64. doi:10.1016/B978-0-08-011275-6.50009-4.

Rao G, Lin AM, Yan B, et al. Tectonic activity and structural features of active intracontinental normal faults in the Weihe Graben, central China. Tectonophysics. 2014;636(1):270–85. doi:10.1016/j.tecto.2014.08.019.

Réda SZ. Fracture density estimation from core and conventional well logs data using artificial neural networks: the Cambro-Ordovician reservoir of Mesdar oil field, Algeria. J Afr Earth Sci. 2013;83:55–73. doi:10.1016/j.jafrearsci.2013.03.003.

Ren JH, Zhang L, Ezekiel J, et al. Reservoir characteristics and productivity analysis of tight sand gas in Upper Paleozoic Ordos Basin China. J Nat Gas Sci Eng. 2014;19:244–50. doi:10.1016/j.jngse.2014.05.014.

Santosh M, Liu SJ, Tsunogae T, et al. Paleoproterozoic ultrahigh-temperature granulites in the North China Craton: implications for tectonic models on extreme crustal metamorphism. Precambr Res. 2012;222–223:77–106. doi:10.1016/j.precamres.2011.05.003.

Savage MH, Shackleton JR, Cooke LM, et al. Insights into fold growth using fold-related joint patterns and mechanical stratigraphy. J Struct Geol. 2010;32(10):1466–76. doi:10.1016/j.jsg.2010.09.004.

Schellart WP, Lister GS. The role of the East Asian active margin in widespread extensional and strike-slip deformation in East Asia. J Geol Soc. 2005;162:959–72.

Smart KJ, Ferrill DA, Morris AP. Impact of interlayer slip on fracture prediction from geomechanical models of fault-related folds. AAPG Bull. 2009;93(11):1447–58. doi:10.1306/05110909034.

Song SG, Niu YL, Su L, et al. Tectonics of the North Qilian orogen, NW China. Gondwana Res. 2013;23(4):1378–401. doi:10.1016/j.gr.2012.02.004.

Sun YJ, Dong SW, Zhang H, et al. Numerical investigation of the geodynamic mechanism for the late Jurassic deformation of the Ordos Block and surrounding orogenic belts. J Asian Earth Sci. 2014;114:623–33. doi:10.1016/j.jseaes.2014.08.033.

Tang X, Zhang JC, Shan YS, et al. Upper Paleozoic coal measures and unconventional natural gas systems of Ordos Basin, China. Geosci Front. 2012;3(6):863–73. doi:10.1016/j.gsf.2011.11.018.

Tong HM, Yin A. Reactivation tendency analysis: a theory for predicting the temporal evolution of preexisting weakness under uniform stress state. Tectonophysics. 2011;503:195–200. doi:10.1016/j.tecto.2011.02.012.

Velázquez SM, Vicente G, Elorza FJ. Intraplate stress state from finite element modeling: the southern border of the Spanish Central System. Tectonophysics. 2009;473(3–4):417–27. doi:10.1016/j.tecto.2009.03.024.

Wan TF. Intraplate deformation, tectonic stress field and their application for Eastern China in Meso-Cenozoic. Beijing: Geological Publishing House; 1994. p. 230.

Wan TF, Zeng HL. The distinctive characteristics of the Sino-Korean and the Yangtze plates. J Asian Earth Sci. 2002;20(8):881–8. doi:10.1016/S1367-9120(01)00068-2.

Wang J, Ye ZR, He JK. Three-dimensional mechanical modeling of large-scale crustal deformation in China constrained by the GPS velocity field. Tectonophysics. 2008;446(1–4):51–60. doi:10.1016/j.tecto.2007.11.006.

Wang LJ, Wang HC, Wang W, et al. Relation among three dimensional tectonic stress field, fracture and migration of oil and gas in oil field. Chin J Rock Mech Eng. 2004;23(23):4052–7 (in Chinese).

Wang CL, Zhou W, Li HB, et al. Characteristics and distribution of multiphase fractures in Yanchang Formation of Zhenjing Block in Ordos Basin, China. J Chengdu Univ Technol (Sci Technol Ed). 2014a;41(5):596–603 (in Chinese).

Wang CY, Sandvol E, Zhu L, et al. Lateral variation of crustal structure in the Ordos block and surrounding regions, North China, and its tectonic implications. Earth Planet Sci Lett. 2014b;387:198–211. doi:10.1016/j.epsl.2013.11.033.

Wang W, Wang DJ, Zhao B, et al. Horizontal crustal deformation in Chinese Mainland analyzed by CMONOC GPS data from 2009–2013. Geod Geodyn. 2014c;5(3):41–5. doi:10.3724/SP.J.1246.2014.03041.

Xie FR, Zhang HY, Cui XF, et al. The modern tectonic stress field and strong earthquakes in China. Recent Dev World Seismol. 2011;1:4–12 (in Chinese).

Xie HP, Ju Y, Li LY, et al. Energy mechanism of deformation and failure of rock masses. Chin J Rock Mech Eng. 2008;27(9):1729–40 (in Chinese).

Yang H, Deng XQ. Deposition of Yanchang Formation deep-water sandstone under the control of tectonic events in the Ordos

Basin. Pet Explor Dev. 2013;40(5):549–57. doi:10.1016/S1876-3804(13)60072-5.

Yang MH, Li L, Zhou J, et al. Segmentation and inversion of the Hangjinqi fault zone, the northern Ordos basin (North China). J Asian Earth Sci. 2013;70–71:64–78. doi:10.1016/j.jseaes.2013.03.004.

Yang SX, Huang LY, Xie FR, et al. Quantitative analysis of the shallow crustal tectonic stress field in China mainland based on in situ stress data. J Asian Earth Sci. 2014;85:154–62. doi:10.1016/j.jseaes.2014.01.022.

Yao JL, Deng XQ, Zhao YD, et al. Characteristics of tight oil in Triassic Yanchang Formation, Ordos Basin. Pet Explor Dev. 2013;40(2):161–9. doi:10.1016/S1876-3804(13)60019-1.

Yuan YS, Hu SB, Wang HJ, et al. Meso-Cenozoic tectonothermal evolution of Ordos basin, central China: insights from newly acquired vitrinite reflectance data and a revision of existing paleothermal indicator data. J Geodyn. 2007;44(1–2):33–46. doi:10.1016/j.jog.2006.12.002.

Zahm KC, Zahm CL, Bellian AJ. Integrated fracture prediction using sequence stratigraphy within a carbonate fault damage zone, Texas, USA. J Struct Geol. 2010;32(9):1363–74. doi:10.1016/j.jsg.2009.05.012.

Zeng LB, Zhao JY, Zhu SX, et al. Impact of rock anisotropy on fracture development. Prog Nat Sci. 2008;18:1403–8. doi:10.1016/j.pnsc.2008.05.016.

Zhang C, Zhou W, Xie RC, et al. Fracture prediction of the Ma 5_{1-2} tight carbonate reservoir in gentle structure zone, Daniudi Gas Field. J Northeast Pet Univ. 2014;38(3):9–17 (in Chinese).

Zhang YQ, Liao CZ, Shi W, et al. Jurassic deformation in and around the Ordos Basin, North China. Earth Sci Front. 2007;14(2):182–96. doi:10.1016/S1872-5791(07)60016-5.

Zhang YQ, Ma YS, Yang N, et al. Cenozoic extensional stress evolution in North China. J Geodyn. 2003;36(5):591–613. doi:10.1016/j.jog.2003.08.001.

Zhao GC, Sun M, Wilde SA, et al. Late Archean to Paleoproterozoic evolution of the North China Craton: key issues revisited. Precambr Res. 2005;136(2):177–202. doi:10.1016/j.precamres.2004.10.002.

Zhao WT, Hou GT, Hari KR. Two episodes of structural fractures and their stress field modeling in the Ordos Block, northern China. J Geodyn. 2016;97:7–21. doi:10.1016/j.jog.2016.02.005.

Zhao WT, Hou GT, Sun XW, et al. Influence of layer thickness and lithology on the fracture growth of clastic rock in east Kuqa. Geotecton Metallog. 2013;4:603–10 (in Chinese).

Zheng RG, Wu TR, Zhang W, et al. Late Paleozoic subduction system in the northern margin of the Alxa block, Altaids: geochronological and geochemical evidences from ophiolites. Gondwana Res. 2014;25(2):842–58. doi:10.1016/j.gr.2013.05.011.

Zhou XG, Zhang LY, Huang CJ, et al. Paleostress judgement of tectonic fractures in Chang 61 low permeable reservoir in Yanhewan area, Ordos Basin in main forming period. Geoscience. 2009a;23(5):843–51 (in Chinese).

Zhou XG, Zhang LY, Qu XF, et al. Characteristics and quantitative prediction of distribution laws of tectonic fractures of low-permeability reservoirs in Yanhewan area. Acta Pet Sin. 2009b;30(2):195–200 (in Chinese).

Zhu SB, Shi YL. Estimation of GPS strain rate and its error analysis in the Chinese continent. J Asian Earth Sci. 2011;40(1):351–62. doi:10.1016/j.jseaes.2010.06.007.

11

Sensitivity-based upscaling for history matching of reservoir models

Saad Mehmood[1] · Abeeb A. Awotunde[2]

Abstract Simulation of reservoir flow processes at the finest scale is computationally expensive and in some cases impractical. Consequently, upscaling of several fine-scale grid blocks into fewer coarse-scale grids has become an integral part of reservoir simulation for most reservoirs. This is because as the number of grid blocks increases, the number of flow equations increases and this increases, in large proportion, the time required for solving flow problems. Although we can adopt parallel computation to share the load, a large number of grid blocks still pose significant computational challenges. Thus, upscaling acts as a bridge between the reservoir scale and the simulation scale. However as the upscaling ratio is increased, the accuracy of the numerical simulation is reduced; hence, there is a need to keep a balance between the two. In this work, we present a sensitivity-based upscaling technique that is applicable during history matching. This method involves partial homogenization of the reservoir model based on the model reduction pattern obtained from analysis of the sensitivity matrix. The technique is based on wavelet transformation and reduction of the data and model spaces as presented in the 2Dwp–wk approach. In the 2Dwp–wk approach, a set of wavelets of measured data is first selected and then a reduced model space composed of important wavelets is gradually built during the first few iterations of nonlinear regression. The building of the reduced model space is done by thresholding the full wavelet sensitivity matrix. The pattern of permeability distribution in the reservoir resulting from the thresholding of the full wavelet sensitivity matrix is used to determine the neighboring grids that are upscaled. In essence, neighboring grid blocks having the same permeability values due to model space reduction are combined into a single grid block in the simulation model, thus integrating upscaling with wavelet multiscale inverse modeling. We apply the method to estimate the parameters of two synthetic reservoirs. The history matching results obtained using this sensitivity-based upscaling are in very close agreement with the match provided by fine-scale inverse analysis. The reliability of the technique is evaluated using various scenarios and almost all the cases considered have shown very good results. The technique speeds up the history matching process without seriously compromising the accuracy of the estimates.

Keywords Upscaling · Inverse analysis · History matching · Sensitivity · Wavelets

1 Introduction

Upscaling is the process of reducing a large number of the fine-scale grid blocks to a smaller number of coarse-scale grid blocks. This is required because it is often impractical to perform simulation at finest scale of the reservoir. Therefore, upscaling is one of the most important components of reservoir simulation. The last few decades have seen significant advancements in upscaling which include development of single-phase and multiphase upscaling as well as upscaling in the near wellbore and away from wellbore regions. Single-phase upscaling involves the

✉ Saad Mehmood
mehmood.saad@hotmail.com

[1] United Energy Pakistan, Bahria Complex-1, M.T. Khan Road, Karachi 74000, Sindh, Pakistan

[2] Department of Petroleum Engineering, King Fahd University of Petroleum and Minerals, Dhahran 31261, Saudi Arabia

Edited by Yan-Hua Sun

upscaling of permeability distribution only. The technique is simple and can be used for structurally complex reservoirs but it neglects the multiphase flow effects (Durlofsky 1991; Ringrose 2007). In multiphase flow upscaling, relative permeability curves are also upscaled in addition to absolute permeability upscaling. This approach is computationally expensive and as such its use is limited to simple reservoir models (Ekrann and Dale 1992; Ringrose 2007).

One method of upscaling involves averaging the parameters and imputing the averaged values directly into the simulation flow grid. Most of the averaging techniques (arithmetic, harmonic, geometric, power law, pressure solver) are only appropriate under the circumstances of perfectly layered or heterogeneous distributions that are perfectly random and seldom observed in realistic reservoir descriptions. Another method of upscaling is an averaging technique that first computes the lower and upper bounds of the effective properties, based on geology, and then uses a new correlation and scaling technique to estimate the effective properties for the upscaled grid (Li et al. 2001). Purely local upscaling methods consider only those fine-scale grids that are combined in the target coarse-scale grid (Durlofsky 1991; King and Mansfield 1999). The hydraulic conductivity upscaling method (Wen and Gómez-Hernández 1996) involves upscaling of hydraulic conductivities at the scale of measurements to a coarser grid of block conductivity tensors. The extended local procedure includes few of the adjacent grids in the local problems (Gómez-Hernández and Journel 1994; Wu et al. 2002). In global upscaling methods, the flow solution utilized to calculate the upscaled parameters is performed over the entire domain (White and Horne 1987; Pickup et al. 1992; Holden and Nielson 2000). This technique can provide high level of accuracy, but it has a disadvantage of requiring global fine-scale solutions.

An adaptive local–global procedure has also been proposed for multiphase near-well problems (Nakashima 2009). Adaptive means that the actual boundary conditions are applied for global coarse-scale simulations rather than the generic set of boundary conditions. The adaptive local–global upscaling technique involves global coarse-scale simulation with initial estimates for wellblock parameters which provides the coarse-block pressure and saturations. The resultant pressure and saturation distributions are then interpolated onto the local well model to obtain boundary conditions for the near-well upscaling computations.

History matching has long been used to estimate reservoir parameters from dynamic production history data. However, a limitation of this procedure is that it is often the case that the information content of the production histories is not enough to resolve the model parameters at the finest scale. Thus, different methods of model space reduction have been proposed in the literature to reduce the number

of model parameters to be estimated from the production history, thereby reducing the non-uniqueness associated with the inverse modeling. One such model reduction method is the wavelet multiscale inverse analysis (Lu and Horne 2000; Sahni and Horne 2005, 2006a, b; Awotunde and Horne 2011a, b, 2012, 2013). The various methods of model space reduction often automatically produce some level of smoothening (homogenization) of the reservoir model parameters such as grid block permeabilities. This smoothening creates a scenario in which several adjacent grid blocks have the same permeability values. However, during such history matching procedures, forward simulation runs are still performed at the finest scale (Lu and Horne 2000; Sahni and Horne 2005; Awotunde and Horne 2012, 2013). In this way, a huge amount of time is spent on forward simulations. Another permeability upscaling procedure using the fast marching method was implemented by Sharifi and Kelkar (2014). The purpose of this work is to utilize the pattern of smoothening in the permeability field created by the model space reduction during history matching, to upscale the forward simulation model with an ultimate goal of reducing the total time required for history matching.

We propose and evaluate an upscaling procedure based on wavelet sensitivity thresholding. Sensitivity-based thresholding has been reported in the literature to reduce model parameter space during history matching (Sahni and Horne 2006a, b; Awotunde and Horne 2012). Sensitivity computation is required for computing the Hessian matrix in the Gauss–Newton and LM algorithms (Gill et al. 1981; Nocedal and Wright 2006; Griva et al. 2009). In addition, a sensitivity matrix has been used to reduce the model space. One of the features of such model reduction is the emergence of several neighboring grids with similar values of the reservoir model parameter. For example, in the 2Dwp–wk approach presented in Awotunde and Horne (2013), the thresholding of wavelet sensitivity matrix computed in early iterations of the nonlinear regression is used to determine the reduced model space.

The back-transformation of the model space coefficients into the real permeability field would identify the regions of homogeneity in the upscaled reservoir model. Combining the grid blocks based on the pattern obtained from the model reduction would result in a coarse-scale unstructured grid system. The method is expected to be more consistent as it predetermines the areas of the reservoir with homogeneous permeability distribution based on sensitivity analysis. To improve the accuracy of simulation results, grid blocks having wells completed in them were not combined with any neighboring grid blocks. Further, two scenarios were tested; one in which all neighboring grid blocks with equal permeabilities but without any well were combined, and the second in which wellblocks and their neighbors were not combined. The second scenario was done to improve the accuracy of the variables obtained from the wells. If the neighbors of a wellblock are much larger in size than the wellblock, the accuracy of variables such as wellbore pressure and water cut may be compromised.

To properly investigate the effectiveness of the sensitivity-based upscaling approach, the methodology was applied to history match data from two synthetic reservoir models. The reliability of the technique was evaluated by comparing the results of history matching performed using coarse-scale forward simulations to those obtained from history matching performed using a fine-scale forward simulation model. In both, the history matching was used to obtain a reduced model parameter space.

2 Reservoir parameter estimation

The process of determining the spatial distribution of reservoir properties, particularly porosity and permeability, is known as reservoir characterization. History matching is the process of modifying the reservoir model by fitting simulation results to actual field data. Originally, history matching was done manually, then progress was made and the industry shifted to automated history matching. Automated history matching often relies on nonlinear regression of the observed dataset. Nonlinear regression comprises the class of inverse analysis techniques used in minimizing the l_2—norm of errors between modeled data and the measured data. In a nonlinear regression, the objective function is often given by

$$\Phi(\vec{\alpha}) = \frac{1}{2\xi} \left\| \vec{d}_{\text{cal}} - \vec{d}_{\text{meas}} \right\|_2^2, \tag{1}$$

where $\vec{\alpha}$ is the vector of the unknown parameters (to be estimated by optimization or nonlinear regression), $\vec{d}_{\text{cal}}$ is the vector of modeled pressure data, $\vec{d}_{\text{meas}}$ is the vector of measured data and ξ is a scaling factor. Most nonlinear regression algorithms follow the Newton–Raphson approach in which the parameters of the model are iteratively estimated by repeatedly finding an optimum direction (first-order optimality) and step-length with which to move the current iterate. This is achieved by computing the gradient $\vec{g}$ and the Hessian H (or its approximation) at each iteration of the nonlinear regression. The optimum direction of descent $\delta\vec{\alpha}$, in the minimization algorithm, is then computed from

$$H\delta\vec{\alpha} = -\vec{g} \tag{2}$$

and the subsequent iterate is

$$\vec{\alpha}^{\kappa+1} = \vec{\alpha}^\kappa + \delta\vec{\alpha}^\kappa, \tag{3}$$

where κ represents the iteration index. In the standard Newton–Raphson approach, the exact Hessian is computed and used to calculate the direction of descent. Although the Newton method gives fast convergence (fewer number of iterations) relative to other gradient-based methods, the computation of the Hessian matrix can be time-consuming when the problem dimension is large. Also, a simple analytic expression for the first and/or second derivative of the objective function may not be obtainable. Thus, an approximation to the Hessian is often used and this forms the basis of the different nonlinear regression algorithms such as the steepest descent, the conjugate gradient, the quasi-Newton methods, the Gauss–Newton approach, and the Levenberg–Marquardt (LM) method (Levenberg 1944; Marquardt 1963). For small- and medium-sized problems, the LM approach is often the nonlinear regression technique of choice. Thus, we use the LM approach to estimate the parameters of the well test problems considered and compare its results to the results from the global optimization techniques. In the LM approach, the Hessian matrix is approximated by

$$H = S^{\mathrm{T}}S + \dot{\lambda}I, \tag{4}$$

where S is the sensitivity matrix computed from

$$S = \frac{\partial \vec{d}_{\text{cal}}}{\partial \vec{\alpha}}, \tag{5}$$

and $\dot{\lambda}$ is a small positive number that ensures the algorithm remains stable.

3 Reservoir simulator

The reservoir model is usually solved by a numerical approach due to its complex nature. The fine-scale simulator used in this work is a three-dimensional, oil–water, black oil, finite-difference reservoir simulator. The upscaling simulator is also purposely built for this work using the same governing equations for the reservoir model. Both simulators have a built-in functionality of computing sensitivity of data to reservoir parameters using the Adjoint-State approach.

3.1 Fine-scale simulator

A three-dimensional reservoir system with a total number of M grid blocks is considered with the total number of wells to be N_{well}. The general residual equation can be given as

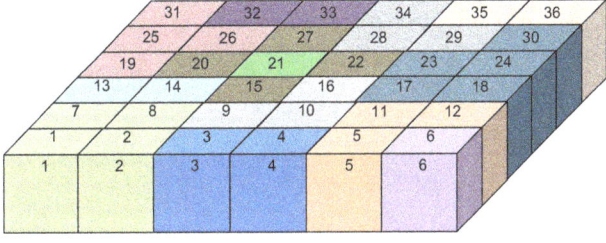

Fig. 1 A reservoir system with some homogeneous patches

$$\vec{f}^{n+1}\left(\vec{u}^{n+1}, \vec{u}^n, \vec{v}, \Delta t; \vec{\alpha}\right) = \vec{0}, \qquad (6)$$

where $\vec{f}$ represents the vector of residual for flow equations; $\vec{\alpha}$ is the reservoir parameters; vector $\vec{v}$ consists of known reservoir properties and vector $\vec{u}$ contains state variables and can be written as

$$\vec{u} = [p_{o,1}, S_{w,1}, \ldots, p_{o,M}, S_{w,M}, p_{wf,1}, \ldots, p_{wf,N_{well}}]^{T}, \quad (7)$$

where p_o is the pressure of the oil phase, S_w is the water saturation, p_{wf} is the wellbore pressure $\vec{f}_{blk}^{n+1}$ contains the residual due to flow in and out of reservoir grid blocks and is given as

$$\vec{f}_{blk}^{n+1} = \left[f_{w,1}^{n+1}, f_{o,1}^{n+1}, f_{w,2}^{n+1}, f_{o,2}^{n+1}, \ldots, f_{w,M}^{n+1}, f_{o,M}^{n+1}\right]^{T} \qquad (8)$$

whereas $\vec{f}_{well}^{n+1}$ represents the residual due to flow into or out of the wells in the reservoir and can be presented as

$$\vec{f}_{well}^{n+1} = \left[f_{well,1}^{n+1}, f_{well,2}^{n+1}, \ldots, f_{well,N_{well}}^{n+1}\right]^{T}. \qquad (9)$$

$\vec{f}_{blk}^{n+1}$ and $\vec{f}_{well}^{n+1}$ both combine to form $\vec{f}^{n+1}$ as

$$\vec{f}^{n+1} = \begin{bmatrix} \vec{f}_{blk}^{n+1} \\ \vec{f}_{well}^{n+1} \end{bmatrix}, \qquad (10)$$

Now, Eq. (8) indicates that $\vec{f}_{blk}^{n+1}$ comprises the residuals of the two phases existing in the reservoir system which can be given as

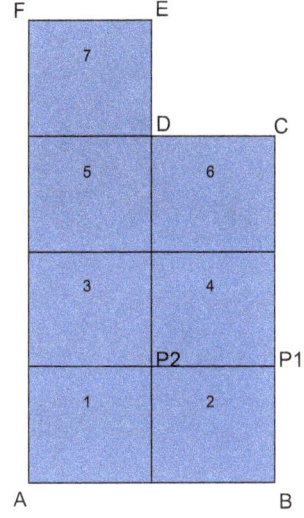

Fig. 2 Example of an upscaled grid block

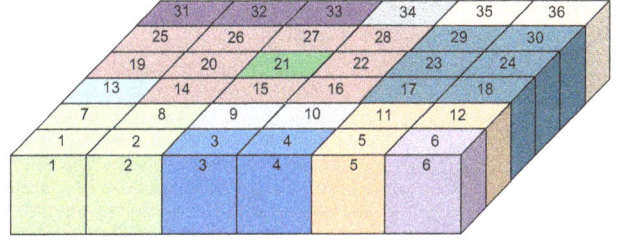

Fig. 3 An unstructured grid system showing a large grid enveloping a small grid

$$\vec{f}_w\left(\vec{p}_o^{n+1}, S_w^{n+1}, \vec{p}_{wf}^{n+1}, \vec{p}_o^n, S_w^n, \vec{\phi}_{ini}, \Delta t; \vec{k}\right) = \vec{0}, \qquad (11)$$

and

$$\vec{f}_o^{n+1}\left(\vec{p}_o^{n+1}, S_w^{n+1}, \vec{p}_{wf}^{n+1}, \vec{p}_o^n, S_w^n, \vec{\phi}_{ini}, \Delta t; \vec{k}\right) = \vec{0}. \qquad (12)$$

whereas the well residual for $\vec{f}_{well}^{n+1}$ of Eq. (9) can be written as

$$\vec{f}_{well}^{n+1}\left(\vec{p}_o^{n+1}, S_w^{n+1}, \vec{p}_{wf}^{n+1}, \vec{p}_o^n, S_w^n, \vec{\phi}_{ini}, \Delta t; \vec{k}\right) = \vec{0}, \qquad (13)$$

Table 1 Combination of fine-scale grid blocks in the upscaled system	Upscaled grid blocks	Fine-scale grids merged	Upscaled grid blocks	Fine-scale grids merged
	1	1, 2, 7, 8	9	19, 25, 26, 31
	2	3, 4	10	20
	3	5, 11, 12	11	21 (well)
	4	6	12	22
	5	9, 10, 16	13	27
	6	13, 14	14	28, 29, 34
	7	15	15	32, 33
	8	17, 18, 23, 24, 30	16	35, 36

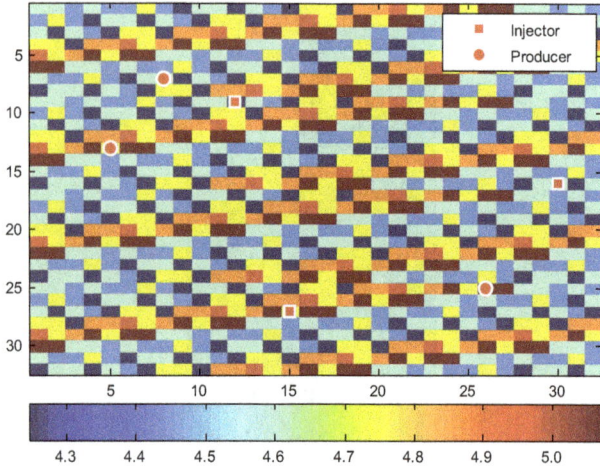

Fig. 4 Log permeability distribution and well locations for the 32×32 reservoir model

where $\vec{\phi}_{\mathrm{ini}}$ represents initial porosity distribution and $\vec{k}$ represents the permeability distribution in the reservoir in all the residual equation presented above.

Consider the constraint of the total production rate,

$$f_{\mathrm{well},i}^{n+1} = \sum_{\mathrm{ph=o,w}} \sum_{j}^{N\mathrm{comp}} q_{\mathrm{ph},j}^{\mathrm{well}} - q_{\mathrm{t,i}} = 0. \qquad (14)$$

In Eq. (14), $q_{\mathrm{ph},j}^{\mathrm{well}}$ represents the flow rate of the phases (oil or water, denoted by ph) at the jth completion and it can be defined as

$$q_{\mathrm{ph},j}^{\mathrm{well}} = \lambda_{\mathrm{ph},j}^{n+1} W I_j \left(p_{\mathrm{ph},j}^{n+1} - p_{\mathrm{wf}}^{n+1} - \gamma_{\mathrm{ph},j}^{n+1} \Delta z_j \right) \qquad (15)$$

There is no p_c (capillary pressure) in Eq. (15) because p_{ph} represents both p_o and p_w, so the capillary pressure will be incorporated in p_w in the case of the water phase.

The mobility ratio and specific gravity of any phase ph at the completion j are represented by $\lambda_{\mathrm{ph},j}^{n+1}$ and $\gamma_{\mathrm{ph},j}^{n+1}$, respectively, while WI_j denotes the well index at the jth completion. The Newton–Raphson iterative method is used in order to solve the nonlinear system of equations at every iteration; so we have at any iteration κ,

$$J^{n+1,\kappa} \delta \vec{u}^{n+1,\kappa} = -\vec{f}^{n+1,\kappa}, \qquad (16)$$

where $J^{n+1,\kappa}$ is known as the Jacobian matrix and can be written as

$$J^{n+1,\kappa} = \frac{\partial \vec{f}^{n+1,\kappa}}{\partial \vec{u}^{n+1,\kappa}}. \qquad (17)$$

the solution is then updated as

$$\vec{u}^{n+1,\kappa+1} = \vec{u}^{n+1,\kappa} + \delta \vec{u}^{n+1,\kappa}. \qquad (18)$$

3.2 Upscaled simulator

The basic governing equations in the upscaled simulator are the same as those used in the fine-scale simulator. The principal difference is in the manner of calculating the transmissibility between the upscaled grid blocks. The central idea in this work involves the upscaling of the reservoir grid blocks based on homogenization of the system parameters. This would result in upscaled grid blocks that have different structures. Also, the neighboring grid blocks will not be structured as in the rectangular grid system. Thus, we need to find an appropriate way of computing the transmissibility between the interacting grid blocks. The detailed description of the upscaling procedure and its calculations is presented in Sect. 4.

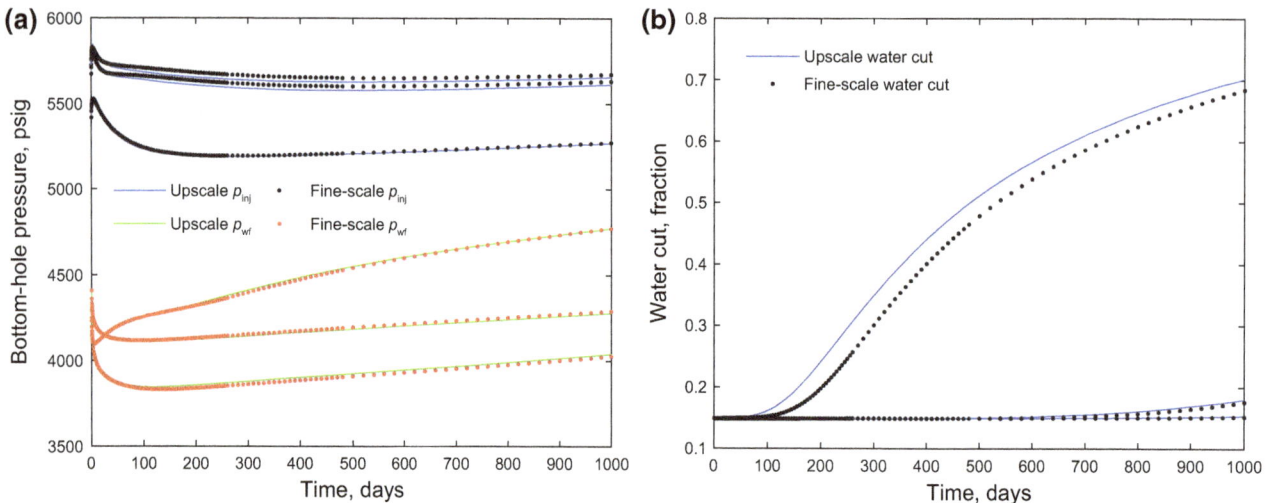

Fig. 5 Production data from the fine-scale and upscale reservoir models for the 32×32 system. **a** Bottom-hole pressures. **b** Water cut (Constraint 1)

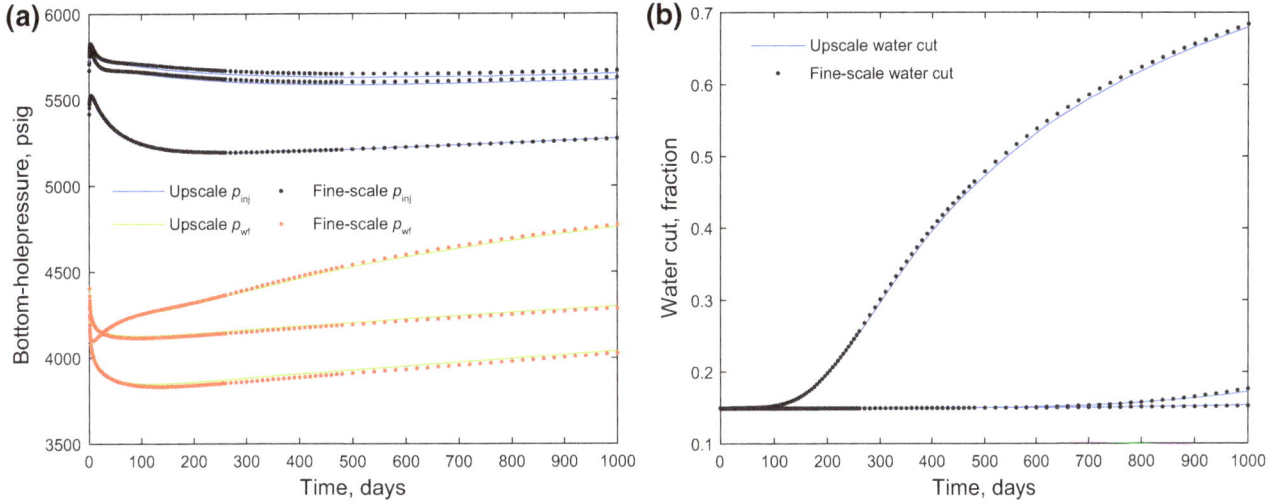

Fig. 6 Production data from the fine-scale and upscale reservoir models for the 32×32 system. **a** Bottom-hole pressures. **b** Water cut (Constraint 2)

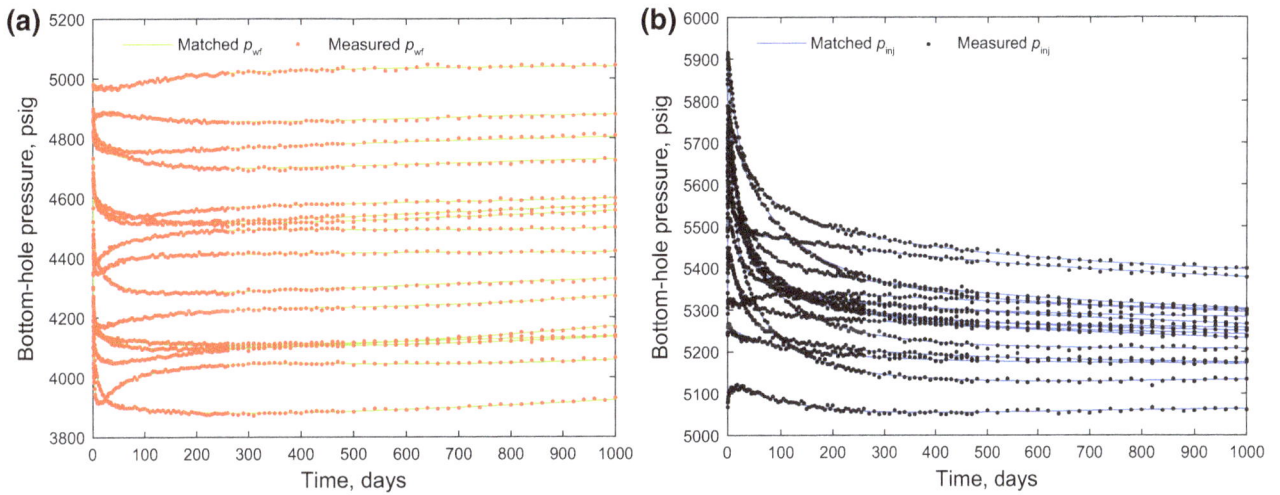

Fig. 7 Match of production data for the 64×64 reservoir system (wavelet fraction 0.60, no upscaling performed during history matching). **a** Bottom-hole pressures in producers. **b** Bottom-hole pressures in injectors

4 Upscaling based on homogenization of reservoir model during history matching

This work involves upscaling of the reservoir system based on the sensitivity of production data to model parameters. Sensitivity computation is a part of some inverse analysis methods such as the Gauss–Newton and the LM algorithms. Sensitivities provide us information on the grid block parameters that have similar effects on calculated production data. During history matching and at any particular nonlinear iteration to estimate the unknown reservoir parameters (e.g., in the LM approach), adjacent grid blocks that exhibit almost similar values of sensitivities are considered as a homogenous patch and may then be

combined to form an upscaled grid block. The pattern of upscaling is thus based on the pattern of homogenous patches obtained through sensitivity analysis during history matching. The adjacent grid blocks whose parameters ($\ln k$) have almost similar effects on the production data are expected to have similar permeability trends and are merged to form an upscaled grid block. The combination of fine-scale grid blocks into larger ones, based on permeability distribution, ultimately reduces the number of grid blocks for simulation and this in turn reduces the required computational resources. The upscaling of grid blocks with similar permeability values is performed subject to one of two different constraints. The first constraint involves ensuring that grid blocks having wells in them (wellblocks)

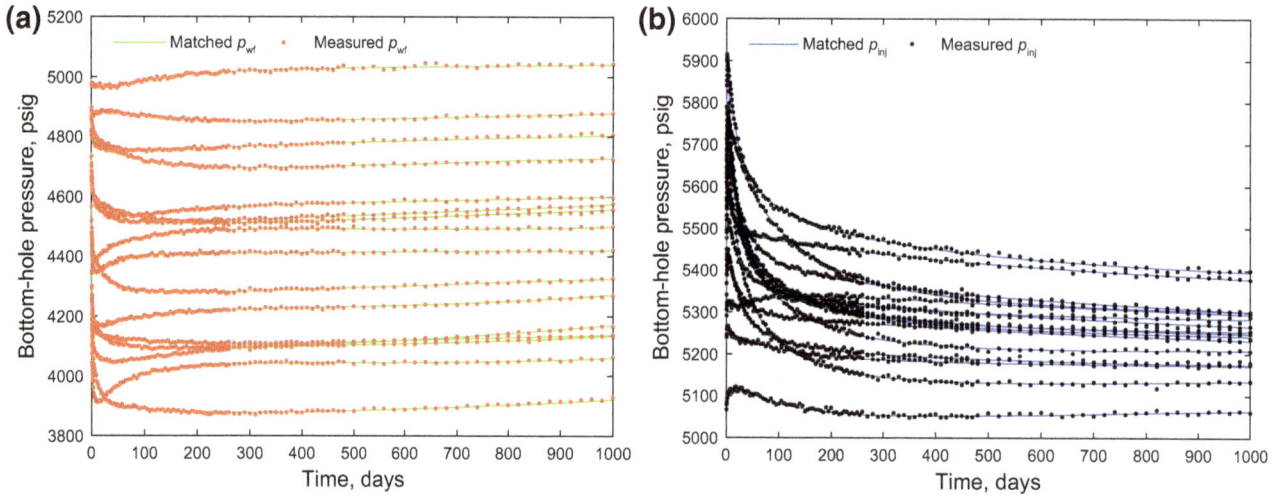

Fig. 8 Match of production data for the 64×64 reservoir system (wavelet fraction 0.40, no upscaling performed during history matching). **a** Bottom-hole pressures in producers. **b** Bottom-hole pressures in injectors

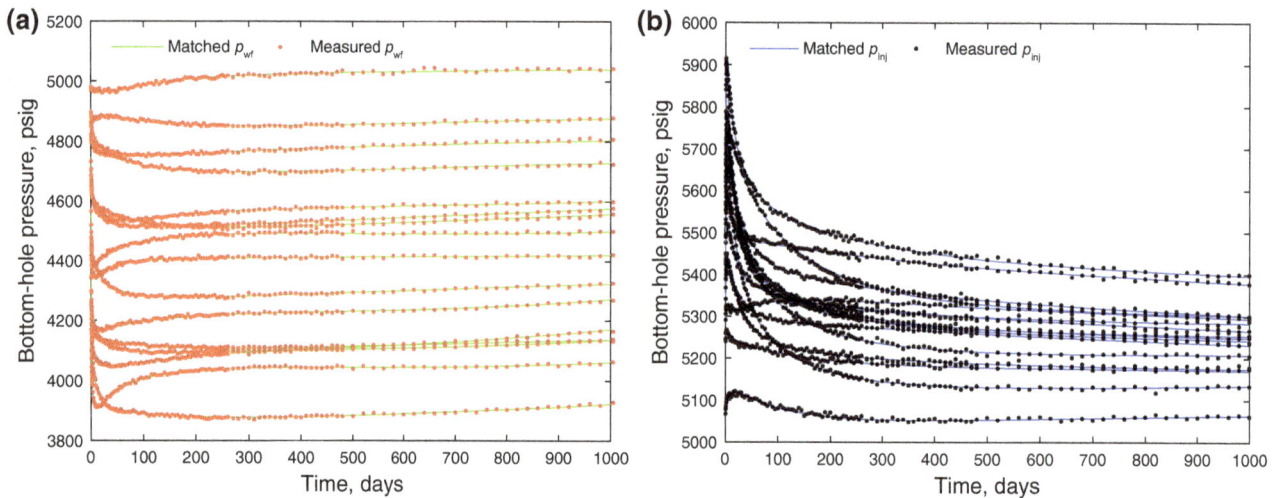

Fig. 9 Match of production data for the 64×64 reservoir system (wavelet fraction 0.25, no upscaling performed during history matching). **a** Bottom-hole pressures in producers. **b** Bottom-hole pressures in injectors

are not combined with any other grid. The second constraint involves ensuring that any wellblock and the grid blocks adjacent to it are not combined with one another or with any other block in their neighborhood.

Figure 1 illustrates an example of a 6×6 reservoir system with the implementation of the second constraint. The figure shows that fine-scale grid blocks are combined based on homogeneity of the system. The grid blocks having the same color have the same permeability value; therefore, these fine-scale grid blocks would be combined to form the upscaled system. The transformation from fine-scale to upscale grid blocks is explained using Table 1.

Table 1 shows the fine-scale grid blocks in the 6×6 reservoir system that are combined due to homogeneity to

form the upscaled system. Figure 1 and Table 1 also illustrate that the grid block having a well in it (i.e., Grid block 21) and the grids adjacent to this wellblock (Grid blocks 15, 20, 22, and 27) are not combined with any other grid block. However, because we combine the grid blocks based on permeability distribution, the resulting upscaled grid blocks do not necessarily form well-defined shapes, resulting often in unstructured gridding systems. This is observed in Fig. 1. This poses a challenge in the calculation of transmissibilities between any pairs of grid blocks. The transmissibility is performed by first locating the centroid of each upscaled grid block. Thus, the algorithm, used to upscale the fine-scale system, based on the presence of homogeneous patches is described below:

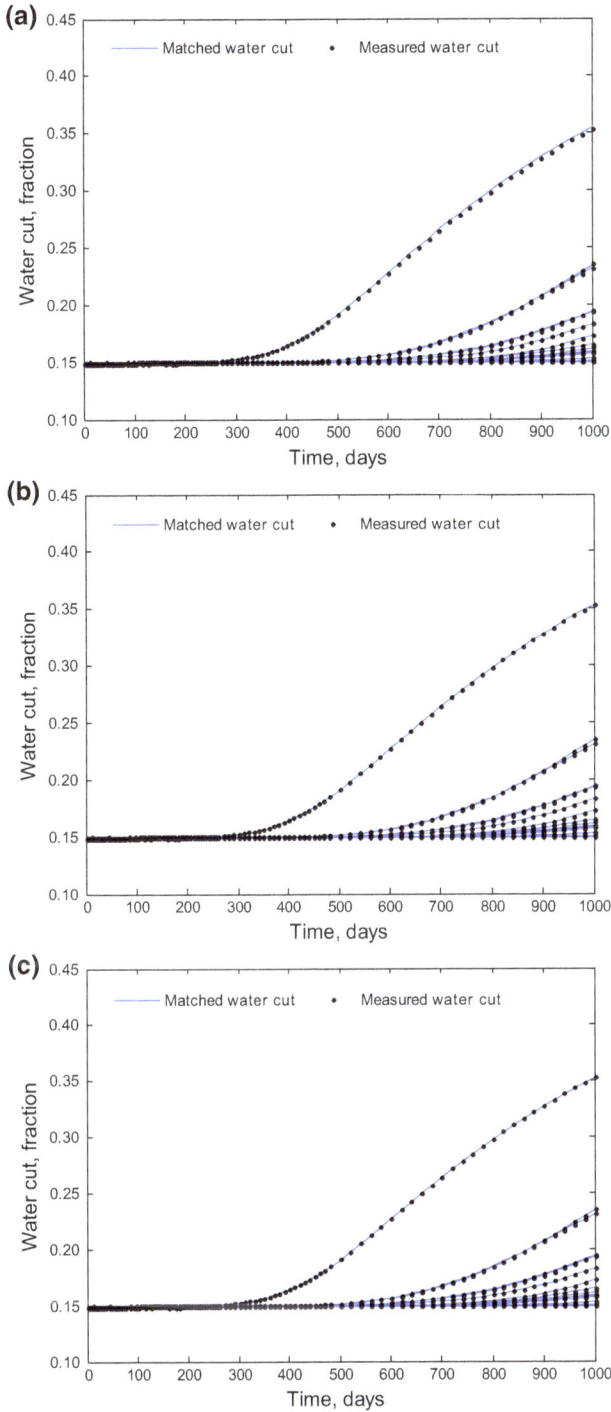

(a)

(b)

(c)

Fig. 10 Match to water cut in all producers for the 64×64 reservoir system with no upscaling performed during history matching. **a** Wavelet fraction 0.60. **b** Wavelet fraction 0.40. **c** Wavelet fraction 0.25

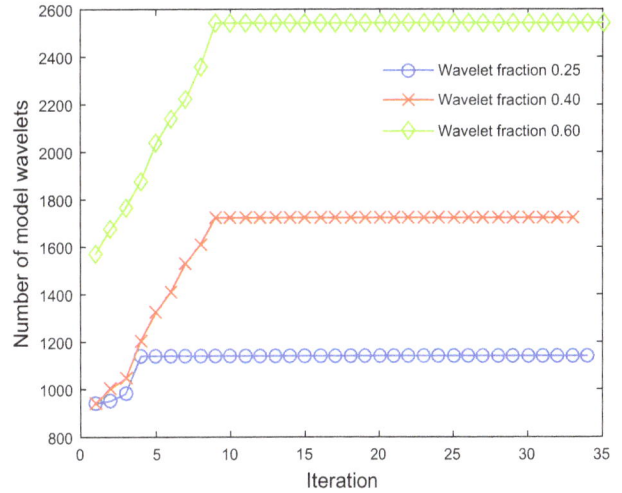

Fig. 11 Trend of wavelet coefficients for the 64×64 reservoir system (all fractions, no upscaling performed during history matching)

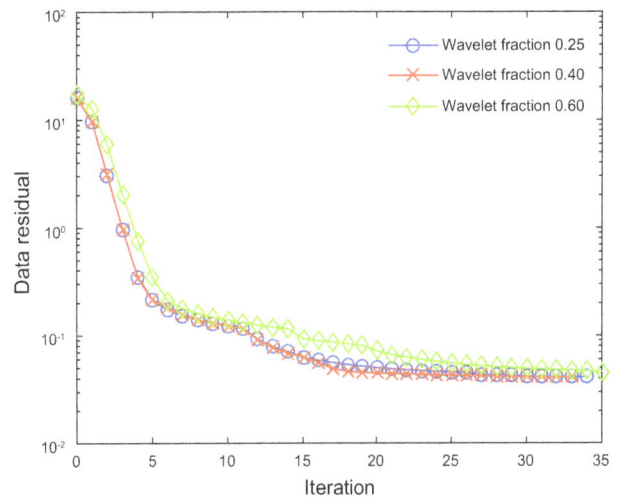

Fig. 12 Data residual for the 64×64 reservoir system (all fractions, no upscaling performed during history matching)

(1) The fine-scale grid blocks are combined to form the upscaled grid blocks based on the presence of homogenous patches in the permeability distribution.

(2) The centroid of each upscaled grid block is calculated.

(3) All adjacent grids to every upscaled grid block are assembled and stored; to be used in transmissibility calculations.

(4) At each simulation time step, the Newton–Raphson iteration is performed.

(5) During each Newton-iteration of each simulation time-step, the transmissibility between pairs of upscaled grid blocks is calculated based on their centroids.

(6) The procedure is repeated in each iteration, and each time step.

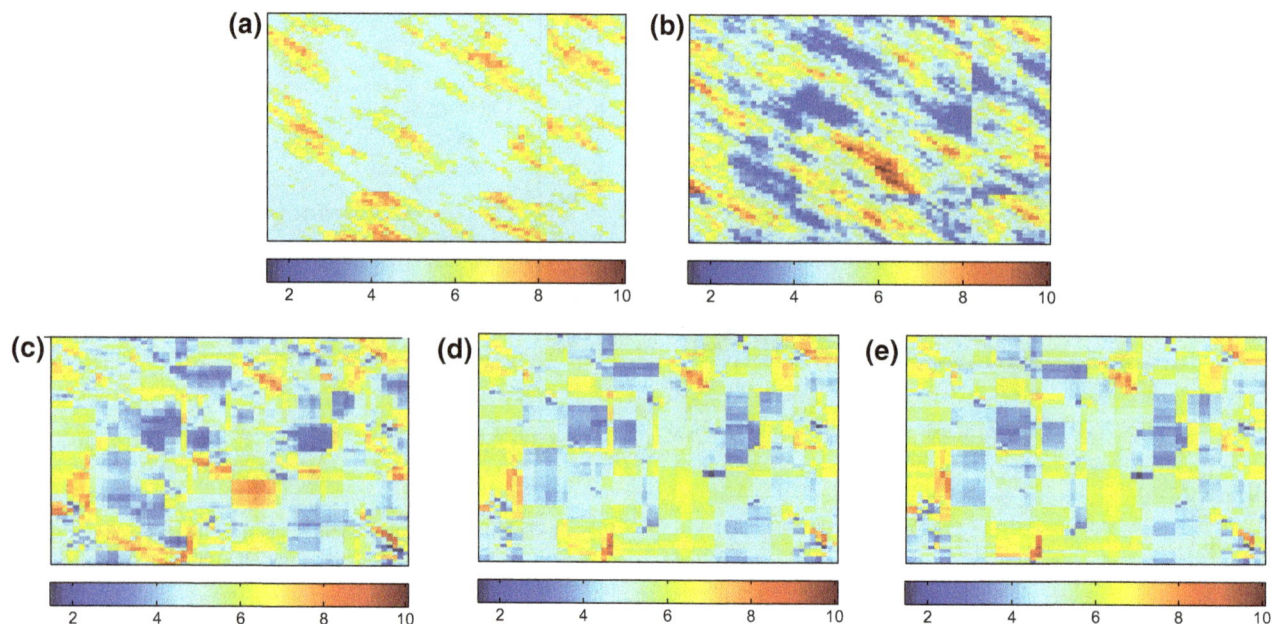

Fig. 13 Log permeability distribution for the 64 × 64 reservoir system. **a** True. **b** Initial guess. **c** Estimate of wavelet fraction 0.60. **d** Estimate of wavelet fraction 0.40. **e** Estimate of wavelet fraction 0.25 (no upscaling performed during history matching)

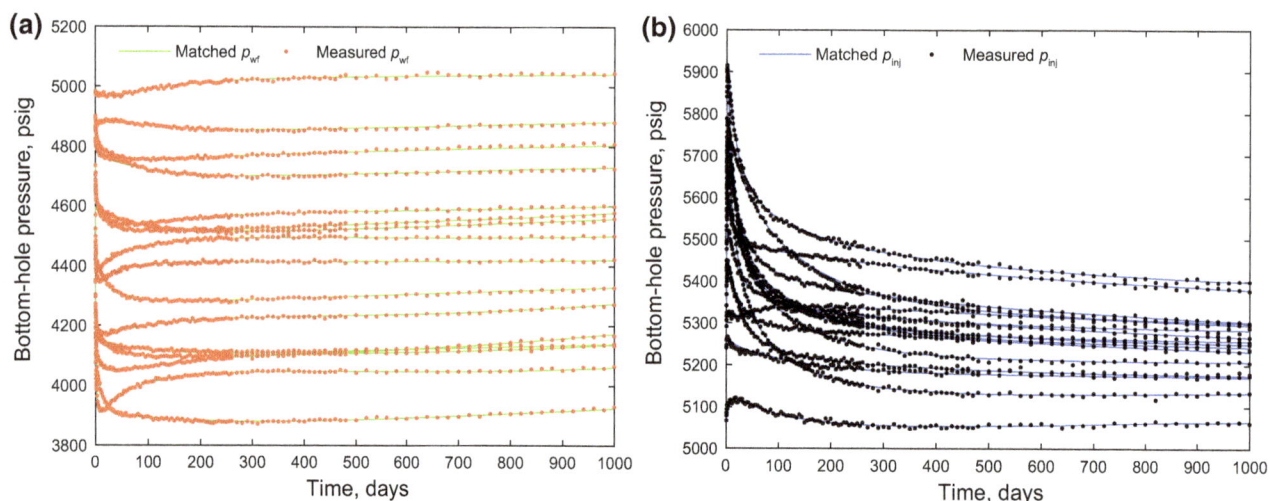

Fig. 14 Match of production data for the 64 × 64 reservoir system (wavelet fraction 0.60, upscaling performed during history matching). **a** Bottom-hole pressures in producers. **b** Bottom-hole pressures in injectors

5 Transmissibility and centroid calculations for the upscaled system

Transmissibility is the property calculated at the interface of the grid blocks. However, the properties used in its calculation are known at the grid center. In a structured gridding system in rectangular coordinates, the size of a grid block and its center are appropriately defined. However, in the type of upscaled system shown in Fig. 1, the shape of the resulting upscaled grid blocks may not be regular, and for calculating transmissibility we need to

define the centers of these grids. Thus, the centroids of the upscaled grid blocks are evaluated and used for transmissibility computation. In a two-dimensional system, the faces of the resulting upscaled grid blocks are polygons. Figure 2 shows one of such grid blocks. In order to calculate the centroid of any object, its vertices should be arranged in either the clockwise or counterclockwise direction. The first step is to define the vertices of the new grid block. An algorithm is developed that finds the vertices of an upscaled grid from all the vertices of fine-scale grids contained in it. This is better

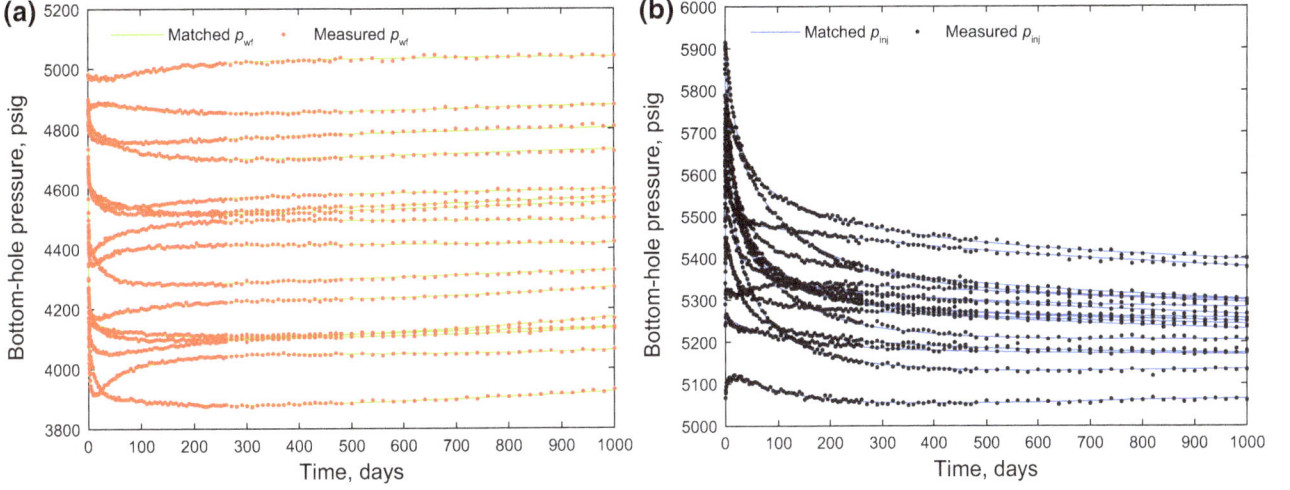

Fig. 15 Match of production data for the 64 × 64 reservoir system (wavelet fraction 0.40, upscaling performed during history matching). **a** Bottom-hole pressures in producers. **b** Bottom-hole pressures in injectors

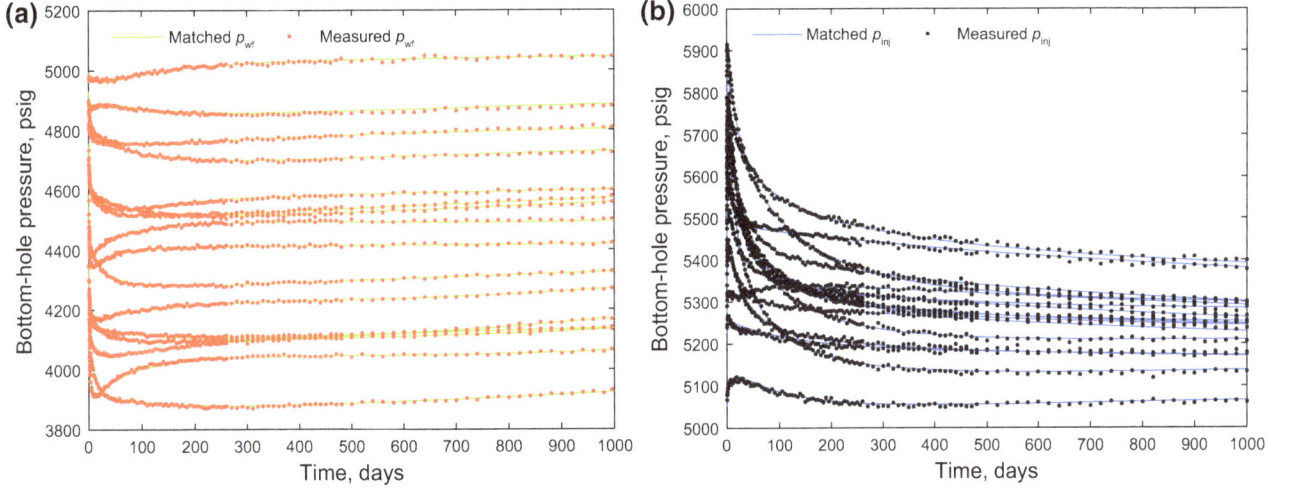

Fig. 16 Match of production data for the 64 × 64 reservoir system (wavelet fraction 0.25, upscaling performed during history matching). **a** Bottom-hole pressures in producers. **b** Bottom-hole pressures in injectors

understood from Fig. 2 in which vertices A, B, C, D, E, and F are the vertices of an upscaled grid, and the algorithm developed in this work finds these vertices. Vertices P1, P2, etc. of the fine-scale grids are excluded as these are not vertices of the upscaled grid. Once these vertices have been found, they are arranged in a clockwise or counter-clockwise order.

A separate algorithm is written to arrange the vertices. This algorithm is based on the structure of x- and y-coordinates and it arranges the vertices in counter-clockwise order. Subsequently, the arranged points are used to determine the centroid using the following equations:

$$C_x = \frac{1}{6A} \sum_{i=0}^{n-1} (x_i + x_{i+1})(x_i y_{i+1} - x_{i+1} y_i) \qquad (19)$$

$$C_y = \frac{1}{6A} \sum_{i=0}^{n-1} (y_i + y_{i+1})(x_i y_{i+1} - x_{i+1} y_i) \qquad (20)$$

where C_x is the x-coordinate of the centroid; C_y is the y-coordinate of the centroid; A is the area of the polygon and is given as

$$A = \frac{1}{2} \sum_{i=0}^{n-1} (x_i y_{i+1} - x_{i+1} y_i) \qquad (21)$$

6 Usefulness and limitations of the sensitivity-based upscaling

The method proposed in this work can be performed only when there is a sensitivity matrix to indicate the response of the well data to changes in grid block parameters. Thus,

(a)

(b)

(c)

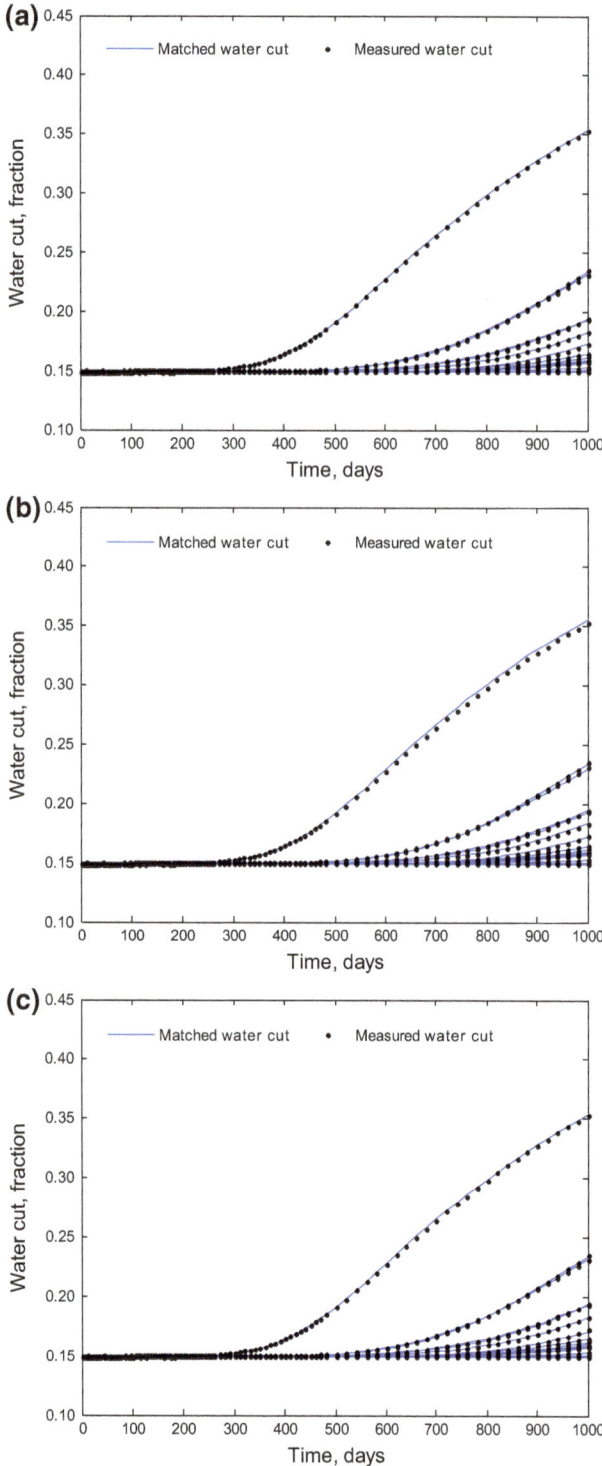

Fig. 17 Match to water cut in all producers for the 64 × 64 reservoir system with upscaling of grid blocks during history matching. **a** Wavelet fraction 0.60. **b** Wavelet fraction 0.40. **c** Wavelet fraction 0.25

the method is useful during history matching only if the method of history matching used involves the computation of a sensitivity matrix at each iteration. For instance, if the conjugate gradient method or the quasi-Newton method is

used, sensitivities are not computed and such sensitivity-based upscaling of grid blocks is not possible. Also, we observed that the method has a severe limitation in the acceptable pattern of homogeneous patches. The problem that arises here is that sensitivity-based upscaling often results in unstructured grid blocks. In the case of such unstructured gridding, the centroid of all grid blocks in the new system must be located. To compute the centroid of a block, all the vertices of the block must form a continuous connection. That is, there must not be a void inside the block. However, in certain cases of unstructured gridding, cases arise in which one large grid block entirely envelopes a smaller grid block so that the large grid block has an opening within it and thus its centroid cannot be computed. A scenario of this type is illustrated in Fig. 3. In this figure, a large-grid formed by combining smaller blocks 14, 15, 16, 19, 20, 22, 25, 26, 27, and 28 (all shown in pink color) envelops a smaller grid, block 21 (shown in green color). In this case, the centroid of the larger grid cannot be computed. This situation is more likely to occur when the reduction in the model size is large. As a result, the model reduction that was achieved in this work is limited by this problem.

7 Comparison of fine-scale and upscale forward simulation

First, we considered a 32 × 32 reservoir model with such permeability distributions that result in some homogenous regions in the reservoir. This reservoir (Fig. 4) has three producers and three injectors. The reservoir was upscaled based on the homogeneous patches indicated by its inherent permeability distribution. Then a flow simulation was performed on both the fine-scale and the upscaled reservoir model. In upscaling the reservoir sample, the two constraints were imposed. The constraints are that all neighboring grid blocks having equal permeability values are allowed to be merged during upscaling except

(1)   those having wells; and
(2)   those having wells and their respective adjacent grid blocks.

The fine-scale and upscaled simulators were run to obtain bottom-hole pressure and water cut data, and the match between the results from the upscaled reservoir model and those from the fine-scale model was used to determine which constraint produced better results. Figure 5 shows the bottom-hole pressure and water cut matches with Constraint 1, while Fig. 6 shows the matches obtained with Constraint 2. We obtained very good matches of the bottom-hole pressure with both constraints, but the water cut match was not good with any of the

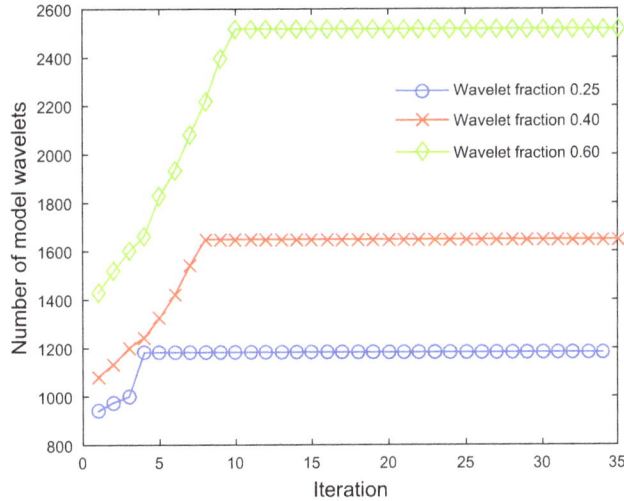

Fig. 18 Trend of wavelet coefficients for the 64 × 64 reservoir system (all fractions, upscaling performed during history matching)

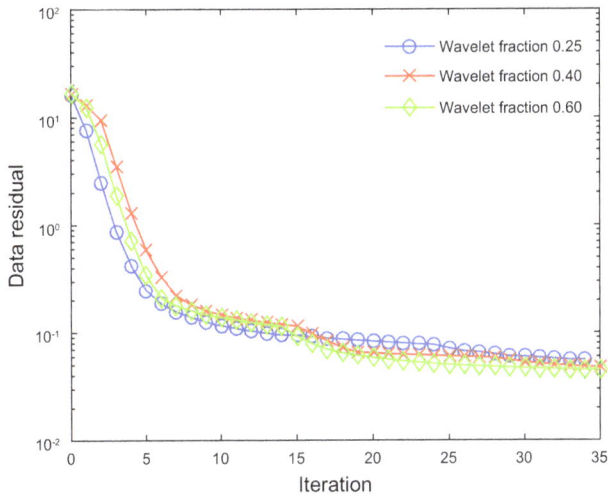

Fig. 19 Data residual for the 64 × 64 reservoir system (all fractions, upscaling performed during history matching)

constraints. However, the results from Constraint 2 were better than those from Constraint 1. Therefore, we chose only Constraint 2 for the history matching.

8 Example application of sensitivity-based upscaling to inverse analysis

In this section, we present history matching results for a reservoir model discretized into 64 × 64 grid blocks. The 64 × 64 reservoir has 16 producers and 16 injectors. Three thresholding values were used separately to select the wavelets that make up the model space. The upscaling procedure used in the forward simulations was based on the homogeneity-pattern created by thresholding the model

space using the wavelet sensitivity matrix. Furthermore, history matching was performed for the upscaled model (this work) and the fine-scale model (Awotunde 2010; Awotunde and Horne 2012, 2013) and the results from both systems were compared. We transformed the model space into wavelets and then perform thresholding to reduce the number of model parameters used for describing the system. This is done to reduce the computation time and also the non-uniqueness associated with the estimated results. The three fractions used in thresholding the model space are 0.60, 0.40, and 0.25. Each of these fractions determines the number of wavelets of the parameters we retain for history matching. The wavelet fraction of 0.60 indicates that the problem dimension is reduced to 60 % of its original size. That is 60 % of the total number of wavelets (of reservoir parameters) are selected. Upon inversion, the selected wavelets result in heterogeneous permeability distribution with some homogeneous patches. These number and size of homogenous patches tend to increase as we reduce the wavelet fraction.

8.1 Fine-scale inverse analysis

Fine-scale history matching was performed for the three wavelet fractions as mentioned above. The match of bottom-hole pressure (producers and injectors) for all the fraction of wavelets considered are shown in Figs. 7, 8, and 9, respectively. The matches to the water cut for fractions 0.60, 0.40, and 0.25 are presented in Fig. 10.

Good matches are obtained to all pressure and water cut histories, in all the cases considered. The trend of the number of wavelet coefficients selected using each fraction of 0.60, 0.40, and 0.25 is shown in Fig. 11. The figure shows that the number of wavelet coefficients increases

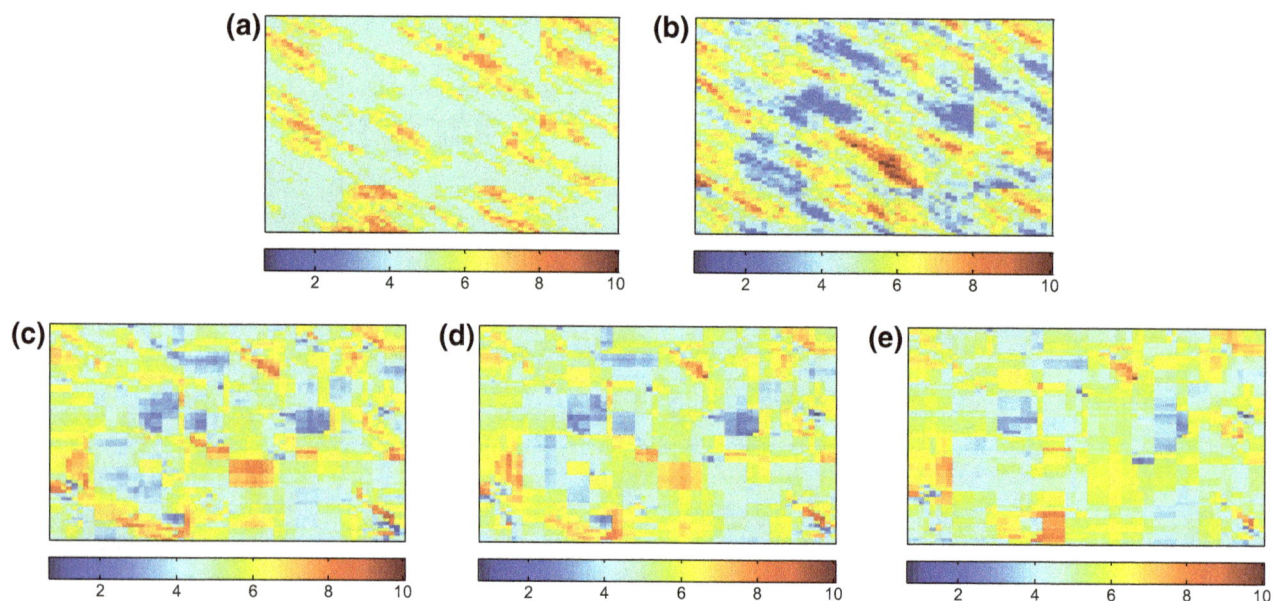

Fig. 20 Log permeability distribution for the 64 × 64 reservoir system. **a** True. **b** Initial guess. **c** Estimate of wavelet fraction 0.60. **d** Estimate of wavelet fraction 0.40. **e** Estimate of wavelet fraction 0.25 (upscaling performed during history matching)

Table 2 Important statistics for history matched 64 × 64 reservoir system

Details	Fine-scale			Upscale		
Number of measured data	6144			6144		
Number of wavelets of data	496			496		
Compression ratio	12.92			12.92		
Number of reservoir parameters	4096			4096		
Fraction	0.60	0.40	0.25	0.60	0.40	0.25
Number of wavelets of parameters	1573–2545	942–1725	942–1144	1432–2520	1082–1650	942–1185

in the early iterations and reaches the maximum preset number of wavelets (based on wavelet fraction used). Figure 12 shows the reduction of error as the iteration proceeds. The true permeability distribution, initial guess, and the final distribution obtained for the three scenarios are shown in Fig. 13. None of the estimates in Fig. 13c–e is close enough to the true permeability map (Fig. 13b). However, the estimate obtained from the fraction 0.60 gives a better representation of the heterogeneity than the other estimates.

8.2 Upscaled inverse analysis

In this case, simulation models are upscaled based on the analysis of sensitivity matrix at each iteration of the LM algorithm. The matches obtained to measured pressure data (producers and injectors), for the three fractions considered, are presented in Figs. 14, 15, and 16 while the matches to the water cut data are presented in Fig. 17. We observe that the pressure histories are adequately matched.

However, matches obtained with data fractions of 0.60 and 0.40 are better than those obtained with 0.25. Similarly, the matches to the water cut data for all fractions are acceptable. The selection of wavelet coefficients for each fraction is shown in Fig. 18 while the reductions in the mismatch error for all the fractions are presented in Fig. 19. All the fractions exhibit similar performances. Figure 20 shows the true permeability, the initial guess, and the permeability maps estimated from the three fractions. We observe that a better estimate of the permeability map is obtained when a 0.60 fraction of all the model wavelets is used.

A summary of performance information from all the cases is presented in Table 2. It represents that we have total of 6144 production data points (pressure and water cut), out of which only 496 have been selected for history matching which gives a compression ratio of 12.92. It is a 64 × 64 grid system that results in 4096 reservoir parameter (permeability) values. As discussed earlier that as the wavelet fraction is reduced the number of parameters considered is also reduced. Thus, a fraction of 0.60 has maximum parameters of 2545 which reduce to 1144 in the case of fraction 0.25. The parameters considered are almost the same in both fine-scale and upscale history matching.

9 Conclusions

The following conclusions can be drawn from the results obtained during this work:

(1) A good match is obtained for all the fine-scale history matching cases, as it was established in earlier work (Awotunde 2010; Awotunde and Horne 2012, 2013).

(2) During upscaling, combining the grid blocks that are adjacent to the well grid blocks does not provide good results. This is evident from the fact the Constraint 2 provides better results than Constraint 1. The upscaling of grid blocks can be achieved by analyzing the transformed sensitivity matrix and implemented during the iterations of the LM algorithm.

(3) Sensitivity-based upscaling during history matching provides reasonable results with a sufficient reduction in computation time as compared to the fine-scale inverse analysis.

(4) It is observed that the results from all three fractions (0.60, 0.40, and 0.25) are reasonably good. However, results from 0.25 are beginning to show some slight deviation from the true results, indicating that further reduction may lead to larger deterioration in the performance of the algorithms.

(5) The reduction of the simulation model size due to sensitivity-based upscaling can be limited by the emergence of an upscaled grid block that envelopes a smaller grid block.

Acknowledgments The authors acknowledge the support received from King Fahd University of Petroleum & Minerals through the DSR research Grant IN111046.

References

Awotunde A. Relating time series in data to spatial variation in the reservoir using wavelets. Ph.D. dissertation, Stanford University, Stanford, California; 2010.

Awotunde A, Horne R. A multiresolution analysis of the relationship between spatial distribution of reservoir parameters and time distribution of well-test data. SPE Res Eval Eng. 2011a;14(3):345–56. doi:10.2118/115795-PA.

Awotunde A, Horne R. A wavelet approach to adjoint state sensitivity computation for steady state differential equations. Water Resour Res. 2011b;47:W03502. doi:10.1029/2010WR009165.

Awotunde A, Horne R. An improved adjoint-sensitivity computations for multiphase flow using wavelets. SPE J. 2012;17(2):402–17. doi:10.2118/133866-PA.

Awotunde A, Horne R. Reservoir description with integrated multiwell data using two-dimensional wavelets. Math Geosci. 2013;45(2):225–52. doi:10.1007/s11004-013-9440-y.

Durlofsky L. Numerical calculation of equivalent grid block permeability tensors for heterogeneous porous media. Water Resour Res. 1991;27(5):699–708. doi:10.1029/91WR00107.

Ekrann S, Dale M. Averaging of relative permeability in heterogeneous reservoirs. In: King P, editor. The mathematics of oil recovery. Cambridge: Cambridge University Press; 1992.

Gill P, Murray W, Wright H. Practical optimization. New York: Academic Press; 1981.

Gómez-Hernández J, Journel A. Stochastic characterization of grid block permeabilities. SPE Form Eval. 1994;9(2):93–9. doi:10.2118/22187-PA.

Griva I, Nash S, Sofer A. Linear and nonlinear optimization. Philadelphia: SIAM Press; 2009.

Holden L, Nielsen BF. Global upscaling of permeability in heterogeneous reservoirs; the output least squares (ols) method. Transp Porous Media. 2000;40(2):115–43. doi:10.1023/A:1006657515753.

King MJ, Mansfield M. Flow simulation of geologic models. SPE Reservoir Eval Eng. 1999;2(4):351–67.

Levenberg K. A method for the solution of certain non-linear problems in least squares. Q Appl Math. 1944;2:164–8.

Li D, Beckner B, Kumar A. A new efficient averaging technique for scaleup of multimillion-cell geologic models. SPE Reserv Eval Eng. 2001;4(4):297–307. doi:10.2118/72599-PA.

Lu P, Horne R. A multiresolution approach to reservoir parameter estimation using wavelet analysis. In: SPE annual technical conference and exhibition, 1–4 October, Dallas; 2000. doi:10.2118/62985-MS.

Marquardt DW. An algorithm for least-squares estimation of nonlinear parameters. J Soc Ind Appl Math. 1963;11(2):431–41.

Nakashima T. Near-well upscaling for two and three-phase flows. Ph.D. dissertation, Stanford University, Stanford, California; 2009.

Nocedal J, Wright S. Numerical optimization. New York: Springer Science & Business Media; 2006.

Pickup GE, Jensen JL, Ringrose PS, Sorbie KS. A method for calculating permeability tensors using perturbed boundary conditions. In: 3rd European conference on the mathematics of oil recovery, 17 June; 1992.

Ringrose PS. Myths and realities in upscaling reservoir data and models. In: EUROPEC/EAGE conference and exhibition, 11–14 June, London, United Kingdom; 2007.

Sahni I, Horne RN. Multiresolution wavelet analysis for improved reservoir description. SPE Reserv Eval Eng. 2005;8(1):53–69. doi:10.2118/87820-PA.

Sahni I, Horne RN. Generating multiple history-matched reservoir-model realizations using wavelets. SPE Reserv Eval Eng. 2006a;9(3):217–26. doi:10.2118/89950-PA.

Sahni I, Horne RN. Stochastic history matching and data integration for complex reservoirs using a wavelet-based algorithm. In: SPE annual technical conference and exhibition, 24–27 September, Texas; 2006b.

Sharifi M, Kelkar M. Novel permeability upscaling method using fast marching method. Fuel. 2014;117:568–78. doi:10.1016/j.fuel.2013.08.084.

White CD, Horne RN. Computing absolute transmissibility in the presence of fine-scale heterogeneity. In: SPE symposium on reservoir simulation, 1–4 February, San Antonio, Texas; 1987. doi: 10.2118/16011-MS.

Wen XH, Gómez-Hernández JJ. Upscaling hydraulic conductivities in heterogeneous media: an overview. J Hydrol. 1996;183(1–2):ix–xxxii. doi:10.1016/S0022-1694(96)80030-8.

Wu XH, Efendiev Y, Hou TY. Analysis of upscaling absolute permeability. Discret Contin Dyn Syst Ser B. 2002;2(2):185–204. doi:10.3934/dcdsb.2002.2.185.

12

Combination and distribution of reservoir space in complex carbonate rocks

Lun Zhao[1] · Shu-Qin Wang[1] · Wen-Qi Zhao[1,2] · Man Luo[1] · Cheng-Gang Wang[1] · Hai-Li Cao[1] · Ling He[1]

Abstract This paper discusses the reservoir space in carbonate rocks in terms of types, combination features, distribution regularity, and controlling factors, based on core observations and tests of the North Truva Oilfield, Caspian Basin. According to the reservoir space combinations, carbonate reservoirs can be divided into four types, i.e., pore, fracture–pore, pore–cavity–fracture, and pore–cavity. Formation and distribution of these reservoirs is strongly controlled by deposition, diagenesis, and tectonism. In evaporated platform and restricted platform facies, the reservoirs are predominately affected by meteoric fresh water leaching in the supergene–para-syngenetic period and by uplifting and erosion in the late stage, making both platform facies contain all the above-mentioned four types of reservoirs, with various pores, such as dissolved cavities and dissolved fractures, or structural fractures occasionally in favorable structural locations. In open platform facies, the reservoirs deposited continuously in deeper water, in an environment of alternative high-energy shoals (where pore–fracture-type reservoirs are dominant) and low-energy shoals (where pore reservoirs are dominant).

Keywords Caspian Basin · Carbonate rock of platform facies · Reservoir space type · Reservoir type · Controlling factor · Distribution regularity

✉ Shu-Qin Wang
  wshuqin@petrochina.com.cn

[1] Research Institute of Petroleum Exploration and Development, CNPC, Beijing 100083, China

[2] School of Energy Resources, China University of Geosciences, Beijing 100083, China

Edited by Jie Hao

1 Introduction

Carbonate reservoirs contribute about 60 % of the world's oil and gas production. The Caspian Basin is one of the major petroliferous basins in the world, where over 90 % of the oil and gas production is from carbonate reservoirs. Compared with sandstone reservoirs, carbonate reservoirs often have stronger heterogeneity due to more complex genesis (Jiang et al. 2014a, b). Many researchers have recognized that the carbonate reservoirs in the Caspian Basin are very complex in diagenesis and fracture formation (Wang et al. 2012a, b; Zhao et al. 2010, 2012; Xu 2011). The authors think the heterogeneity of the carbonate reservoirs is the combined result of complex distribution and combination of reservoir space, i.e., pore, cavity, and fracture. Thus, finding out distribution features of such reservoir space is essential for evaluating the heterogeneity of carbonate reservoirs. Taking the North Truva Oilfield in the eastern margin of the Caspian Basin as an example, based on core observations and tests, this paper discusses the features and genesis of pores, fractures, and dissolved cavities in carbonate reservoirs of open platform—evaporated platform facies. It gives a classification of the carbonate reservoirs according to the features of their reservoir space, describes their physical property features, and analyzes the distribution and combination regularity and controlling factors of different types of carbonate reservoirs.

2 Overview

The North Truva Oilfield is located in the eastern margin of the Caspian Basin (Fig. 1). Its main oil-bearing formations are Carboniferous carbonate reservoirs of platform facies,

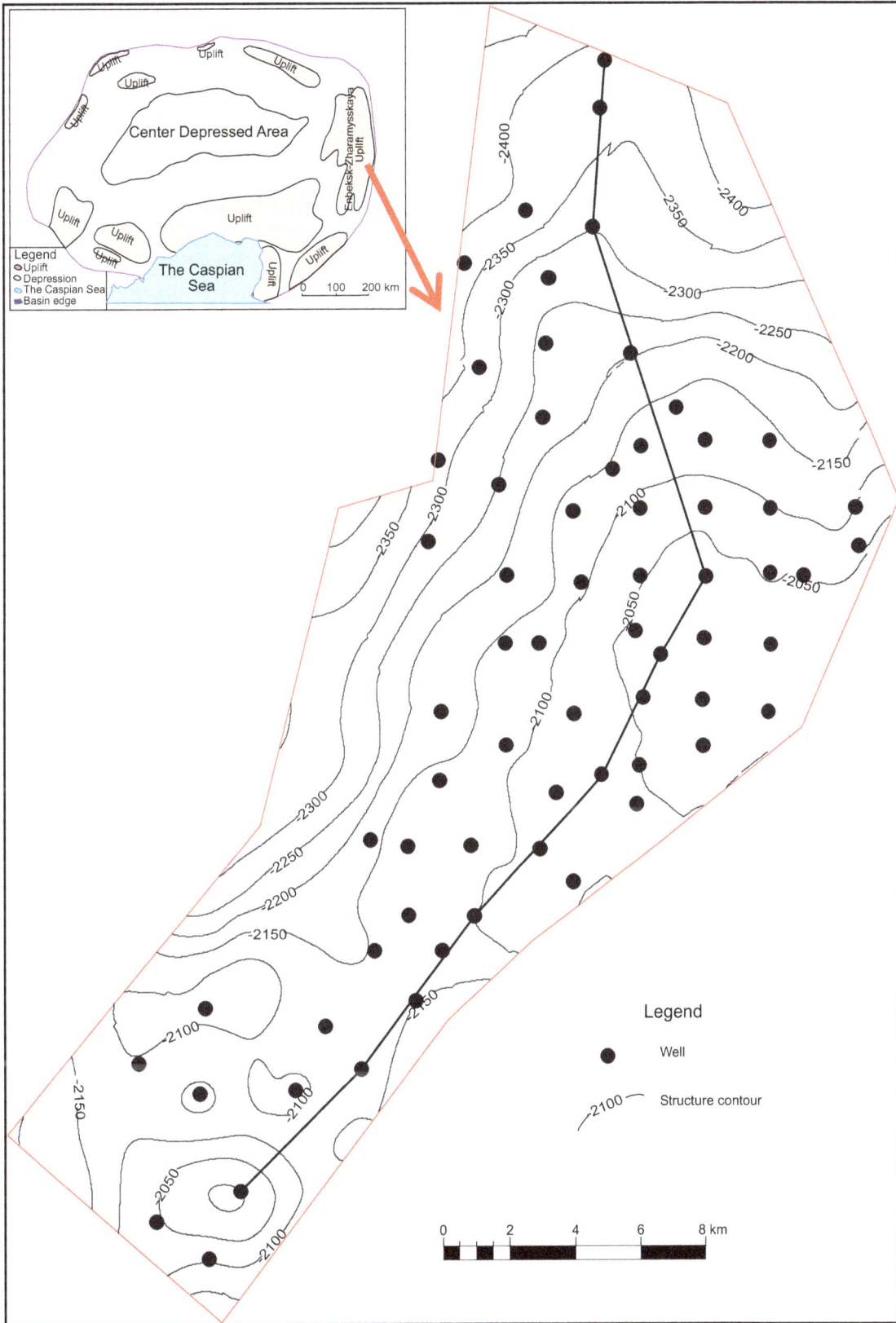

Fig. 1 Location and structural map of the study area

including KT-I and KT-II; two sets of oil/gas-bearing zones vertically. KT-I, composed of deposits of evaporated platform-restricted platform facies at the burial depth of 2300–2800 m, is further divided into A, Б, and B oil layers. Layer A presents evaporated platform facies in the north part, and transits to restricted platform facies southwards, from top to bottom, with water depth during deposition gradually increasing. It is composed of calcareous dolomite, dolomitic limestone and anhydrite-bearing dolomite, anhydrite dolomite, with dolomitic and gypsum content decreasing but calcareous content increasing from the evaporated platform to restricted platform facies (Fig. 2a) (Esrafili-Dizaji and Rahimpour-Bonab 2009). A1 and part of A2 are missing in the north of the oilfield, indicating that the area experienced uplifting and erosion after the deposition of Carboniferous and before deposition of Permian. Core analysis of KT-I shows that its porosity is 7.2 %– 39.2 %, with an average of 16.9 %, and permeability is 0.025–2170 mD, with an average of 107.3 mD. KT-II represents the carbonate rocks of open platform facies at a burial depth of 3100–3400 m, with depositional water depth decreasing and energy enhancing from bottom to top, since it is dominated by low-energy shoal deposits in the lower part and high-energy shoal deposits in the upper part. KT-II is mostly composed of pure bioclastic limestone, with calcareous content of 99.6 % (Fig. 2b). Core analysis shows that its porosity is 8 %–20.1 %, with an average of 12.1 %, and permeability is 0.003–415 mD, with an average of 35.9 mD.

3 Types and genesis of reservoir space

9600 core samples taken from 23 wells were analyzed to ascertain the features and genesis of the major reservoir space of KT-I and K-II, and the reservoir space is classified according to their genesis and shape. The main types of reservoir space in the study area are pores, fractures, and dissolved cavities. For KT-I and KT-II, pores account for 84 % and 97.8 %, dissolved cavities account for 6.7 % and 0 %, and fractures account for 9.3 % and 2.2 %, respectively.

3.1 Pores

Pores are defined as reservoir space less than 2 mm in diameter, which are the most widespread type of reservoir space in the study area, and also the most complicated. By genesis, pores in the study area can be divided into 7 subcategories under 2 categories (Table 1), i.e., primary pores (syngenetic or para-syngenetic) and mainly secondary pores formed during burial diagenesis (Zheng et al. 2010). Primary pores are the pores formed in syngenetic or para-syngenetic periods during deposition, with shapes and

sizes changing during diagenesis. In the study area, the most prominent primary pores are intergranular pores preserved after the compaction between detrital grains, visceral foramen preserved after soft body decomposition, and the pores preserved between biological frameworks (Fig. 3a, b, c). Secondary pores refer to the pores formed by dissolution during the process of burial diagenesis, which are dominated by intergranular dissolved pores, intragranular dissolved pores, and intercrystalline dissolved pores (Fig. 3d, e, f). The distribution of pore types is significantly influenced by the depositional–diagenetic environment (Yue et al. 2005; Gao et al. 2013; Haq and Al-Qahtani 2005).

KT-I is composed of deposits of evaporated platform-restricted platform facies. Its deposits were affected by sabkha dolomitization and reflux-seepage dolomitization during syngenetic, para-syngenetic, and eogenetic periods and further transformed due to dissolution and cementation during burial diagenesis. In Late Carboniferous, the entire study area was uplifted and eroded by meteoric leaching (Huang et al. 2009; López-Horgue et al. 2010; Wang et al. 2012a, b). Due to the effect of multiple factors, the reservoirs have various pores, including primary (including syngenetic and para-syngenetic) pores, interframework pores and visceral foramen, and secondary pores, such as intergranular dissolved pores, intragranular dissolved pores, intercrystalline dissolved pores, crystal moldic pores, etc. Meanwhile, the average surface porosity is almost the same for all pores. Primary visceral foramen, intergranular dissolved pores, intragranular dissolved pores, and intercrystalline dissolved pores are 2.15 %– 3.45 % in average surface porosity (Fig. 4), not differing much.

KT-II is dominated by deposits of open platform facies, including bioclastic shoals, algal reef, and intershoal deposits developed in deeper water, which are continuous and free from uplifting and erosion in later stages. The formation of reservoir pores was controlled by dissolution and cementation during diagenesis, predominately including various primary (syngenetic or para-syngenetic) visceral foramen remaining after soft tissue decomposition and intergranular dissolved pores, and intragranular dissolved pores formed due to selective dissolution of soluble components during burial diagenesis. The average surface porosity of visceral foramen, intergranular dissolved pores, and intragranular dissolved pores are 1.8 %, 5.24 %, and 0.46 % respectively, and less than 0.1 % for other types of pores (Fig. 4).

3.2 Fracture

Fracture, another kind of reservoir space commonly found in the study area, is also an important oil and gas flow

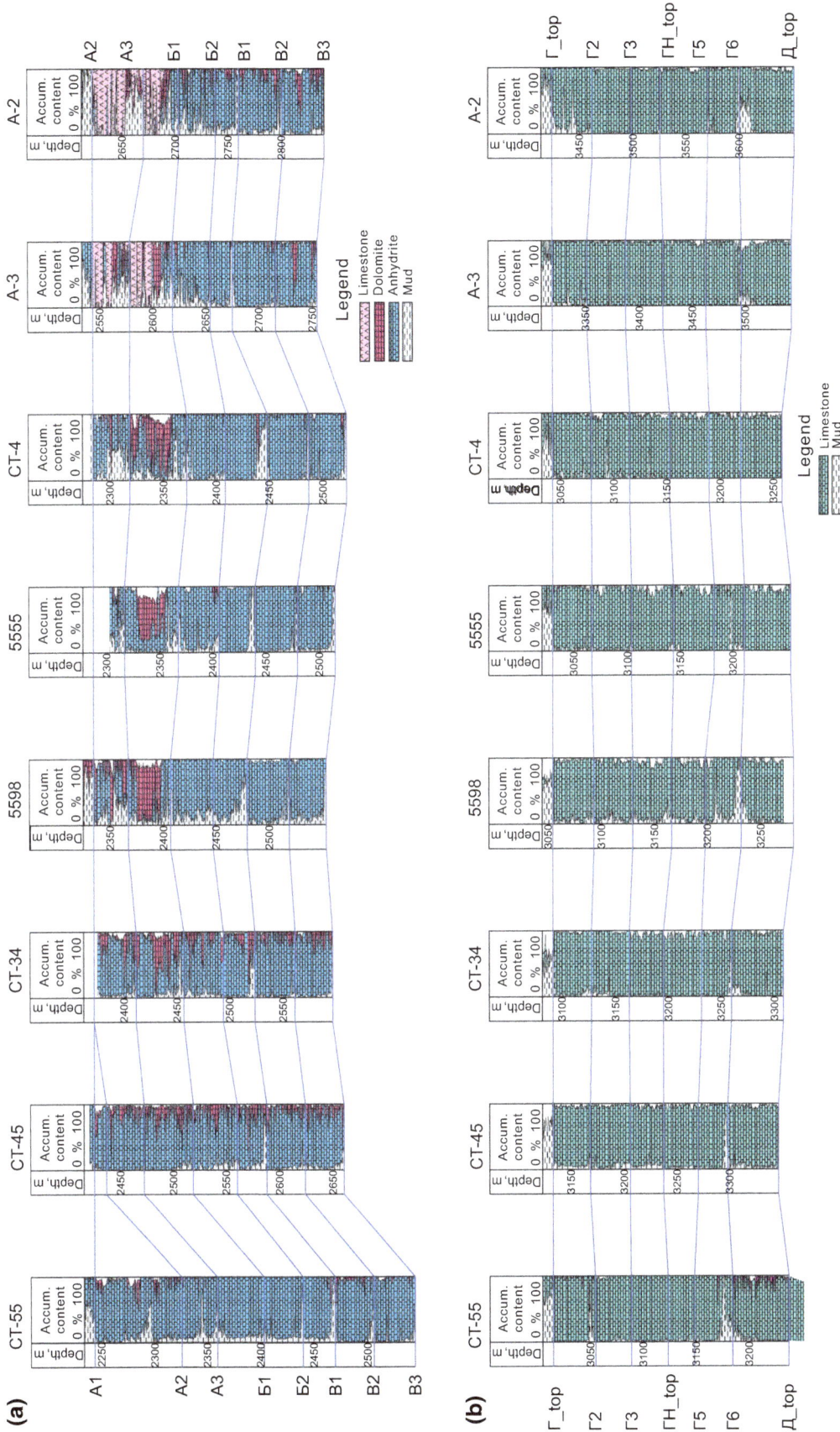

Fig. 2 Stratum and lithological profile of the study area. **a** KT-I, **b** KT-II

Table 1 Classification and definition of pore types in the study area

Classification		Definition
Category	Subcategory	
Primary pores	Intergranular pores	The space not filled by marl or cement between grains or the partially filled residual pores
	Interframework pores	The framework space formed due to reef-building, the unfilled or partially filled pores
	Visceral foramen	The visceral foramen remaining after the decomposition of soft tissues, unfilled or partially filled residual pores
Secondary pores	Intergranular dissolved pores	The pores formed by partial or complete dissolution of the micrite or sparry cements previous filled between particles
	Intragranular dissolved pores	Pores inside bioclastic, oolitic, and sand clast grains formed due to partial dissolution
	Intercrystalline dissolved pores	Enlarged dissolved pores between powder crystalline and fine crystalline
	Crystal moldic pores	Formed by complete dissolution of gypsum or salt crystals

channel. Various types of fractures are observed in cores from the study area. These can be divided into structural fractures and nonstructural fractures according to their genesis (Davies and Smith Jr. 2006; Wang et al. 2012a, b; Zheng et al. 2009). Structural fractures, formed due to fracturing of rocks under stress fields (Figs. 3h, 5b–d), arrange in a certain direction. Nonstructural fracture including dissolved fracture and pressure dissolved fracture (sutures), has no apparent directionality (Figs. 3i, 5e, f). The fractures can also be divided into macrofractures and microfractures according to their extended length. The former refer those in cores visible to the naked eye, while the latter can be only identified under a microscope.

During the depositional period of KT-II, the open platform water energy changed intermittently, leading to the alternate deposition of high-energy shoals, low-energy shoals, and inter shoals (the low-lying land between shoals). Nonreservoirs deposited between shoals are tight and strong in resistance to pressure, while reservoirs between shoals are loose and weak in resistance to pressure. Under tectonic stress, the reservoirs of high-energy shoals and low-energy shoals would release pressure first since their particles were breaking up, resulting in the formation of microfractures; then, the tight nonreservoirs between shoals would form macrofractures. Thus, structural macrofractures of open platform facies carbonate deposition are mostly present in tight nonreservoirs, accounting for 84.4 %, and seldom in low-energy shoals. High-energy shoals are dominated by granular fracturing microfractures.

The reservoirs deposited in the evaporated platform—restricted platform facies are mostly calcareous dolomite and dolomitic limestone with strong brittleness, while nonreservoirs are composed of argillaceous and gypsum rocks with strong plasticity. Thus, under tectonic stress, the

reservoir interval is more likely to form structural fractures due to fracturing (Aqrawi et al. 1998; Carnell and Wilson 2004; He et al. 2012; Moutaz et al. 2010). Fractures account for 71 % of the reservoir space in the reservoir interval but 29 % of the space in the nonreservoir interval. During the syngenetic–para-syngenetic period, dissolved fractures were likely to form due to strong leaching by meteoric fresh water. The uplifting and erosion after Carboniferous deposition and before Permian deposition leached the formations with meteoric water again, making the dissolved fractures and structural fractures formed earlier enlarge further. Therefore, dissolved fractures are quite abundant in these reservoirs. But they are small in scale, mostly dissolved microfractures, and bigger ones are only found in local areas (Fig. 5f).

The statistical results of microfractures observed from thin sections of reservoir intervals show that there are more structural microfractures and dissolved fractures in KT-I than KT-II, with a surface fracture ratio of 0.04 % and 0.28 %, respectively, that is 2–10 times that of KT-II, but the granular fracturing microfractures of KT-II have a higher surface fracture ratio than that of KT-I (Fig. 4). KT-II is tighter and more brittle than KT-I, so macrofractures are richer in KT-II. The statistical results of cores and imaging logging reveal that the linear fracture density is 6.49/m for KT-II and 3.38/m for KT-I.

3.3 Dissolved cavities

Dissolved cavities refer to reservoir space more than 2 mm in diameter. They are only found in the evaporated platform facies of KT-I, but are absent in KT-II. The formation of dissolved cavities in KT-I is mainly due to dissolution of carbonate components, which is directly related to meteoric leaching by weathering and erosion resulted from the

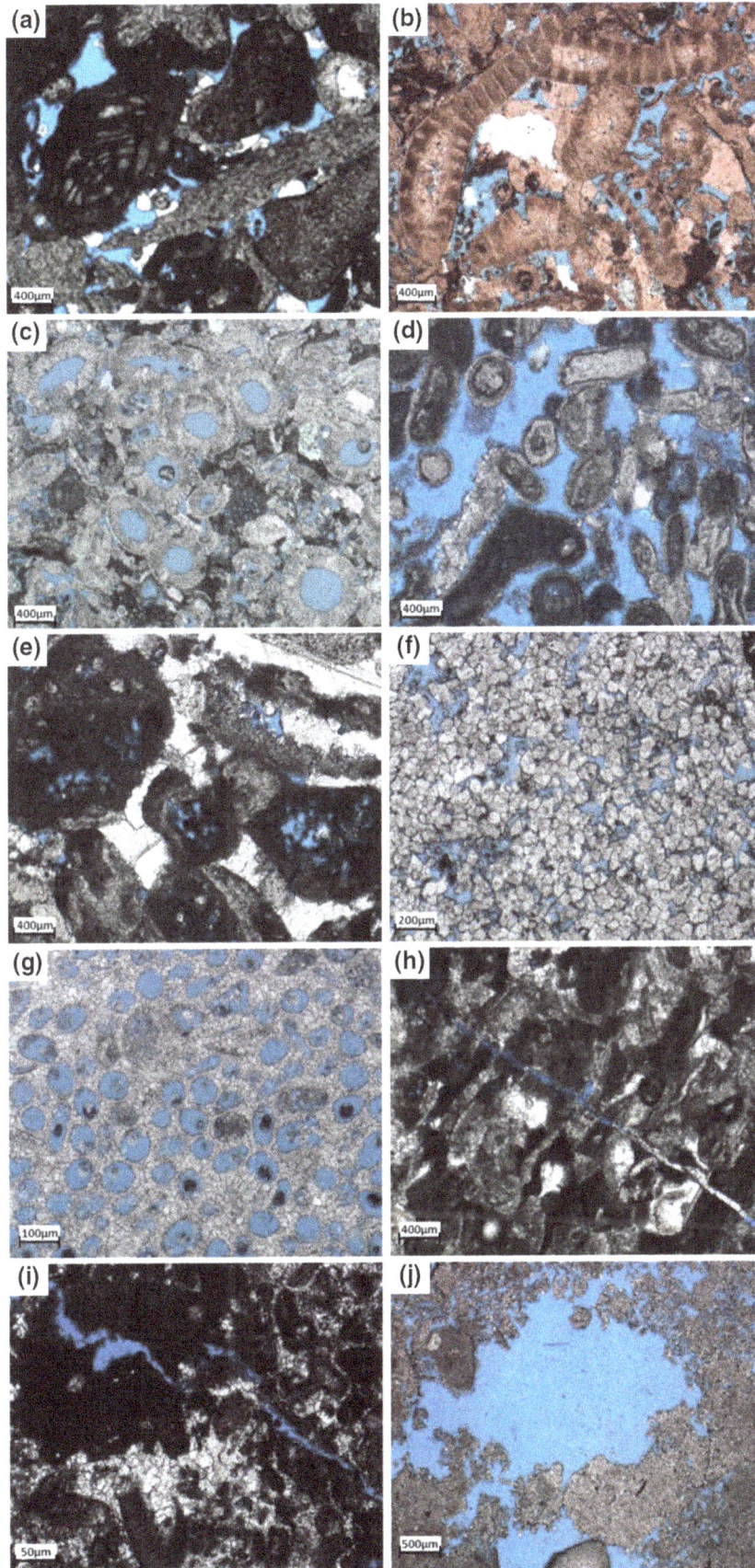

◀**Fig. 3** Microscopic features of reservoir space. **a** Residual intergranular pores after cementation and compaction, Well CT-22, 3170.28 m, sparry algal limestone; **b** green algae framework pores, Well 5555, 3123.16 m, micrite green framework limestone; **c** visceral foramen within derbesia neglecta, Well 5598, 3155.13 m, micrite foraminiferal red algae limestone; **d** superficial oolith intergranular dissolved pores, Well 5598, 3201.37 m, sparry algae superficial oolitic limestone; **e** dissolved pores within algae aggregate, Well CT-22, 2339.72 m, micrite foraminiferal red algal limestone; **f** dolomitic intercrystalline dissolved pores, Well CT-4, 2335.52 m, coarse powdery crystalline dolomite; **g** oolite moldic pores, pore diameter: 0.05–0.1 mm, Well A2, 2887.27 m, sparry moldic oolitic limestone; **h** structural fracture, half-filled by calcite, Well CT-22, 3148.82–3148.85 m, sparry algae foraminiferal limestone; **i** dissolved fracture, Well CT-22, 2301.15 m, micrite foraminiferal limestone; **j** large dissolved cavity, pore diameter: 2–5 mm, Well CT-4, 2342.48 m, powdery crystalline residual bioclastic dolomite

late-stage uplifting (Guo 2011; Müller et al. 1990; Yao et al. 2008). Core observations and image logging data show there are apparent dissolved cavities at the top of KT-I, which are 3–5 mm in diameter generally, and up to 25 m at maximum (Fig. 5g–h). Dissolved cavities, pores, and fractures generally coexist. Microanalysis of core thin sections shows the dissolved cavity porosity in KT-1 is 0.45 %.

4 Features of reservoir space combination

The analysis above reveals that the reservoir space in evaporated platform—restricted platform facies is dominated by pores, fractures, and dissolved cavities, while the reservoir space in open platform facies has pores and fractures, without dissolved cavities. Different types of reservoir space are not evenly distributed in different layers and positions, which manifests in differences in reservoir space combinations. Based on core observations, and

combination of pores, fractures, and dissolved cavities, the reservoirs in the study area can be divided into four types, i.e., pore, fracture–pore, pore–fracture–cavity, and pore–cavity (Borkhataria et al. 2005; Huang 1997; Mahdi and Aqrawi 2014; Whitaker et al. 2004). The porosity and permeability of these different types of reservoirs were determined from logging data to compare their physical properties.

4.1 Pore-type reservoir

This type of reservoir contains reservoir space of pores of different genesis, with an average porosity of 9.8 % and average permeability of 5.7 mD (Table 2). Pore-type reservoirs are most widespread in all intervals, but differ widely in physical properties in different intervals. KT-I has various types of pores, and higher porosity and permeability than KT-II. The porosity and permeability are 10.9 % and 6.1 mD for KT-I, and 9.7 % and 5.2 mD for KT-II. Moreover, their porosity and permeability show a gradually decreasing trend from top to bottom. For KT-I, this variation in physical properties was controlled by meteoric fresh water leaching and uplifting erosion, but for KT-II, the variation in physical properties is controlled by the sedimentary environment, the Г1–Г5 layers are high-energy shoal facies with an average porosity of 9.9 %–10.7 %, and average permeability of 4.7–14.8 mD; while the Г6–Д3 layers are low-energy shoal facies with an average porosity of 8.5 %–9.9 % and average permeability of 0.6–2.2 mD. It is obvious that the reservoirs of upper high-energy shoal facies have better physical properties than those of low-energy shoal facies.

4.2 Fracture–pore-type reservoir

This type of reservoir mainly contains pores and fractures of varying genesis. Second in extensiveness, it has an

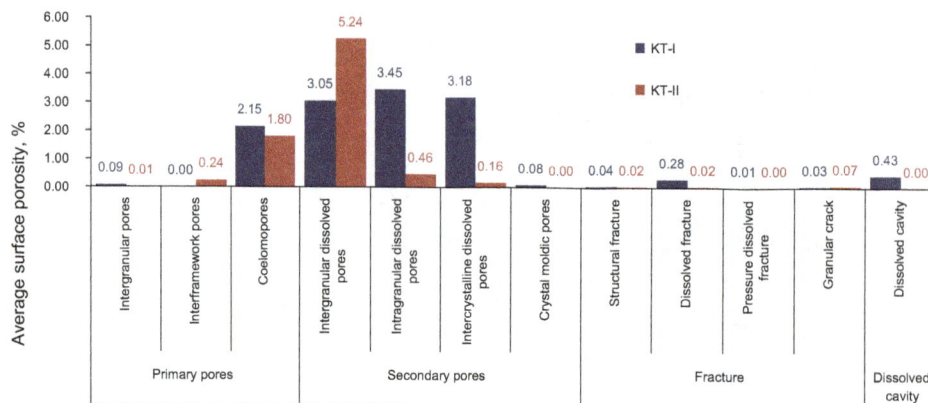

Fig. 4 Surface porosity of reservoir space (microscope) obtained from the statistics of core thin sections

Fig. 5 Macroscopic features of reservoir space. **a** Poorly connected pores, Well A2, 3466.21–3466.24 m, limestone; **b** netted fracture, Well CT10, 2343.06–2343.17, dolomitic limestone; **c** high-angle structural fracture, Well 5555, 2332.15–2333.49 m, dolomitic limestone; **d** low-angle structural fracture, Well CT-4, 2294.90–2395.02 m, dolomitic limestone; **e** suture line, Well CT10, 2350.46–2350.58 m, dolomitic limestone; **f** dissolved fracture, Well CT4, 2346.72–2346.96 m, dolomitic limestone; **g** dissolved cavity, Well 5555, 2341.03–2341.26 m, dolomite with oil patch and dissolved cavity; **h** dissolved cavity, Well CT4, 2341.54–2341.66 m, dolomitic limestone

Table 2 Physical properties of various types of reservoirs

Horizon	Porosity, %					Permeability, mD				
	Pore–cavity	Pore	Pore–cavity–fracture	Fracture–pore	Average	Pore–cavity	Pore	Pore–cavity–fracture	Fracture–pore	Average
A1		11.6	12.1	10.1	11.4		5.1	3.0	1.8	4.3
A2	14.3	10.8	14.4	11.5	12.1	49.0	15.1	54.5	8.2	25.4
A3	13.1	10.2	13.0	10.6	11.7	18.3	5.3	17.9	7.0	12.2
Б1	14.2	11.0		10.2	11.5	43.2	2.9		4.6	11.0
Б2		10.0		9.0	9.7		3.1		4.2	3.3
B1		10.0		8.9	9.7		1.0		2.9	1.6
B2		9.8		10.0	9.9		3.3		9.6	5.5
B3		13.2		10.8	13.0		2.1		3.4	2.2
B4		12.4		12.9	12.5		0.4		2.7	0.8
B5		10.1			10.1		1.3			1.3
KT-I	13.9	10.9	13.2	10.7	11.7	40.1	6.1	22.8	6.9	13.4
Г1		9.9		8.6	9.0		4.7		4.3	4.4
Г2		10.4		10.8	10.5		7.6		13.2	9.9
Г3		10.4		10.1	10.3		8.3		6.3	7.3
Г4		10.7		11.2	10.9		8.4		15.6	12.4
Г5		10.2		10.3	10.3		14.8		25.4	19.2
Г6		9.0		9.1	9.0		2.2		8.1	3.5
Д1		8.5		8.1	8.5		0.6		1.5	0.6
Д2		9.4		8.4	9.3		1.6		1.8	1.6
Д3		9.7		8.2	9.7		0.8		1.0	0.8
KT-II		9.7		10.4	9.9		5.2		12.3	7.8
Oilfield	13.9	10.1	13.2	10.5	10.8	40.1	5.2	22.8	9.8	10.5

average porosity of 10.5 % and average permeability of 9.8 mD. This type of reservoir in KT-I and KT-II are similar in porosity, but as KT-II is tighter and more brittle, structural macrofractures and microfractures are more likely to form under tectonic stress in KT-II, which can significantly improve reservoir permeability, therefore, this type of reservoir in KT-II has higher permeability than that in KT-I (12.3 mD for the former and 6.9 for the latter) (Table 2).

4.3 Pore–fracture–cavity-type reservoir

With pores, fractures, and cavities of various genesis as reservoir space, this type of reservoir mainly occurs in Layer A at the top of KT-I, and is composed of calcareous dolomite with minor dolomitic limestone. Reservoir space in this type of reservoir is most diverse due to the combined effects of meteoric leaching and uplifting and erosion, including pores, fractures, and dissolved cavities of various origins. The fractures are dominated by diagenetic dissolved fractures, followed by structural fractures. This type

of reservoir largely occurs in A1, A2, and A3 layers in the upper part of KT-I 30–80 m below the KT-I top. With an average porosity of 13.2 % and average permeability of 22.8 mD (Table 2), it is one type of reservoir with fairly good physical properties in the study area. The A2 layer in the central north of the oilfield, high in structural position, and suffering strong deformation, is best in physical properties, with an average porosity of 14.4 % and an average permeability of 54.5 mD.

4.4 Pore–cavity-type reservoir

Often associated with pore–fracture–cavity reservoirs, this type of reservoir has pores and cavities of various origins as reservoir space, and occurs in A2, A3, and B1 layers. It is composed of dolomitic limestone, the dissolved cavities are mostly unfilled or partially filled, and connected by pores of various sizes. With an average porosity of 13.1 %–14.3 % and an average permeability of 40.1 mD, it is the reservoir type with the best physical properties in the study area (Table 2).

5 Distribution regularity of reservoir types

The thicknesses of various types of reservoirs in KT-I of 156 wells and KT-II of 146 wells were compiled and the thickness ratio of various types of reservoirs was analyzed by sublayer to find out the thickness variation pattern of different types of reservoirs in the vertical direction and horizontally.

The different types of carbonate reservoirs follow different variation patterns in the vertical direction (Fig. 6) strongly controlled by deposition and diagenesis. KT-I has all four types of reservoir, due to its complicated sedimentary–diagenetic environment, thus it has diverse reservoir space and reservoir space combinations. Of which, layer A with all four types of reservoirs is most diverse in reservoir type. Layer Б has three types of reservoirs, pore, fracture–pore, and pore–cavity type (Fig. 6). Layer B has only two types of reservoirs, pore, and fracture–pore types. Pore–fracture–cavity reservoirs mainly occur in sublayers A1, A2, and A3 at the top of KT-I, its thickness ratio in three sublayers are 13.9 %, 22.1 %, and 49 %, respectively. Pore–cavity-type reservoirs mainly occur in A2–Б2 sublayers, at a thickness ratio of 1.6 %–19.5 %. Found in all sublayers, pore-type reservoirs increase in thickness ratio from top to bottom, and reach 100 % in the B5 sublayer. Fracture–pore-type reservoirs occur in all sublayers except B5, with a thickness ratio of 9.8 %–28.7 %. In general, the thicknesses of pore–cavity–fracture, pore, and fracture–pore-type reservoirs account for 32.6 %, 28.7 %, and 27.0 %, respectively, while pore–cavity type reservoirs account for 4.7 % in thickness. The above changes in reservoir type have a direct effect on the reservoir physical properties, as a result, the permeability of reservoirs in A1–Б1 is much higher than the permeability of reservoirs in the Б2–B5 sublayer (Table 2).

Deposited in a simple depositional–diagenetic environment, thus monotonous in lithology, KT-II only has two types of reservoir space, pore and fracture, and only two reservoir types, pore and fracture–pore. These two types of reservoirs show patterns in thickness variation, the thickness ratio of fracture–pore reservoirs drops down from the upper sublayer Г1 which is 70 % to the underneath sublayer Д3 which is 1.5 %; while the thickness ratio of pore-type reservoirs increases up from the upper sublayer Г1 which is 30 % to the underneath sublayer Д3 which is 98.5 %. The change in reservoir type directly affects the reservoir physical properties. As a result, the permeability of reservoirs in Г1–Г5 in the upper part is much higher than those in Г6-Д3 (Table 2).

Horizontally, pore-type and fracture–pore-type reservoirs are widespread, but thinner in relatively higher part of structures in K-I (Fig. 7a, b). Pore–cavity–fracture-type reservoirs and pore–cavity-type reservoirs are predominately distributed in structural high positions because tectonic stress is stronger there, fractures are more developed, and dissolution is more likely to occur when fluid flows into the reservoirs via the fractures formed in the early stage. Furthermore, the structural high positions often have been uplifted by large amounts, so leaching and erosion are more intense in the late stage. Under this tectonic and diagenetic environment, a large quantity of dissolved fractures, dissolved pores, and dissolved cavities are formed in structural high positions, resulting in concentrated distribution of pore–cavity–fracture-type reservoirs and pore–cavity-type reservoirs there.

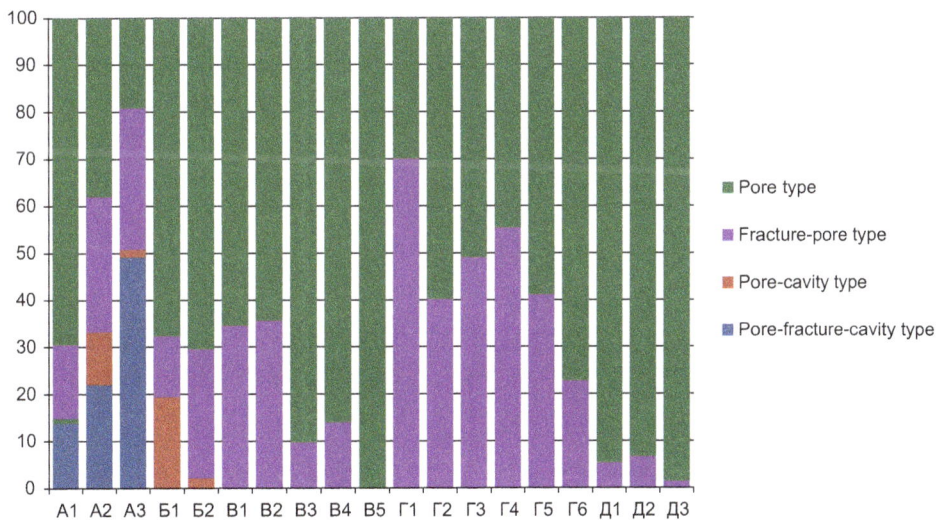

Fig. 6 Thickness ratio of various types of reservoirs in the study area

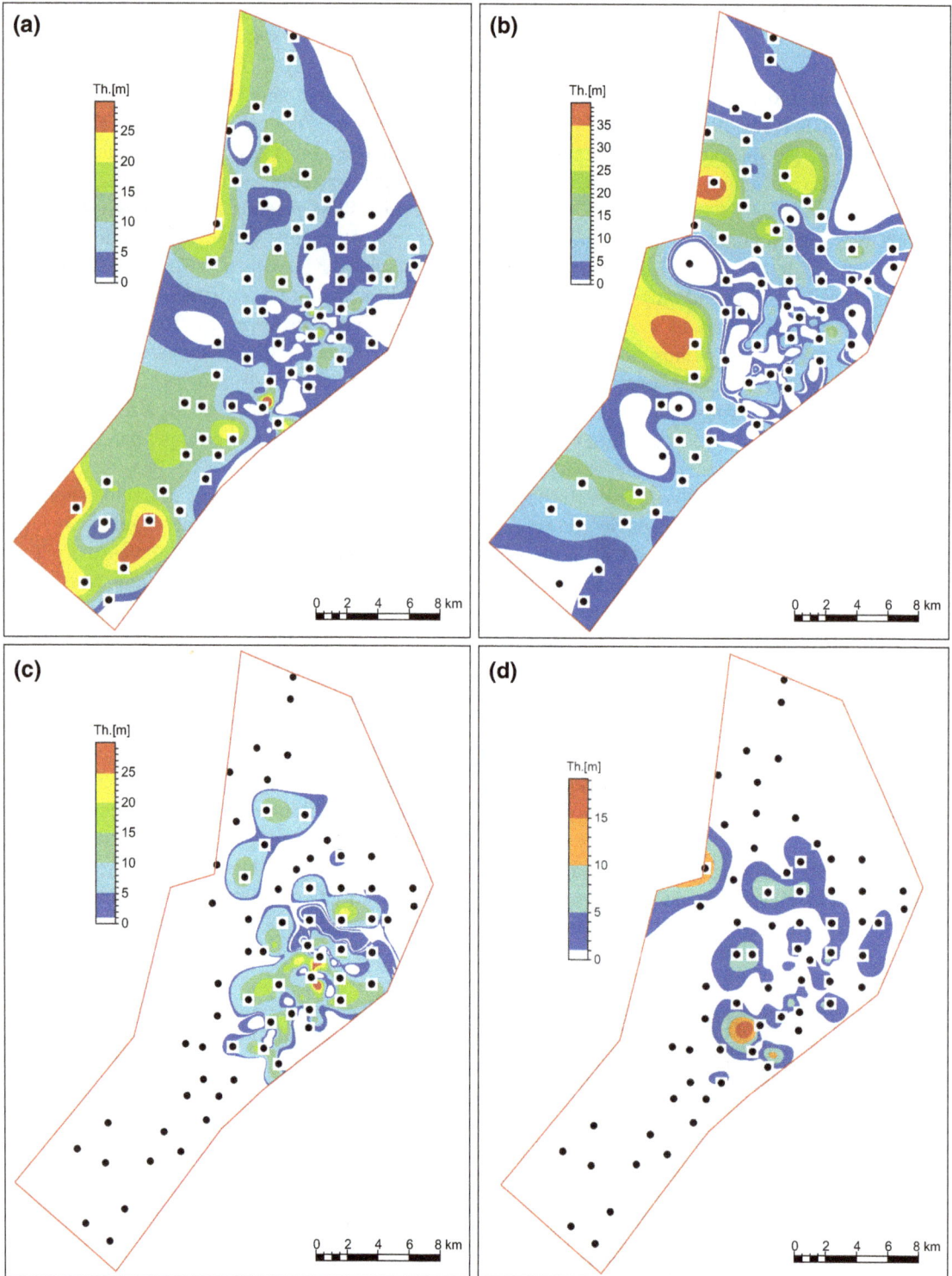

Fig. 7 Reservoir-type distribution in KT-I. **a** Pore type, **b** fracture–pore type, **c** pore–fracture–cavity type, **d** pore–cavity type

Horizontally, pore-type and fracture–pore-type reservoirs are widespread in KT-II (Fig. 8), fracture–pore-type reservoirs are mainly distributed in structural high positions on an NE trend. This is because structural stress is stronger at structural highs, and under the effect of structural stress, fractures are likely to be created in the reservoir. In contrast, pore-type reservoirs mainly controlled by deposition and diagenesis, have no apparent regularity in distribution.

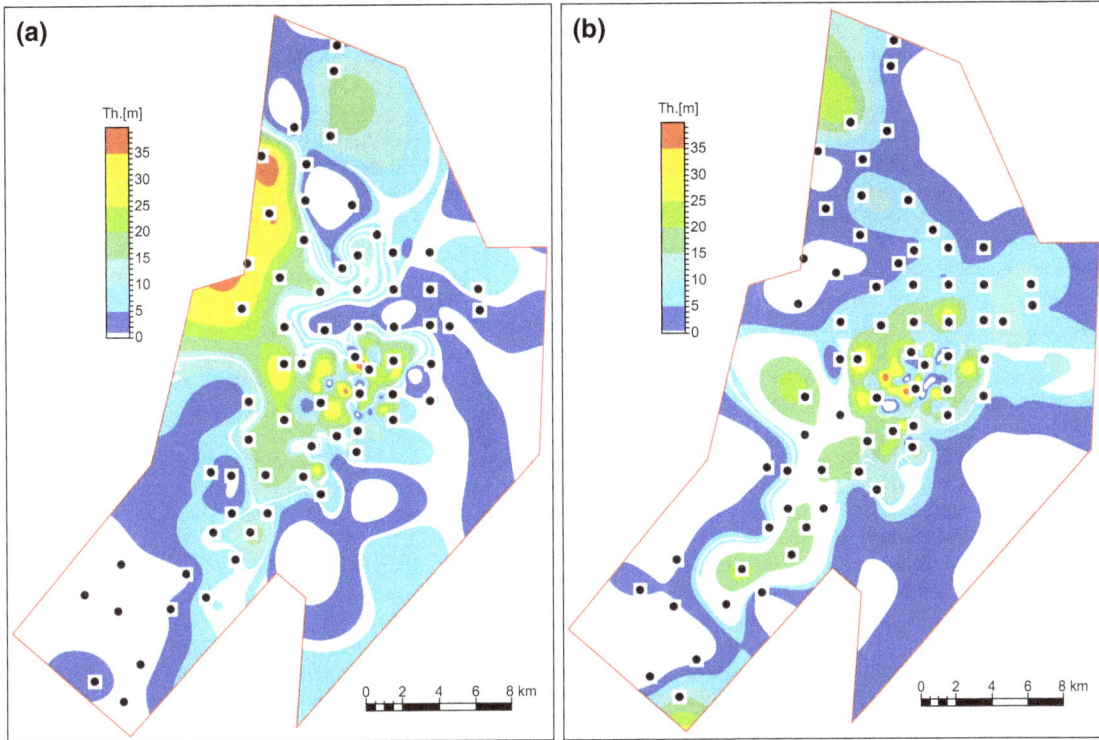

Fig. 8 Reservoir-type distribution in KT-II. **a** Pore type, **b** fracture–pore type

6 Conclusions

Observation of a large quantity of core samples and core analysis data of the North Truva Oilfield in the eastern margin of the Caspian Basin reveals that in the carbonate reservoirs there are many kinds of reservoir space, including pores, fractures, and dissolved cavities of different genesis, and complex reservoir space combinations. Accordingly, the reservoirs can be divided into four types, i.e., pore, fracture–pore, pore–cavity–fracture, and pore–cavity types. The distribution of these reservoirs is controlled by deposition, diagenesis, and tectonism. In evaporated platform and restricted platform facies, affected by meteoric fresh water leaching in supergene–para-syngenetic periods and uplifting and erosion in later stages, there developed pores of various types, including dissolved cavities and fractures, and structural fractures at favorable structural positions. So the reservoirs are versatile in type, including all the above four types. In open platform deeper water facies, the deposition was continuous, the depositional environment of alternative high-energy shoals and low-energy shoals directly controlled the degree of fracture development, so pore-type reservoirs dominate in low-energy shoals, while pore–fracture reservoirs dominate in high-energy shoals.

Acknowledgments This paper is supported by the National Major Science and Technology Project (No. 2016ZX05030002).

References

Aqrawi AAM, Thehni GA, Sherwani GH, et al. Mid-Cretaceous rudist-bearing carbonates of the Mishrif Formation: an important reservoir sequence in the Mesopotamian Basin. J Pet Geol. 1998;21(1):57–82.

Borkhataria R, Aigner T, Poppelreiter M, et al. Characterisation of epeiric "layer-cake" carbonate reservoirs: upper Muschelkalk (Middle Triassic), the Netherlands. J Pet Geol. 2005;28(2):119–46.

Carnell AJH, Wilson MEJ. Dolomites in SE Asia—varied origins and implications for hydrocarbon exploration. Geol Soc Lond. 2004;235:255–300.

Davies GR, Smith LB Jr. Structurally controlled hydrothermal dolomite reservoir facies: an overview. AAPG Bull. 2006;90(11):1641–90.

Esrafili-Dizaji B, Rahimpour-Bonab H. Effects of depositional and diagenetic characteristics on carbonate reservoir quality: a case study from the South Pars gas field in the Persian Gulf. Pet Geosci. 2009;15(4):325–44.

Gao JX, Tian CB, Zhang WM, et al. Characteristics and genesis of carbonate reservoir of the Mishrif Formation and Rumaila oil field, Iraq. Acta Pet Sin. 2013;34(5):843–52.

Guo F. Carbonate sedimentology. Beijing: Petroleum Industry Press; 2011.

Haq BU, Al-Qahtani AM. Phanerozoic cycles of sea-level change on the Arabian platform. Geoarabia. 2005;10(2):127–60.

He YL, Fu XY, Liu B, et al. Control of oolitic beaches sedimentation and diagenesis on reservoirs in Feixianguan Formation, northeastern Sichuan Basin. Pet Explor Dev. 2012;39(4):434–43 (in Chinese).

Huang SJ. Carbon and strontium isotopes of Late Palaeozoic marine carbonates in the Upper Yangtze platform, Southwest China. Acta Geol Sin. 1997;71(3):45–53 (in Chinese).

Huang SJ, Tong HP, Liu LH, et al. Petrography, geochemistry and dolomitization mechanisms of Feixianguan dolomites in Triassic, NE Sichuan, China. Acta Pet Sin. 2009;25(10):1363–72 (in Chinese).

Jiang L, Worden RH, Cai CF, et al. Dolomitization of gas reservoirs: The Upper Permian Changxing and Lower Triassic Feixianguan Formations, Northeast Sichuan Basin, China. J Sediment Res. 2014a;84(10):792–815.

Jiang L, Worden RH, Cai CF, et al. Thermochemical sulfate reduction and fluid flow evolution of the Lower Triassic Feixianguan Formation sour gas reservoirs; the Northeast Sichuan Basin, China. AAPG Bull. 2014b;98(5):947–73.

López-Horgue MA, Iriarte E, Schröder S, et al. Structurally controlled hydrothermal dolomites in Albian carbonates of the Asón valley, Basque Cantabrian Basin, Northern Spain. Mar Pet Geol. 2010;27(5):1069–92.

Mahdi TA, Aqrawi AAM. Sequence stratigraphic analysis of the mid-Cretaceous Mishrif Formation, Southern Mesopotamian Basin, Iraq. J Pet Geol. 2014;37(3):287–312.

Moutaz AD, Jassim AJ, Saad AJ. Depositional environments and porosity distribution in regressive limestone reservoirs of the Mishrif Formation, Southern Iraq. Arab J Geosci. 2010;3(1):67–78.

Müller DW, McKenzie JA, Mueller PA. Abu Dhabi sabkha, Persian Gulf, revisited: application of strontium isotopes to test an early dolomitization model. Geology. 1990;18(7):618–21.

Wang J, Liu HQ, Xu J, et al. Formation mechanism and distribution law of remaining oil in fracture-cavity reservoirs. Pet Explor Dev. 2012a;39(5):585–90 (in Chinese).

Wang SQ, Zhao L, Cheng XB, et al. Geochemical characteristics and genetic model of dolomite reservoirs in the eastern margin of the Pre-Caspian Basin. Pet Sci. 2012b;9:161–9.

Whitaker FF, Smart PL, Jones GD. Dolomitization: from conceptual to numerical models. Geol Soc. 2004;235:99–139.

Xu KQ. Characteristics of hydrocarbon migration and accumulation and exploration practice in the Eastern Margin of the Pre-Caspian Basin. Beijing: Petroleum Industry Press; 2011. p. 37–50.

Yao GS, Shen AJ, Pan WQ, et al. Carbonate reservoirs. Beijing: Petroleum Industry Press; 2008.

Yue DL, Wu SH, Lin CY, et al. Remaining oil distribution controlled by intercalation in reef limestone reservoir. Pet Explor Dev. 2005;32(4):113–8.

Zhao L, Li JX, Li KC, et al. Fracture development and formation mechanism for complex carbonate reservoirs—a case study on Kazakhstan Zahnanor oilfield. Pet Explor Dev. 2010;3:304–9 (in Chinese).

Zhao WZ, Shen AJ, Hu SY, et al. Geological conditions and distributional features of large-scale carbonate reservoirs onshore China. Pet Explor Dev. 2012;39(1):1–12 (in Chinese).

Zheng RC, Dang LR, Zheng C, et al. Diagenetic system of carbonate reservoirs in Huanglong Formation from the east Sichuan to north Chongqing area. Acta Pet Sin. 2010;31(2):237–45 (in Chinese).

Zheng RC, Wen HG, Zheng C, et al. Genesis of dolostone of the Feixianguan Formation, Lower Triassic in the NE Sichuan Basin: evidences from rock structure and strontium content and isotopic composition. Acta Pet Sin. 2009;25(10):2459–68 (in Chinese).

13

Reservoir stress path and induced seismic anisotropy: results from linking coupled fluid-flow/geomechanical simulation with seismic modelling

D. A. Angus[1,7] · Q. J. Fisher[1] · J. M. Segura[2,6] · J. P. Verdon[3] · J.-M. Kendall[3] · M. Dutko[4] · A. J. L. Crook[5]

Abstract We present a workflow linking coupled fluid-flow and geomechanical simulation with seismic modelling to predict seismic anisotropy induced by non-hydrostatic stress changes. We generate seismic models from coupled simulations to examine the relationship between reservoir geometry, stress path and seismic anisotropy. The results indicate that geometry influences the evolution of stress, which leads to stress-induced seismic anisotropy. Although stress anisotropy is high for the small reservoir, the effect of stress arching and the ability of the side-burden to support the excess load limit the overall change in effective stress and hence seismic anisotropy. For the extensive reservoir, stress anisotropy and induced seismic anisotropy are high. The extensive and elongate reservoirs experience significant compaction, where the inefficiency of the developed stress arching in the side-burden cannot support the excess load. The elongate reservoir displays significant stress asymmetry, with seismic anisotropy developing predominantly along the long-edge of the reservoir. We show that the link between stress path parameters and seismic anisotropy is complex, where the anisotropic symmetry is controlled not only by model geometry but also the nonlinear rock physics model used. Nevertheless, a workflow has been developed to model seismic anisotropy induced by non-hydrostatic stress changes, allowing field observations of anisotropy to be linked with geomechanical models.

Keywords Coupled fluid-flow/geomechanics · Reservoir characterization · Seismic anisotropy · Stress path

1 Introduction

Extraction and injection of fluids within hydrocarbon reservoirs alters the in situ pore pressure leading to changes in the effective stress field within the reservoir and surrounding rocks. However, changes in pore pressure do not necessarily lead to a hydrostatic change in effective stress. For instance, a reduction in fluid pressure within a reservoir is often accompanied by a slower increase in the minimum effective horizontal stress with respect to the vertical effective stress change (e.g., Segura et al. 2011). This asymmetry can result in the development of stress anisotropy that may promote elastic failure within the rock, such as fault reactivation and borehole deformation. From the perspective of seismic monitoring, changes in the stress field can lead to microseismicity as well as nonlinear changes in seismic velocity and, in cases where stress anisotropy develop, to stress-induced seismic anisotropy. This has important implications on the interpretation of time-lapse (4D) seismic as well as microseismic data, where stress anisotropy can result in anisotropic perturbations in the velocity field, offset and azimuthal variations in reflection amplitudes and shear-wave splitting.

✉ D. A. Angus
    D.Angus@leeds.ac.uk

[1] School of Earth and Environment, University of Leeds, Leeds, UK

[2] Formerly School of Earth and Environment, University of Leeds, Leeds, UK

[3] Department of Earth Sciences, University of Bristol, Bristol, UK

[4] Rockfield Software Ltd., Swansea, UK

[5] Three Cliffs Geomechanics, Swansea, UK

[6] Present Address: Repsol, Madrid, Spain

[7] Present Address: ESG Solutions, Kingston, Canada

Edited by Jie Hao

Reservoir stress path, expressed as the ratio of change of effective horizontal to effective vertical stress from an initial stress state, is a useful concept in characterizing the evolution of stress anisotropy due to production (e.g., Aziz and Settari 1979; Sayers 2007; Alassi et al. 2010). The stress path of a reservoir during production is sensitive to the geometry of the reservoir system, pore pressure and the material properties of the reservoir and surrounding rock mass (e.g., Segura et al. 2011). In the field, stress path can be measured from borehole pressure tests, and so maps of reservoir stress path are extrapolated out into the reservoir via limited discrete measurements. History matching borehole reservoir stress path parameters with seismic anisotropy measurements may provide a more reliable prediction of reservoir stress path throughout the reservoir volume. However, linking seismic anisotropy measurements with stress path requires a better understanding of the link between geomechanical deformation (evolution of stress and strain), fluid-flow, rock physical properties and seismic attributes.

Recent studies focusing on linking numerical coupled fluid-flow and geomechanical simulation with seismic modelling have improved our understanding of the relationship between seismic attributes, fluid properties and mechanical deformation due to reservoir fluid extraction and injection (e.g., Rutqvist et al. 2002; Dean et al. 2003; Herwanger and Horne 2009; Alassi et al. 2010; Herwanger et al. 2010; Verdon et al. 2011; He et al. 2015; Angus et al. 2015). Analytic and semi-analytic approaches using poroelastic formulations have previously been used to understand surface subsidence (e.g., Geertsma 1973) and seismic travel-time shifts (e.g., Fuck et al. 2009; Fuck et al. 2010) due to pore pressure changes. Coupled fluid-flow and geomechanical numerical simulation algorithms integrate the influence of multi-phase fluid-flow as well as deviatoric stress and strain to provide more accurate models of the spatial and temporal behaviour of various rock properties within and outside the reservoir (e.g., Herwanger et al. 2010). Linking changes in reservoir physical properties, such as porosity, permeability and bulk modulus, to changes in seismic attributes is accomplished via rock physics models (e.g., Prioul et al. 2004) to generate so-called dynamic (high strain rate and low strain magnitude suitable for seismic frequencies) elastic models.

In this paper, we investigate the sensitivity of seismic anisotropy to reservoir stress path using the micro-structural nonlinear rock physics model of Verdon et al. (2008). This work follows the coupled fluid-flow and geomechanical characterization of reservoir stress path of Segura et al. (2011), who explore the influence of reservoir geometry and material property contrast on stress path for poroelastic media. We present results from coupled fluid-flow and geomechanical simulations for the same

geometries to investigate stress-induced seismic anisotropy. A major point of departure of our approach to the approaches of Rutqvist et al. (2002), Herwanger and Horne (2009) and Fuck et al. (2009) is that we extend the material behaviour from poroelastic to include plasticity (i.e., so-called poroelastoplastic behaviour). Poroelastoplasticity can incorporate matrix failure during simulation, allowing strain hardening and weakening to develop within the model. This is especially important for modelling reservoir stress path and stress path asymmetry. Furthermore, poroelastoplasticity also enables the prediction of when and where failure occurs in the model, allowing us to model the likely microseismic response of a reservoir (Angus et al. 2010, 2015).

2 Modelling approach

2.1 Coupled fluid-flow and geomechanical simulation

Industry-standard fluid-flow simulation algorithms solve the equations of flow for multi-phase fluids (e.g., Aziz and Settari 1979), but neglect the influence of changing pore pressure on the geomechanical behaviour of the reservoir and surrounding rock. Formulations exist for fully coupled fluid-flow and geomechanical simulation, yet they tend to be computationally expensive (e.g., Minkoff et al. 2003). However, iterative and loose coupling of fluid-flow simulators with geomechanical solvers can be more efficient and yield sufficiently accurate results compare to fully coupled solutions (e.g., Dean et al. 2003; Minkoff et al. 2003). Furthermore, iterative and loosely coupled approaches allow the use of already existing commercial reservoir fluid-flow modelling software. In this paper, the coupled fluid-flow and geomechanical simulations are performed using the finite-element geomechanical solver ELFEN (Rockfield Software Ltd.) linked with the commercial fluid-flow simulation package TEMPEST (Roxar), where the simulations are loosely coupled using a message-passing interface (Muntz et al. 2007).

Predicting the geomechanical response of reservoirs depends on the ability of the geomechanical solver to model the nonlinear behaviour of rocks. The nonlinear dependence of rocks with stress is generally attributed to closure of microcracks and pores, as well as increasing grain boundary contact with increasing confining stress (e.g., Nur and Simmons 1969). Rocks also display stress hysteresis (e.g., Helbig and Rasolofosaon 2000), and this hysteresis has been observed to occur not only at large strains but also small strains (e.g., Johnson and Rasolofosaon 1996). This observation represents a potentially important rock characteristic in explaining the asymmetric behaviour of 4D seismic observations of producing reservoirs (e.g., Hatchell and Bourne 2005). Thus, it is important to incorporate such nonlinear and hysteretic properties within a constitutive model for coupled flow-geomechanical simulation. The constitutive relationships used by ELFEN are derived from laboratory experiments that incorporate linear elastic and plastic behaviour (e.g., Crook et al. 2002) as well as lithology specific behaviour (e.g., Crook et al. 2006). Specifically, the constitutive model used for the simulations within this paper is the so-called SR3 model. This model is defined as a single-surface rate-independent non-associated elastoplastic model that includes geomechanical anisotropy, rate dependence and creep into the basic material characterization (e.g., Crook et al. 2006). In other words, the constitutive model can include the effects of both linear elastic and nonlinear static elastoplastic response.

2.2 Micro-structural nonlinear rock physics model

To model the seismic response due to geomechanical deformation, rock physics model is required to link changes in fluid saturation, pore pressure and triaxial stresses to changes in the dynamic elastic stiffness. Rock physics models should incorporate phenomena observed in both laboratory core experiments and in the field, such as the nonlinear stress-velocity response (e.g., Nur and Simmons 1969; Sayers 2007; Hatchell and Bourne 2005) and the development of stress-induced anisotropy in initially isotropic rocks (e.g., Dewhurst and Siggins 2006; Olofsson et al. 2003).

The model we have developed is based on the approach outlined by Sayers and Kachanov (1995) and Schoenberg and Sayers (1995), where the overall compliance of the rock S_{ijkl} (compliance being the inverse of stiffness) is a function of the background compliance of the rock frame, S_{ijkl}^0, plus additional compliance introduced by the presence of low aspect ratio, highly compliant pore space ΔS_{ijkl} (such as microcracks or grain boundaries),

$$S_{ijkl} = S_{ijkl}^0 + \Delta S_{ijkl}. \tag{1}$$

S_{ijkl}^0 can be estimated from either the mineral composition (e.g., Kendall et al. 2007) or the behaviour at high effective stresses, where it is assumed that the compliant pore space is completely closed (e.g., Sayers 2002). The additional compliance can be modelled using second- and fourth-rank crack density tensors α_{ij} and β_{ijkl}, respectively,

$$\Delta S_{ijkl} = \frac{1}{4}\left(\delta_{ik}\alpha_{jl} + \delta_{il}\alpha_{jk} + \delta_{jk}\alpha_{il} + \delta_{jl}\alpha_{ik}\right) + \beta_{ijkl}. \tag{2}$$

Sayers (2002), Hall et al. (2008) and Verdon et al. (2008) apply this micro-structural formulation to invert for stress-dependent elastic stiffness and observe that the behaviour of sedimentary rock can be modelled adequately using the second-rank crack density tensor α_{ij} and assuming the fourth-rank crack density tensor β_{ijkl} is negligible.

Based on this micro-structural approach, Verdon et al. (2008) incorporate the analytical formulation of Tod (2002) to predict the response of the crack density tensor to changes in effective stress. The crack number density (hereafter referred to as crack density) for each diagonal component of α_{ij} is expressed as a function of the initial crack density at a reference stress state, ε_i^0, and the average initial crack aspect ratio, a_i^0, at this reference stress state.

$$\alpha_{ii} = \frac{\varepsilon_i^0}{h_i}e^{\left(-c_r\sigma_{ii}^e\right)}, \tag{3}$$

where

$$c_r = \frac{\lambda_i + 2\mu_i}{(\pi\mu_i a_i^0)(\lambda_i + \mu_i)} \quad \text{and} \quad h_i = \frac{3E_i(2 - v_i)}{32(1 - v_i^2)} \tag{4}$$

E_i is Young's modulus, v_i is Poisson's ratio, and λ_i and μ_i are the Lame constants of the background material. σ_{ii}^e is the principal effective stress in the ith direction. This derivation yields an expression for the dynamic elastic stiffness that models stress-dependent seismic velocities and seismic anisotropy induced by non-hydrostatic stress fields.

The nonlinear rock physics model is incorporated within an aggregate elastic model (see Angus et al. 2011). The approach has the benefit of allowing us to incorporate the many causes of seismic anisotropy that act on multiple length-scales. Intrinsic anisotropy, caused by alignment of anisotropic mineral crystals (such as clays and micas), is included using an anisotropic background elasticity S_{ijkl}^0. Stress-induced anisotropy is incorporated implicitly within our rock physics model. For instance, even if initial crack density terms are isotropic ($\varepsilon_1^0 = \varepsilon_2^0 = \varepsilon_3^0$), the second-order crack density terms are anisotropic ($\alpha_{11} \neq \alpha_{22} \neq \alpha_{33}$) unless the stress field is hydrostatic. Finally, the influence of larger-scale fracture sets can also be modelled using the Schoenberg and Sayers (1995) effective medium approach, adding the additional compliance of the larger fracture sets to the stress-sensitive compliance computed in Eq. (6). Fluid substitution can also be included into this rock physics model, using either the Brown and Korringa (1975) anisotropic extension to Gassmann's equation, which is appropriate as a low-frequency end member, or incorporating the dispersive effects of squirt flow between pores (e.g., Chapman 2003). In this paper, we focus on the development of stress-induced anisotropy, assuming that the rock has no intrinsic anisotropy, and that large-scale fracture sets are not present. Although squirt flow has been shown to generate observable seismic anisotropy (e.g., Maultzsch et al. 2003; Baird et al. 2013), in this paper, we focus on the influence of stress on seismic anisotropy and so ignore fluid substitution and squirt-flow effects.

The necessary input parameters for the nonlinear analytical model are the background elasticity ($C^0_{ijkl} = 1/S^0_{ijkl}$), effective triaxial stress tensor (σ^e_{ijkl}), and the initial crack density and aspect ratio (ε^0_i and a^0_i). Populating dynamic stress-dependent elastic models from the coupled flow-geomechanical simulation are achieved by passing the background static elastic tensor, rock density, stress tensor and pore pressure for each grid point within the model. The static stiffness is often observed empirically to correlate with dynamic stiffness (e.g., Olsen et al. 2008), providing a potential starting point for seismic modelling when independent estimates of the initial, pre-production seismic velocities are not available. Because we have no independent seismic information for these idealized models, we assume that the initial dynamic stiffness is scaled to the static stiffness. Additionally, there are parameters needed for the rock physics model that are not provided by the geomechanical simulation, in particular the initial crack density and aspect ratio. These parameters are derived from stress-velocity behaviour observed in core samples. Angus et al. (2009, 2012) provide a catalogue of over 200 of such measurements for a range of lithologies, inverting for ε^0 and a^0 to provide constraints for typical values of these parameters.

Before performing the coupled flow-geomechanical simulations, a geomechanical equilibration stage is required for all model geometries. Specifically, the stress state within the models evolves from an initial equilibrium state where the horizontal effective stresses are defined as a function of the vertical effective stress using horizontal stress coefficients. Thus, the initial stress field is non-hydrostatic and controlled by the reservoir geometry, model material properties and initial depth-dependent pore pressure. Application of the analytic stress-dependent rock physics model (Eqs. 1–3) would lead to initially anisotropic elasticity due to the non-hydrostatic effective stresses. To focus solely on the development of stress-dependent anisotropy related to production-induced changes in effective stresses, we include a stress initiation term ΔS^{init}_{ijkl} in Eq. (1), to ensure that the initial overall compliance S_{ijkl} is isotropic and scaled to the inverse of the static geomechanical elastic stiffness:

$$S_{ijkl} = S^0_{ijkl} + \Delta S_{ijkl} - \Delta S^{init}_{ijkl}, \tag{5}$$

where the stress initialization second-rank crack density term is defined

$$\alpha^{init}_{ii} = \frac{\varepsilon^0_i}{h_i} e^{\left(-c_r \sigma^{init}_{ii}\right)} \tag{6}$$

and σ^{init}_{ii} is the initial (baseline) principal effective stress in the ith direction. Including the stress initiation term

prescribes an initially isotropic elastic tensor equal to the static elastic tensor provided from the geomechanical solver. However, there is flexibility to incorporate various forms of anisotropy, which can be due to sedimentary and/or tectonic fabric (e.g., fine layering and fractures) as well as a basin and/or regionally developed stress related anisotropy (e.g., stress disequilibrium related to basin uplift) by adding additional anisotropic compliance terms. It should be noted that we use the static elasticity to compute the seismic velocities, and hence the magnitude of the seismic velocities is lower than typically observed in the field. Given that we are considering simple models, we choose not to perform a static-to-dynamic elasticity conversion (e.g., Angus et al. 2011) as would typically be done for field studies (e.g., He et al. 2016a). As this would involve a constant shift, not performing a static-to-dynamic elasticity conversion will not effect the main conclusions of this paper.

2.3 Stress path

Segura et al. (2011) model the influence of reservoir geometry and material properties on stress path using a more extensive suite of models considered in this paper. Using poroelastic constitutive material behaviour, Segura et al. (2011) observed that the stress arching effect is significant in small, thin reservoirs that are soft compared to the surrounding rock. Under such circumstances, the stresses will not evolve within the reservoir and so stress evolution occurs primarily in the overburden and sideburden. Furthermore, stiff reservoirs do not display any stress arching regardless of the geometry. Stress anisotropy decreases with reduction in bounding material strength (e.g., Young's modulus), and this is especially true for small reservoirs. However, when the dimensions extend in one or two lateral directions the reservoir deforms uniaxially and the horizontal stresses are controlled by the reservoir Poisson's ratio.

To understand the stress path parameters, it is helpful to review the concept of effective stress and the Mohr circle. The concept of stress path is based on Terzhagi (1943) effective stress principle and is expressed assuming compression is positive:

$$\sigma^e = \sigma - bP, \tag{7}$$

where σ^e is the effective stress, σ is the total stress, P is pressure, and b is Biot's coefficient (which we assume is 1 for simplicity in this paper). The Mohr circle is an effective graphical representation of the stress state for a material point (e.g., Jaeger et al. 2007). The Mohr circle allows one to evaluate how close a region is to elastic failure assuming the normal and shear strength is known. The Mohr circle is defined in terms of the principle stresses considering the normal σ^e) and the shear stresses (τ) on a plane at an angle θ:

$$\sigma^e = \frac{\sigma_3^e + \sigma_1^e}{2} + \frac{\sigma_3^e - \sigma_1^e}{2}\cos\theta \qquad (8)$$

$$\tau = \frac{\sigma_3^e - \sigma_1^e}{2}\sin\theta, \qquad (9)$$

where σ_3^e and σ_1^e are the maximum and minimum principle stresses, respectively.

The stress path parameters describe the evolution of the Mohr circle and are defined by three terms, the stress arching parameter γ_3, the horizontal stress path parameter γ_1 and the deviatoric stress path parameter or stress anisotropy parameter K. Since only two of the three parameters are independent we choose γ_3 and K as the reference parameters (e.g., Segura et al. 2011). The stress arching parameter is defined

$$\gamma_3 = \frac{\Delta\sigma_3^e}{\Delta P} \qquad (10)$$

and describes the development of stress arching, where γ_3 high indicates stress arching is occurring with very little stress evolution in the reservoir. The stress anisotropy parameter is defined

$$K = \frac{\Delta\sigma_1^e}{\Delta\sigma_3^e} \qquad (11)$$

and describes the development of stress anisotropy, where K low indicates increase in stress anisotropy with lower changes in horizontal effective stress with respect to changes in vertical effective stress.

3 Numerical examples

3.1 Geomechanical model

Segura et al. (2011) generated a series of 3D numerically coupled poroelastic hydro-mechanical models to investigate the influence of reservoir geometry and material property contrast on the development of reservoir stress path. In their paper, they show the importance of reservoir geometry and material property discontinuities on the development of stress anisotropy. In this paper, we focus on a subset of those reservoir geometries (see Fig. 1) and extended the simulation to include plasticity. The three reservoir geometries are described by a rectilinear sandstone reservoir at depth of 3050 m and having vertical thickness of 76 m. To reduce the computational requirements, the model is reduced to one-quarter geometry based on symmetry arguments. A vertical production well is located in the centre of the reservoir (i.e., at the origin) and produces until the pore pressure declines to 10 MPa within

the reservoir. The surrounding volume is defined laterally 10 km × 10 km and vertically 3220 m, where the non-reservoir rock is shale. The lateral dimensions of the three reservoir geometries are:

- Small reservoir: lateral dimension 190.5 m × 190.5 m
- Elongate reservoir: lateral dimension 4000 m × 200 m
- Extensive reservoir: lateral dimension 4000 m × 4000 m

At reservoir depth, the strength of the overburden and reservoir is equivalent (see Segura et al. 2011) for discussion of geomechanical model parameters). Although ELFEN is capable of incorporating anisotropic elastic material within the geomechanical simulation, we limit the material elasticity to isotropy to allow a clear analysis of geometry related stress-induced anisotropy.

3.2 Stress path evolution

Figure 2 plots the evolution of the stresses during production for several specified points in the reservoir: at the production well and at the edges of the reservoir (the locations of points 1, 2 and 3 can be seen in Fig. 3). The stress path parameters can be estimated from the slopes of these curves. The slope of the curves in the top panels represents the stress arching parameter γ_3 and the slope of the curves in the bottom panels the stress anisotropy parameter K. The stress path development for both the stress arching and stress anisotropy parameters is linear for the small reservoir ($K \approx 0.4$ and $\gamma_3 \approx 0.3$). This stress path development is characteristic of elastic behaviour. However, the stress path development becomes progressively nonlinear for the elongate and extensive geometry, respectively. For the extensive reservoir, the evolution of the stress anisotropy is characteristic of uniaxial compaction (see Fig. 5 in Pouya et al. 1998). Initially, the stress anisotropy has an elastic phase (low K) and then evolves asymptotically into another linear final trend. The transition occurs, while the material undergoes shear-enhanced compaction (stress state intersects the yield surface and plastic consolidation). The final linear trend depends on the plastic potential of the constitutive model (i.e., is a function of the strain hardening). The stress arching is initially low but increases as failure within the reservoir increases and sheds the load onto the side-burden. For the elongate reservoir, the asymmetry of the geometry leads to a behaviour differing from uniaxial compaction. The stress arching is relatively linear and high ($\gamma_3 \approx 0.9$), yet the stress anisotropy transitions from high ($K \approx 0.9$) to moderate ($K \approx 0.4$).

The linear trends of these curves are shown in Table 1 along with the poroelastic predictions (see Segura et al. 2011). There are similarities between the poroelastic and

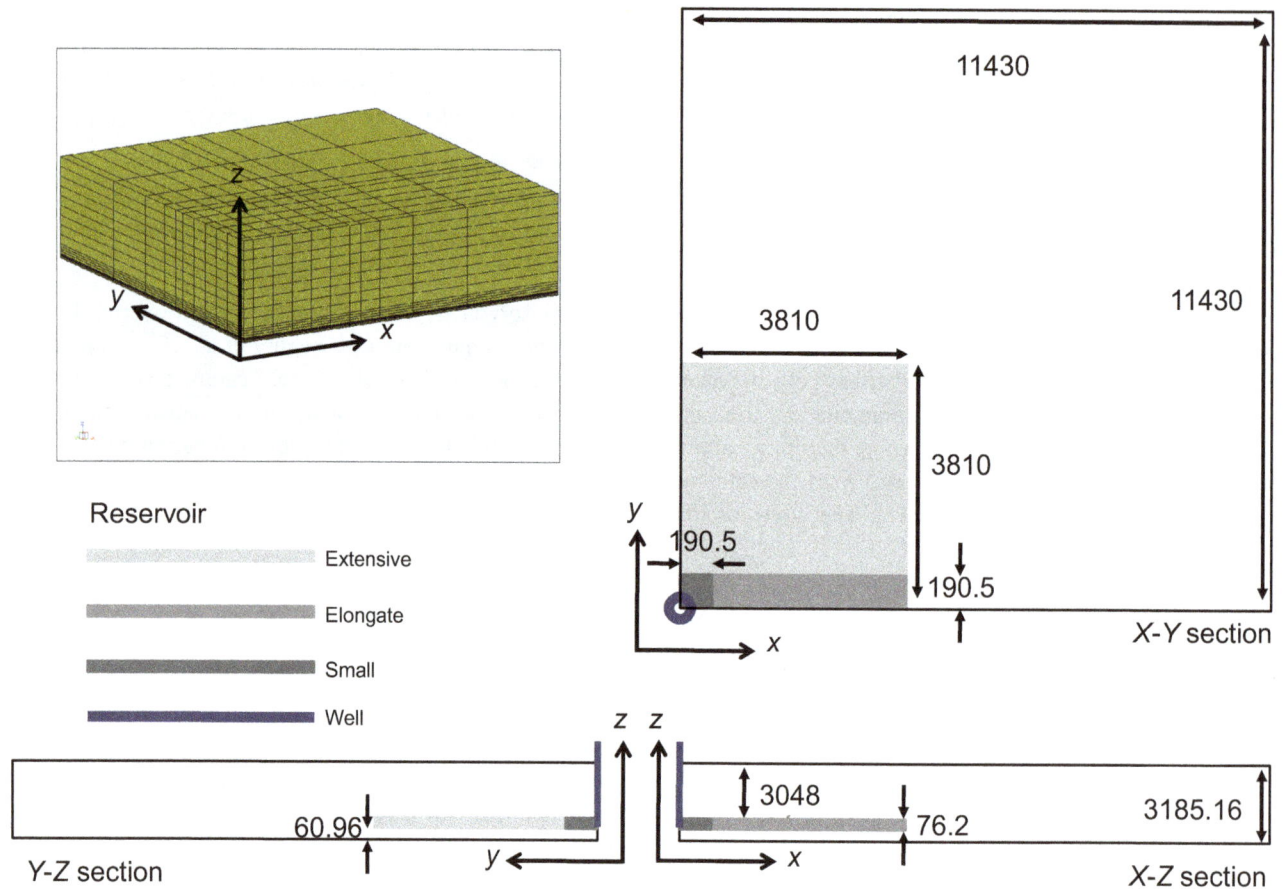

Fig. 1 Geometry of the three simple reservoir models (all spatial units are in metres). The structured finite-element mesh used in the geomechanical simulation is illustrated *top-left*, and the locations of the three reservoir geometries are displayed in *X–Y* (*top-right*), *X–Z* (*bottom-right*) and *Y–Z* (*bottom-left*) sections

poroelastoplastic case for the small geometry, with the exception of more moderate stress arching along the boundaries of the poroelastoplastic case. However, there are noticeable differences between the poroelastic and poroelastoplastic simulation results for the elongate and extensive geometries. The elongate geometry shows greater stress arching with similar moderate stress anisotropy (after the transition from high). The evolution of the stress parameters in the extensive model is more complicated (i.e., nonlinear). The stress arching evolves from low to moderate, whereas the stress anisotropy fluctuates from low to high and then moderate.

3.3 Seismic anisotropy

We use the modelled stress tensors at the end of production to compute the development of stress-induced *P*-wave anisotropy. To do so, we assume the initial crack densities (ε_i^0) and initial aspect ratios (a_i^0) are isotropic (e.g., $\varepsilon_x^0 = \varepsilon_y^0 = \varepsilon_z^0$) for simplicity. This could be relaxed if there

were prior petrophysical information to suggest otherwise; in real field examples this is likely to be the case (e.g., Crampin 2003). Following the calibration studies of Angus et al. (2009, 2011), we choose $\varepsilon_i^0 = 0.25$ and $a_i^0 = 0.001$ for the sandstone reservoir and $\varepsilon_i^0 = 0.125$ and $a_i^0 = 0.005$ for the surrounding shale. These values are taken as representative of the global trend for sandstones and shales observed by Angus et al. (2009, 2011). However, these measurements were biased towards rocks sampled from reservoir depths, with limited data from shallower cores (which are often of less interest commercially). For the extensive reservoir, stress changes are observed to occur as a result of production throughout the overburden even up to the surface. It is unclear whether the trends observed by Angus et al. (2009, 2011) are suitable for softer, poorly consolidated near-surface material. Thus, for the extensive reservoir geometry, an additional simulation is performed using the same initial crack densities, but scaling the initial aspect ratio with depth, with an aspect ratio of 0.001 at the base of the model and increasing to 0.01 at the surface.

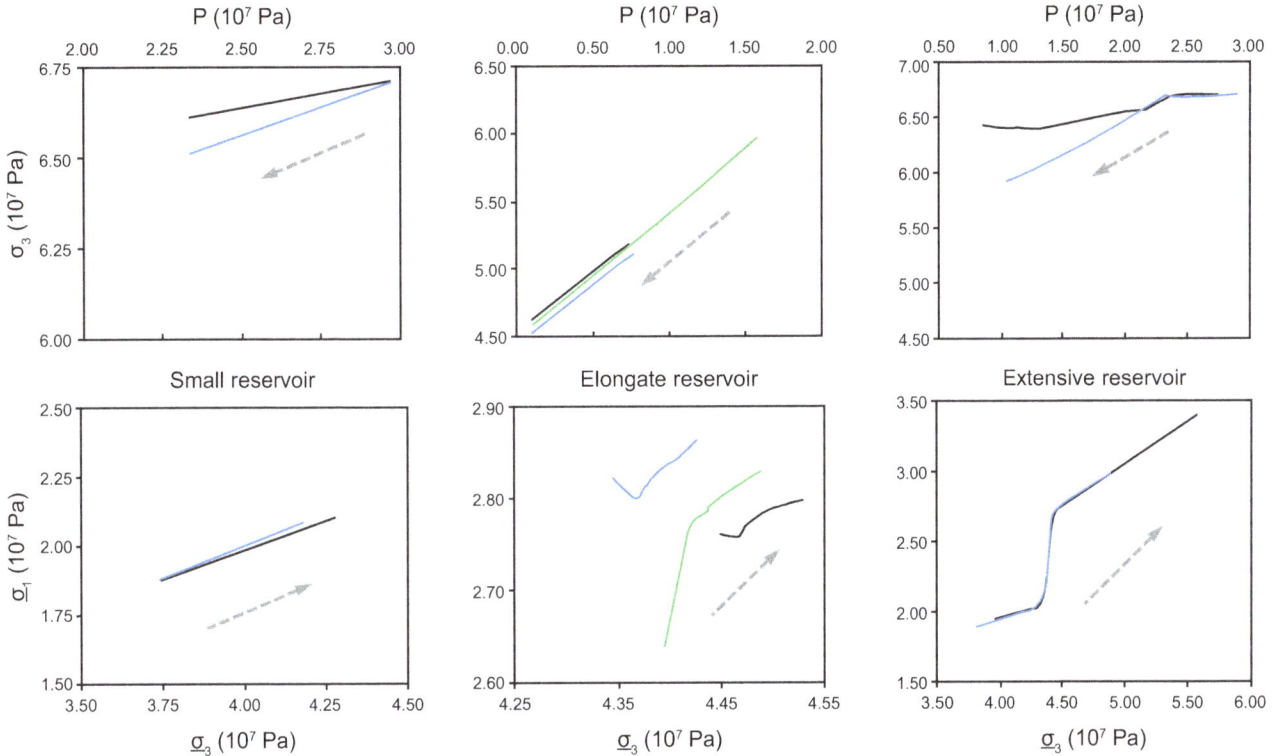

Fig. 2 Evolution of stress path parameters for the three reservoir geometries: (*left column*) small reservoir, (*middle column*) elongate reservoir and (*right column*) extensive reservoir. The slope of the curves in *each panel* represents the stress path parameters: (*top row*) stress arching parameter 3 and (*bottom row*) stress anisotropy parameter K. The *solid lines* are *colour-coded black* for point 1, *green* for point 2 and *blue* for point 3. The *dashed arrow* represents the direction of evolution of the stress path parameter

This is done to replicate shallow core measurements of shale velocity stress dependence (e.g., Podio et al. 1968), where increasing the aspect ratio tends to reduce the overall stress dependence except at very low confining stresses.

3.4 Small reservoir geometry

The small reservoir geometry is characteristic of a highly compartmentalized reservoir, with limited spatial extent. The stress-induced anisotropy that develops during production is plotted in Figs. 3 and 4. These plots show the maximum P-wave anisotropy for near-vertical incidence waves (0°–30°), as well as upper hemisphere plots showing the P-wave velocity at all incidence angles for specified points in and around the reservoir. The P-wave anisotropy is confined to a small volume surrounding the sandstone reservoir with modelled P-wave anisotropy >1 %.

For points within the reservoir, we observe an approximately hexagonal anisotropic symmetry, where the maximum P-wave velocity is vertical. This implies that the reservoir is compacting vertically (closing of microcracks that are oriented horizontally, increasing vertical P-wave velocities). For points outside the reservoir, hexagonal symmetry is again observed, but the vertical P-wave

velocity is now the minimum velocity, implying vertical extension (opening of microcracks that are oriented horizontally, reducing vertical P-wave velocities).

However, there is in fact an observed reduction in both the P- and S-wave velocities throughout the reservoir on the order of 0.5 % or less. For point 1, the maximum P-wave velocity (which is vertical) is 1666 m/s, yet the initial isotropic pre-production P-wave velocity was 1672 m/s. Points 2 and 3 within the reservoir adjacent to the boundary also display sub-vertical maximum P-wave velocity, with minimum P-wave velocity oriented horizontally perpendicular to the reservoir edge. This implies that the maximum horizontal stress is parallel to the reservoir edge. For points 5 and 6, along the borehole above and below the reservoir, the post-production anisotropic symmetries predict minimum vertical velocities equal to the initial pre-production velocities and maximum sub-horizontal P-wave velocities larger than pre-production values.

Table 2 summarizes the anisotropic symmetry decomposition of the stress-induced anisotropy using the approach of Browaeys and Chevrot (2004). Although the stress-induced anisotropy is weak (i.e., isotropic components are all 99 %) for all grid points, there are components of the elastic tensor that require hexagonal and

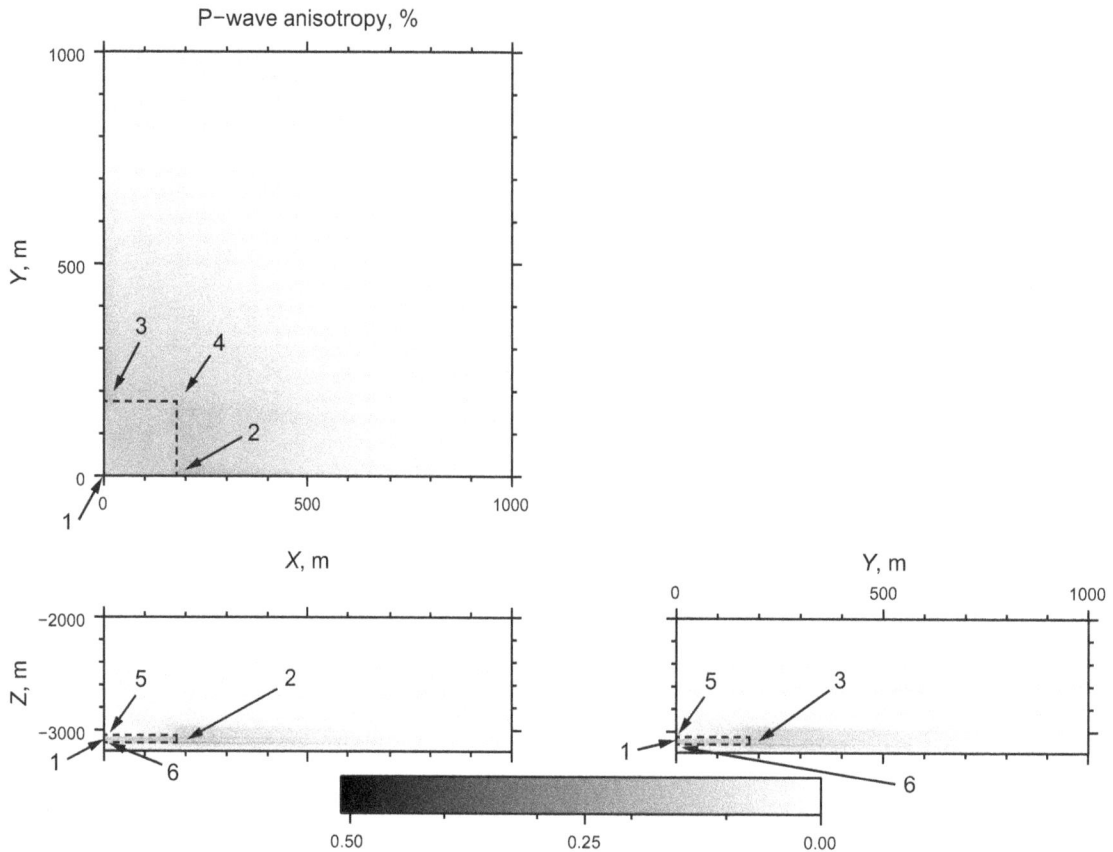

Fig. 3 Contour plot of maximum *P*-wave anisotropy (%) for near offset seismic propagation (0° to 30°) for the small reservoir geometry. (In this and subsequent figures, the *x–z* section is at *y* = 0 m, the *y–z* section is at *x* = 0 m, and the *x–y* section is at *z* = −3000 m. The reservoir is defined as the region within the *dashed lines*)

Table 1 Stress parameters for poroelastic and poroelastoplastic geomechanical simulations. Note that the values for the extensive reservoir are estimated from the initial and final stage and so neglect the nonlinear stress path seen in Fig. 2

Model	Poroelastic			Poroelastoplastic		
	Point	K	γ_3	Point	K	γ_3
Small	1	0.42	0.12	1	0.41	0.16
	2	0.47	0.18	2	0.46	0.31
	3	0.47	0.18	3	0.46	0.31
Elongate	1	0.35	0.10	1	*n to m*	0.88
	2	0.40	0.12	2	*h to m*	0.94
	3	0.40	0.12	3	*n to m*	0.88
Extensive	1	0.33	0.00	1	*l–h–m*	*n to l*
	2	0.37	0.07	2	*l–h–m*	*m to l*
	3	0.37	0.07	3	*l–h–m*	*m to l*

n to *m* refers to transition from negative to moderate, *h* to *m* from high to moderate, *m* to *l* from moderate to low and *l–h–m* transition from low to high and then moderate

orthorhombic symmetry and hence do not fit elliptical anisotropy. These results are not consistent with other studies that suggest that $\Delta\varepsilon = \Delta\delta$ (e.g., Fuck et al. 2010;

Herwanger and Horne 2009). Rasolofosaon (1998) suggests that the lowest order of stress-induced anisotropic symmetry can be at most elliptical anisotropy, $\Delta\varepsilon = \Delta\delta$. However, this is only the case when considering third-order elasticity theory and assuming isotropic third-order elastic tensors (see Fuck and Tsvankin 2009). For the microstructural nonlinear formulation in this paper, we are not limited to elliptical anisotropy.

The results from the small geometry are counter-intuitive and are not consistent with model predictions from other simulations (e.g., Fuck et al. 2009; Herwanger and Horne 2009), where reservoir compaction results in an increase in seismic velocities within the reservoir, and reduction in velocities in the over- and under-burden. Although we would expect the weak reservoir sandstone to deform under the increased effective stress conditions due to pore pressure reduction, we observe instead stress arching occurring within the vicinity of the reservoir. Since the strength of the reservoir and surrounding shale is approximately equal, the shale acts to support deformation occurring within the reservoir. In terms of the rock physics model, the reduction in pore pressure leads to an increase

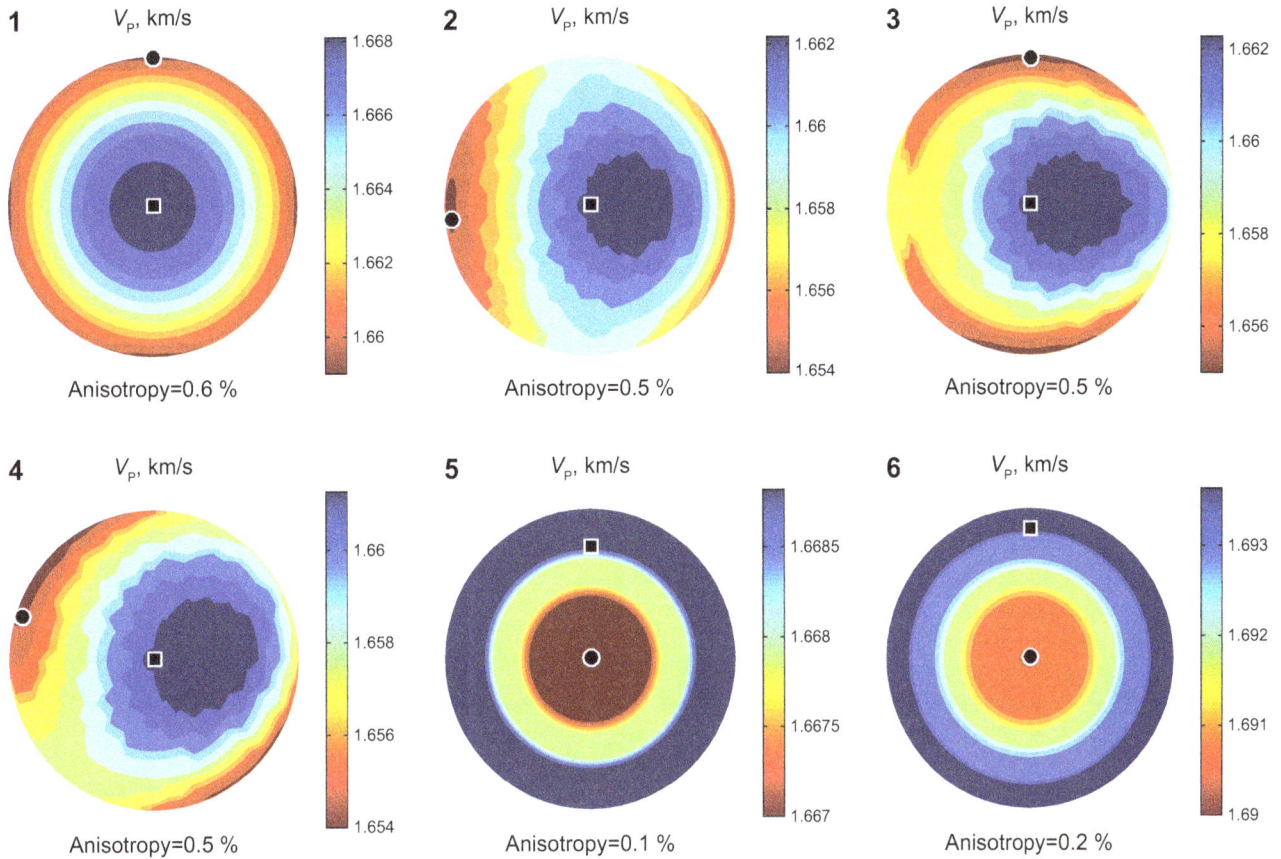

Fig. 4 Upper-hemisphere plots of *P*-wave phase velocity for various points in the small reservoir geometry (see Fig. 3)

Table 2 Decomposition of elastic tensor for all anisotropic symmetries for the small reservoir (labelled points shown in Fig. 3)

Point	Isotropic	Hexagonal	Tetragonal	Orthorhombic	Monoclinic	Triclinic
1	99.57	0.43	0.00	0.00	0.00	0.00
2	99.67	0.16	0.00	0.17	0.00	0.00
3	99.67	0.18	0.00	0.15	0.00	0.00
4	99.63	0.20	0.00	0.17	0.00	0.00
5	99.89	0.01	0.00	0.10	0.00	0.00
6	99.87	0.02	0.00	0.11	0.00	0.00

in microcracks (i.e., opening of existing cracks) with very little reservoir rock compaction and vertical extension above and below in the shale. The changes in seismic attributes suggest that most of the deformation is occurring inside and within the immediate vicinity of the reservoir with minimal influence on the surrounding shale. This is expected due to the small spatial dimensions of the reservoir, where the impact of pressure depletion limits the strength and spatial extend of stress redistribution.

3.5 Elongate reservoir geometry

For this geometry, the elongate reservoir is characteristic of a relatively large compartmentalized reservoir, such as a horst bounded by impermeable faults. In Figs. 5 and 6, the

P-wave anisotropy is no longer confined to a small volume immediately surrounding the sandstone reservoir, but extends laterally away from the long-axis (*y*-axis) of the reservoir by as much as 500 m. There is also a weak increase in *P*-wave anisotropy laterally away from the short-axis, and vertically towards the surface. The largest predicted anisotropy is as large as 3 % and mainly within the side-burden adjacent to the long-axis of the reservoir.

Focusing on the anisotropic symmetries, we observe noticeable differences from the small reservoir. For point 1, in the reservoir adjacent to the well, the maximum *P*-wave velocity is 1840 m/s and vertical, a decrease from the initial pre-production *P*-wave velocity of 1863 m/s similar to the small reservoir geometry. Points 2 and 4 within the reservoir adjacent to the short boundary display

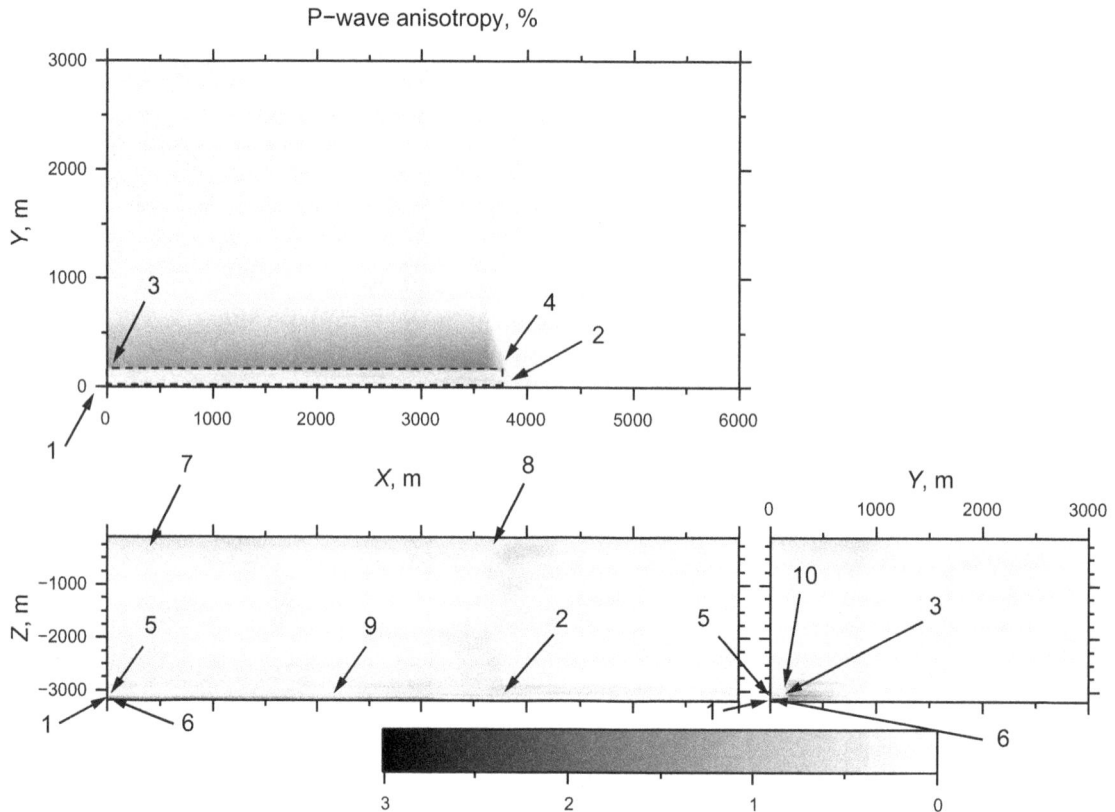

Fig. 5 Contour plot of maximum *P*-wave anisotropy (%) for near offset seismic propagation (0° to 30°) for the elongate reservoir geometry

sub-horizontal maximum *P*-wave velocity (an increase of up to 4 m/s from pre-production) perpendicular to the long-axis, implying a preferred orientation of vertical microcracks oriented parallel to the reservoir short-axis. Point 3 within the reservoir along the long-axis shows sub-horizontal maximum *P*-wave velocity (an increase from pre-production) parallel to the long-axis. The post-production anisotropic symmetry for point 5 predicts maximum *P*-wave velocity slightly greater than pre-production (2 m/s) normal to the long-axis with microcracks oriented sub-vertically perpendicular to the *x*-axis. For point 6, the symmetry appears to be rotated by 90° about the vertical axis with horizontal maximum *P*-wave velocity parallel to the *x*-axis (an increase of approximately 20 m/s) with vertically oriented microcracks parallel to the long-axis. Points 7 and 8 in the near sub-surface indicate slight extension with the opening of horizontal microcracks (decrease of vertical velocity of 1 m/s) with a more prominent horizontal velocity increase (roughly 5 m/s) perpendicular to the long-axis. Points 9 and 10 represent regions adjacent to the reservoir in the overburden some distance from the borehole and show a sub-horizontal increase in velocity along the *y*-axis with microcracks oriented sub-horizontally. Table 3 summarizes the anisotropic symmetry decompositions for the elongate reservoir points.

These results are slightly more intuitive than those of the small geometry. In this model, we still see a velocity reduction near the well within the reservoir, but there is now compaction occurring along the edges of the reservoir, albeit horizontal and not vertical. Thus, we are still observing stress arching above the reservoir, but with some of the load "pushing" into the sides of the reservoir. The changes in seismic attributes suggest that the deformation is no longer confined to the immediate vicinity of the reservoir, where we are observing significant perturbations in the side-burden as well as near the surface.

3.6 Extensive reservoir geometry

The extensive reservoir geometry is characteristic of a non-compartmentalized reservoir. Figures 7 and 8 display the results of the modelled *P*-wave anisotropy based on using the un-scaled initial aspect ratio (i.e., the same rock physics model parameters as was used for the other reservoir geometries). The *P*-wave anisotropy is on the order of 25 % near the surface of the model and hence any anisotropy within the vicinity of the reservoir is overshadowed by the surface perturbations. The anisotropic symmetry for point 7 near the well displays characteristic subsidence pattern with extension vertically (i.e., horizontal microcrack development) and radial horizontal compression.

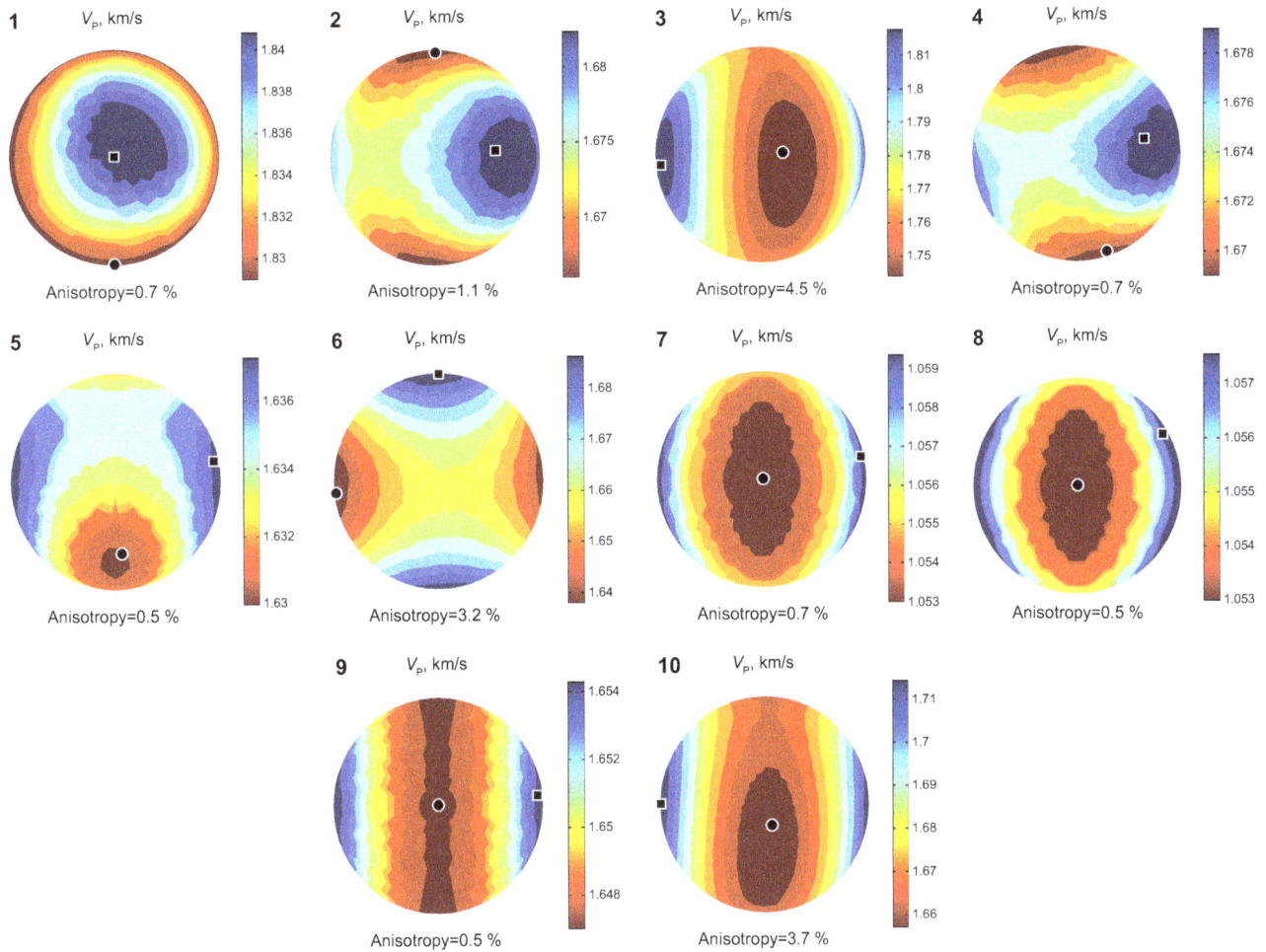

Fig. 6 Upper-hemisphere plots of *P*-wave phase velocity for various points in the elongate reservoir geometry (see Fig. 5)

Table 3 Decomposition of elastic tensor for all anisotropic symmetries for elongate reservoir (labelled points shown in Fig. 4)

Point	Isotropic	Hexagonal	Tetragonal	Orthorhombic	Monoclinic	Triclinic
1	99.45	0.53	0.00	0.01	0.01	0.00
2	99.27	0.37	0.00	0.36	0.00	0.00
3	96.75	2.43	0.00	0.82	0.00	0.00
4	99.51	0.25	0.00	0.24	0.00	0.00
5	99.74	0.13	0.00	0.13	0.00	0.00
6	97.83	1.21	0.00	0.96	0.00	0.00
7	99.49	0.34	0.00	0.17	0.00	0.00
8	99.58	0.24	0.00	0.18	0.00	0.00
9	99.60	0.35	0.00	0.01	0.04	0.00
10	97.23	2.38	0.00	0.39	0.00	0.00

Point 8 is near the edge of the subsiding region and displays sub-horizontal compression in the *y*-direction (or tangential to the subsidence bowl) with sub-vertical microcracks oriented along the *y*-axis (tangential to the subsidence bowl). This result is consistent with fast shear-wave polarization observations and predictions at Valhall (see Fig. 15 of Herwanger and Horne 2009).

Figures 9 and 10 show the results of the anisotropy predictions after defining a depth-dependent initial aspect ratio in order to focus in on perturbations within the vicinity of the reservoir. The results indicate that *P*-wave anisotropy is of the order of 2 % within the side-burden and over-burden. The volume of rock affected extends laterally away from the reservoir boundary by as much as

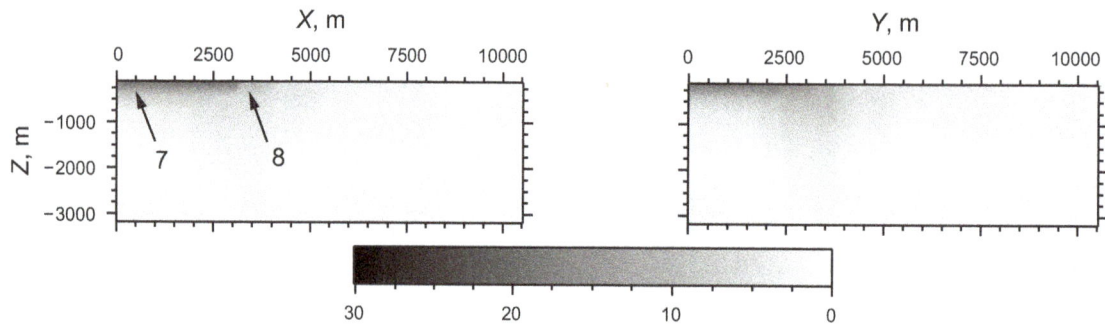

Fig. 7 Contour plot of maximum *P*-wave anisotropy (%) for near offset seismic propagation (0° to 30°) for the extensive reservoir geometry for the *x–z* section. The large magnitude of anisotropy reflects the sensitivity of the elastic model to the rock physics input parameters. In this case, the rock physics parameters are based on core taken from reservoir depths and so are not representative of near-surface rock

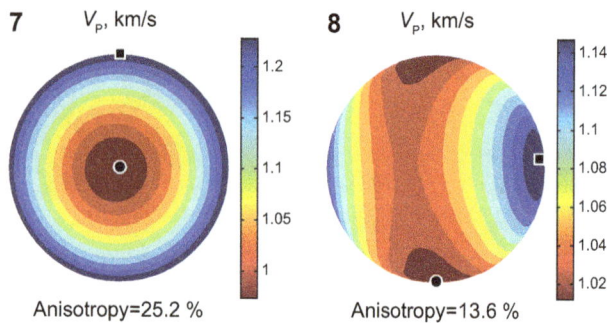

Fig. 8 Upper-hemisphere plots of *P*-wave phase velocity for various points in the extensive reservoir geometry for the *x–z* section (see Fig. 7)

1000 m. Looking at the anisotropic symmetries, we again observe noticeable differences from the other reservoir geometries. For point 1, the maximum *P*-wave velocity is 1685 m/s and vertical, yet the initial isotropic pre-production *P*-wave velocity is 1695 m/s. Points 2 and 3 within the reservoir adjacent to the boundaries display sub-horizontal maximum *P*-wave velocity (increase of up to 5 m/s from pre-production) parallel to the boundary, and with a preferred orientation of vertical microcracks oriented parallel to the reservoir edge. Point 4 at the corner of the reservoir indicates sub-horizontal maximum *P*-wave velocity (greater than the pre-production isotropic velocity) skirting around the reservoir edge with sub-vertical microcracks oriented tangential to the boundary. The post-production anisotropic symmetry for point 5 indicates extension in the overburden, with maximum *P*-wave velocity vertical and slightly less than pre-production (5 m/s), and microcracks oriented vertically and radial. For point 6, the symmetry is VTI maximum P-wave velocity horizontal (increase of approximately 15 m/s) with horizontally oriented microcracks. Table 4 summarizes the anisotropic symmetry decomposition of the stress-induced anisotropy.

In the near surface, these results are consistent with Herwanger and Horne (2009) and Fuck et al. (2010). However, differences can be seen within the vicinity of the reservoir, where the influence of reservoir geometry and the poroelastoplastic constitutive model impacts the development of stress arching.

4 Discussion

The results of the rectilinear reservoir model show that the geometry of the reservoir influences stress path evolution during production, and therefore evolution of seismic anisotropy. For the small reservoir, the geomechanical stress anisotropy is moderate reflecting the influence of the reservoir boundaries on stress redistribution. Yet the developed seismic anisotropy is low due to the limited volumetric influence of the small reservoir as well as the weak development of stress arching in the side-burden. Under such circumstances, it would be reasonable to expect little or no microseismicity. For the extensive reservoir, the geomechanical stress and seismic anisotropy are high resulting from the large size of influence of the producing reservoir. However, the reservoir experiences significant shear-enhanced compaction during production, indicating that the time evolution of anisotropy is necessary for the characterization of compacting reservoirs. There would likely be significant microseismicity occurring within the side-burden due to stress arching leading to larger zones of high shear stress and failure. Also, due to the significant stress redistribution from fluid extraction, microseismicity would likely be observed within the shallow subsurface. This will have important implications for assessing the risk of compaction on production related activities, from the surface down to the reservoir. The elongate reservoir displays the greatest asymmetry, with significant seismic anisotropy (and hence strong potential for shear type

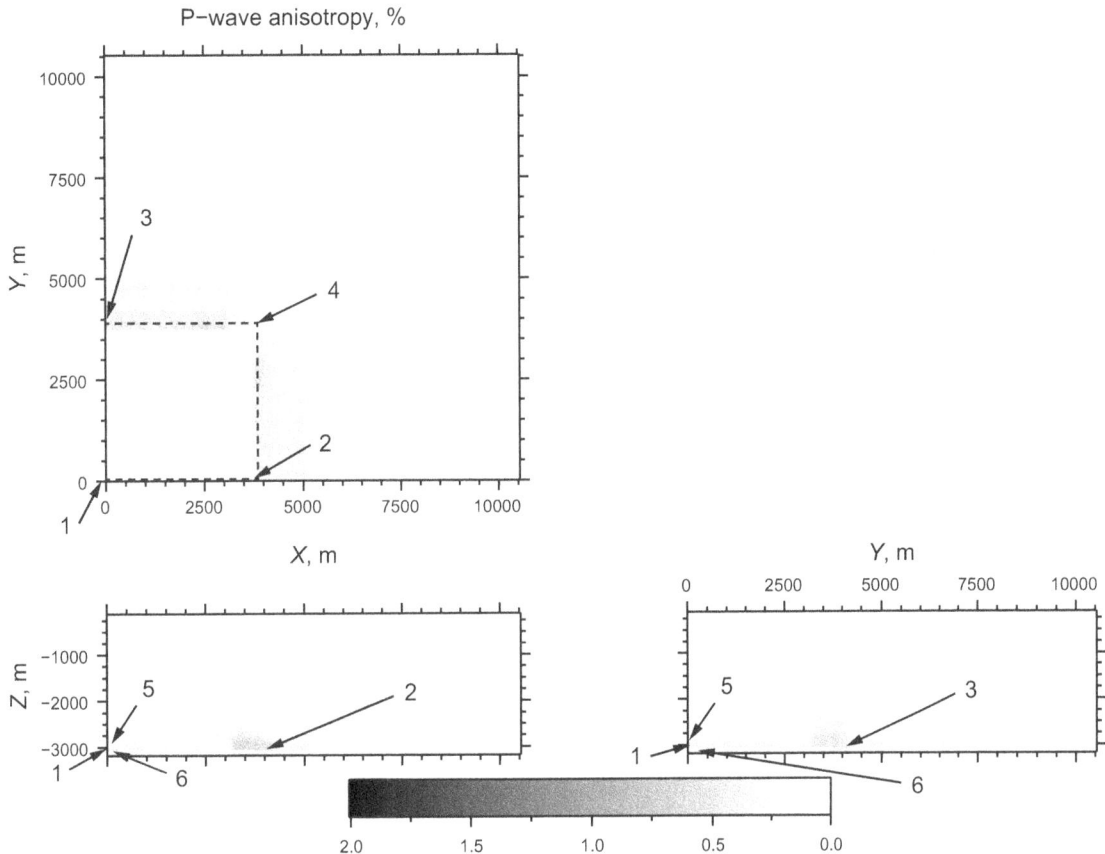

Fig. 9 Contour plot of maximum *P*-wave anisotropy (in percent) for near offset seismic propagation (0° to 30°) for the extensive reservoir geometry after applying an ad hoc depth-dependent scaling of the nonlinear rock physics input parameters

microseismicity) developing along the long-edge of the reservoir. Although the reservoir experiences shear-enhanced compaction within the reservoir, stress arching remains relatively high and suggests that stress arching could have significant influence the fault/fluid-flow behaviour.

In all simulations, it has been shown that elliptical anisotropy is not a prerequisite of stress-induced anisotropy and is controlled not only by model geometry but also the rock physics model used. However, there is no unambiguous diagnostic link between predicted seismic anisotropy and stress path parameters. The most crucial point to note is the dependence of the seismic predictions on the rock physics model to map geomechanical parameters to dynamic elasticity. In particular, the depth dependence of the rock physics model is poorly constrained, which can lead to biases in predicted magnitude of seismic anisotropy. However, in full field simulations the models can be calibrated via history matching (e.g., Kristiansen and Plischke 2010). Certainly a parametric study of seismic attributes to the stress sensitivity of nonlinear rock physics models would be useful to determine the most influential model input parameters.

Nevertheless, this paper demonstrates that detectable amounts of seismic anisotropy can be produced by stress changes in and around a producing reservoir. At present there is a push towards developing methods to image-induced geomechanical deformation (e.g., Verdon et al. 2011; He et al. 2016a; Angus et al. 2015). Normal incidence travel-time shifts characterized by "R-factors" (e.g., Hatchell and Bourne 2005; He et al. 2016b) have been the most common observation used to do so. However, R-factors do not provide a full characterization of the changes in triaxial state, nor do all modes of geomechanical deformation lead to normal incidence travel-time shifts (for example changes in horizontal stresses). Therefore, characterization of seismic anisotropy in and around a producing reservoir can provide a more complete picture of deformation. This paper has outlined a workflow for predicting seismic anisotropy based on geomechanical simulation. By imaging seismic anisotropy around deforming reservoirs, we can begin to match modelled predictions with field observations in order to improve our understanding of production-induced deformation.

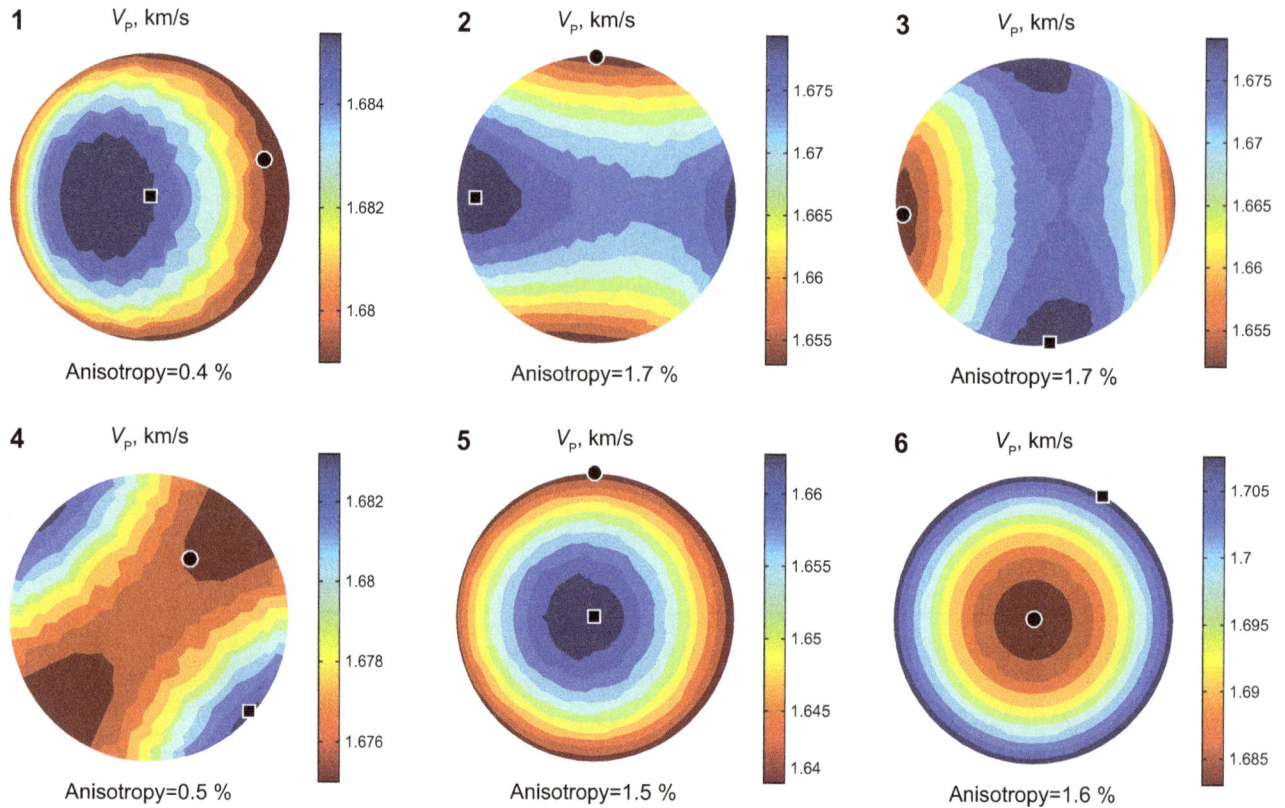

Fig. 10 Upper-hemisphere plots of P-wave phase velocity for various points in the extensive reservoir geometry (see Fig. 9)

Table 4 Decomposition of elastic tensor for all anisotropic symmetries for extensive reservoir (labelled points in Figs. 5, 6)

Point	Isotropic	Hexagonal	Tetragonal	Orthorhombic	Monoclinic	Triclinic
1	99.62	0.34	0.00	0.04	0.00	0.00
2	98.79	0.35	0.00	0.86	0.00	0.00
3	98.79	0.28	0.00	0.93	0.00	0.00
4	99.59	0.34	0.00	0.03	0.04	0.00
5	98.87	1.07	0.00	0.00	0.06	0.00
6	98.77	0.17	0.00	1.06	0.00	0.00
7	81.39	2.56	0.06	15.99	0.00	0.00
8	89.89	8.58	0.00	1.53	0.00	0.00

5 Conclusions

In this paper, we present a workflow that links coupled fluid-flow and geomechanical simulation with seismic modelling. The workflow allows the prediction of seismic anisotropy induced by non-hydrostatic stress changes. The seismic models from the coupled flow-geomechanical simulations for several rectilinear reservoir geometries highlight the relationship between reservoir geometry, stress path and seismic anisotropy. The results confirm that reservoir geometry influences the evolution of stress during production and subsequently stress-induced seismic anisotropy. Although the geomechanical stress anisotropy is high for the small reservoir, the effect of stress arching and

the ability of the side-burden to support the excess load limit the overall change in effective stress resulting in minimal development of seismic anisotropy. For the extensive reservoir, stress anisotropy and induced seismic anisotropy are high. The extensive and elongate reservoirs experience significant shear-enhanced compaction, where the inefficiency of the developed stress arching in the side-burden cannot support the excess load. The elongate reservoir displays significant stress asymmetry, with seismic anisotropy developing predominantly along the long-edge of the reservoir. Although the link between stress path parameters and seismic anisotropy is complex, the results suggest that developments in time-lapse seismic anisotropy analysis will have potential in calibrating geomechanical models. Furthermore, the results of the seismic anisotropy analysis find that elliptical anisotropy is not a prerequisite of stress-induced anisotropy, where the anisotropic symmetry is controlled not only by model geometry but also the nonlinear rock physics model used.

Acknowledgements The authors would like to thank Rockfield Software and Roxar for permission to use their software. We also acknowledge ITF and the sponsors of the IPEGG project, BG, BP, Statoil and ENI. D.A. Angus also acknowledges the Research Council UK (EP/K035878/1; EP/K021869/1; NE/L000423/1) for financial support.

References

Alassi H, Holt R, Landro M. Relating 4D seismics to reservoir geomechanical changes using a discrete element approach. Geophys Prospect. 2010;58:657–68.

Angus DA, Verdon JP, Fisher QJ, Kendall J-M. Exploring trends in microcrack properties of sedimentary rocks: an audit of dry-core velocity-stress measurements. Geophysics. 2009;74:E193–203.

Angus DA, Kendall J-M, Fisher QJ, Segura JM, Skachkov S, Crook AJL, Dutko M. Modelling microseismicity of a producing reservoir from coupled fluid-flow and geomechanical simulation. Geophys Prospect. 2010;58:901–14.

Angus DA, Verdon JP, Fisher QJ, Kendall J-M, Segura JM, Kristiansen TG, Crook AJL, Skachkov S, Yu J, Dutko M. Integrated fluid-flow, geomechanic and seismic modelling for reservoir characterization. Can Soc Explor Geophys Rec. 2011;36(4):18–27.

Angus DA, Verdon JP, Fisher QJ. Exploring trends in microcrack properties of sedimentary rocks: an audit of dry and water saturated sandstone core velocity stress measurements. Int J Geosci Adv Seismic Geophys. 2012;3:822–33.

Angus DA, Dutko M, Kristiansen TG, Fisher QJ, Kendall J-M, Baird AF, Verdon JP, Barkved OI, Yu J, Zhao S. Integrated hydro-mechanical and seismic modelling of the Valhall reservoir: a case study of predicting subsidence, AVOA and microseismicity. Geomech Energy Environ. 2015;2:32–44.

Aziz K, Settari A. Petroleum reservoir simulation. London: Applied Science Publishers Ltd.; 1979.

Baird AF, Kendall J-M, Angus DA. Frequency dependent seismic anisotropy due to fracture: fluid flow versus scattering. Geophysics. 2013;78(2):WA111–22.

Browaeys JT, Chevrot S. Decomposition of the elastic tensor and geophysical applications. Geophys J Int. 2004;159:667–78.

Brown RJS, Korringa J. On the dependence of the elastic properties of a porous rock on the compressibility of the pore fluid. Geophysics. 1975;40:608–16.

Chapman M. Frequency-dependent anisotropy due to meso-scale fractures in the presence of equant porosity. Geophys Prospect. 2003;51:369–79.

Crampin S. The new geophysics: shear-wave splitting provides a window into the crack-critical rock mass. Lead Edge. 2003;22(6):536–49.

Crook AJL, Yu J-G, Willson SM. Development and verification of an orthotropic 3-D elastoplastic material model for assessing borehole stability. In: shales: SPE/ISRM Rock Mechanics Conference. Texas: Irving; 2002, 20–23 Oct, p. 78239.

Crook AJL, Willson SM, Yu J-G, Owen DRJ. Predictive modelling of structure evolution in sandbox experiments. J Struct Geol. 2006;28:729–44.

Dean RH, Gai X, Stone CM, Minkoff SE. A comparison of techniques for coupling porous flow and geomechanics. SPE. 2003;79709:1–9.

Dewhurst DN, Siggins AF. Impact of fabric, microcracks and stress field on shale anisotropy. Geophys J Int. 2006;165:135–48.

Fuck RF, Tsvankin I. Analysis of the symmetry of a stressed medium using nonlinear elasticity. Geophysics. 2009;74(5):WB79–87.

Fuck RF, Bakulin A, Tsvankin I. Theory of traveltime shifts around compacting reservoirs: 3D solutions for heterogeneous anisotropic media. Geophysics. 2009;74(1):D25–36.

Fuck RF, Tsvankin I, Bakulin A. Influence of background heterogeneity on traveltime shifts for compacting reservoirs. Geophys Prospect. 2010;59(1):78–89.

Geertsma J. Land subsidence above compacting oil and gas reservoirs. J Pet Geol. 1973;25(6):734–44.

Hall SA, Kendall J-M, Maddock J, Fisher Q. Crack density tensor inversion for analysis of changes in rock frame architecture. Geophys J Int. 2008;173:577–92.

Hatchell PJ, Bourne S. Rocks under strain: strain-induced time-lapse time shifts are observed for depleting reservoirs. Lead Edge. 2005;24:1222–5.

He Y, Angus DA, Yuan S, Xu YG. Feasibility of time-lapse AVO and AVOA analysis to monitor compaction-induced seismic anisotropy. J Appl Geophys. 2015;122:134–48.

He Y, Angus DA, Blanchard TD, Garcia A. Time-lapse seismic waveform modeling and seismic attribute analysis using hydro-mechanical models for a deep reservoir undergoing depletion. Geophys J Int. 2016a;205:389–407.

He Y, Angus DA, Clark RA, Hildyard MW. Analysis of time-lapse travel-time and amplitude changes to assess reservoir compartmentalisation. Geophys Prospect. 2016b;64:54–67.

Helbig K, Rasolofosaon PNJ. A theoretical paradigm for describing hysteresis and nonlinear elasticity in arbitrary anisotropic rocks. In: Anisotropy 2000: fractures, converted waves and case studies 2000.

Herwanger J, Horne S. Linking reservoir geomechanics and time—lapse seismics: predicting anisotropic velocity changes and seismic attributes. Geophysics. 2009;74(4):W13–33.

Herwanger JV, Schiøtt CR, Frederiksen R, If F, Vejbæk OV, Wold R, Hansen HJ, Palmer E, Koutsabeloulis N. Applying time-lapse seismic to reservoir management and field development planning at South Arne, Danish North Sea. In: Vining BA, Pickering SC (eds) Petroleum geology: from mature Basins to New Frontiers, Proceedings of the 7th Petroleum Geology Conference; 2010.

Jaeger JC, Cook NGW, Zimmerman RW. Fundamentals of rock mechanics. 4th ed. Oxford: Blackwell Publishing; 2007.

Johnson PA, Rasolofosaon PNJ. Manifestation of nonlinear elasticity in rock: convincing evidence over large frequency and strain intervals from laboratory studies. Nonlinear Processes in Geophys. 1996;3:77–88.

Kendall J-M, Fisher QJ, CoveyCrump S, Maddock J, Carter A, Hall SA, Wookey J, Valcke SLA, Casey M, Lloyd G, Ben Ismail W. Seismic anisotropy as an indicator of reservoir quality in siliciclastic rocks, structurally complex reservoirs. In: Jolley SJ, Barr D, Walsh JJ, Knipe RJ, editors. Geological society. London: Special Publication; 2007. pp. 123–36.

Kristiansen TG, Plischke B. History matched full field geomechanics model of the Valhall Field including water weakening and re-pressurization, SPE. 2010;131505.

Maultzsch S, Chapman M, Liu E, Li X-Y. Modelling frequency-dependent seismic anisotropy in fluid-saturated rock with aligned fractures: implication of fracture size estimation from anisotropic measurements. Geophys Prospect. 2003;51(5):381–92.

Minkoff SE, Stone CM, Bryant S, Peszynska M, Wheeler MF. Coupled fluid flow and geomechanical deformation modeling. J Pet Sci Eng. 2003;38:37–56.

Muntz S, Fisher QJ, Angus DA, Dutko M, Kendall JM. A project on the coupling of fluid flow and geomechanics. In: 9th US National Congress on Computational Mechanics, San Francisco, 2007, 23–26 July.

Nur A, Simmons G. Stress-induced velocity anisotropy in rock: an experimental study. J Geophys Res. 1969;74:6667–74.

Olofsson B, Probert T, Kommedal JH, Barkved OI. Azimuthal anisotropy from the Valhall 4C 3D survey. Lead Edge. 2003;22:1228–35.

Olsen C, Christensen HF, Fabricius IL. Static and dynamic Young's moduli of chalk from the North Sea. Geophysics. 2008;73:E41–50.

Podio AL, Gregory AR, Gray KE. Dynamic properties of dry and water-saturated Green River shale under stress. SPE. 1968;8(4):389–404.

Pouya A, Djeran-Maigre I, Lamoureux-Var V, Grunberger D. Mechanical behaviour of fine grained sediments: experimental compaction and three-dimensional constitutive model. Mar Pet Geol. 1998;15:129–43.

Prioul R, Bakulin A, Bakulin V. Nonlinear rock physics model for estimation of 3D subsurface stress in anisotropic formations: theory and laboratory verification. Geophysics. 2004;69:415–25.

Rasolofosaon P. Stress-induced seismic anisotropy revisited. Revue de l'institut français du pet. 1998;53(5):679–92.

Rutqvist J, Wu Y-S, Tsang CF, Bodvarsson G. A modeling approach for analysis of coupled multiphase fluid flow heat transfer, and deformation in fractured porous rock. Int J Rock Mech Min Sci. 2002;39:429–42.

Sayers CM. Stress-dependent elastic anisotropy of sandstones. Geophys Prospect. 2002;50:85–95.

Sayers CM. Asymmetry in the time-lapse seismic response to injection and depletion. Geophys Prospect. 2007;55(5):699–705.

Sayers C, Kachanov M. Microcrack-induced elastic wave anisotropy of brittle rocks. J Geophys Res. 1995;100:4149–56.

Schoenberg M, Sayers CM. Seismic anisotropy of fractured rock. Geophysics. 1995;60(1):204–11.

Segura JM, Fisher QJ, Crook AJL, Dutko M, Yu J, Skachkov S, Angus DA, Verdon J, Kendall J-M. Reservoir stress path characterization and its implications for fluid-flow production simulations. Pet Geosci. 2011;17:335–44.

Terzhagi K. Theoretical soil mechanics. New York: Wiley; 1943.

Tod SR. The effects of stress and fluid pressure on the anisotropy of interconnected cracks. Geophys J Int. 2002;149:149–56.

Verdon JP, Angus DA, Kendall J-M, Hall SA. The effects of microstructure and nonlinear stress on anisotropic seismic velocities. Geophysics. 2008;73(4):D41–51.

Verdon JP, Kendall J-M, White DJ, Angus DA. Linking microseismic event observations with geomechanical models to minimise the risks of storing CO_2 in geological formations. Earth Planet Sci Lett. 2011;305:143–52.

Hydrocarbon charge history of the Paleogene reservoir in the northern Dongpu Depression, Bohai Bay Basin, China

You-Lu Jiang[1] · Lei Fang[2] · Jing-Dong Liu[1] · Hong-Jin Hu[1] · Tian-Wu Xu[3]

Abstract The hydrocarbon charge history of the Paleogene in the northern Dongpu Depression was analyzed in detail based on a comprehensive analysis of the generation and expulsion history of the major hydrocarbon source rocks, fluorescence microscopic features and fluid inclusion petrography. There were two main stages of hydrocarbon generation and expulsion of oil from the major hydrocarbon source rocks. The first stage was the main hydrocarbon expulsion stage. The fluorescence microscopic features also indicated two stages of hydrocarbon accumulation. Carbonaceous bitumen, asphaltene bitumen and colloidal bitumen reflected an early hydrocarbon charge, whereas the oil bitumen reflected a second hydrocarbon charge. Hydrocarbon inclusions also indicate two distinct charges according to the diagenetic evolution sequence, inclusion petrography features combined with the homogenization temperature and reservoir burial history analysis. According to these comprehensive analysis results, the hydrocarbon charge history of the Paleogene reservoir in the northern Dongpu Depression was divided into two phases. The first phase was from the late Dongying depositional period of the Oligocene to the early uplift stages of the late Paleogene. The second phase was from the late Minghuazhen period of the Pliocene to the

Quaternary. Reservoirs formed during the first period were widely distributed covering the entire area. In contrast, reservoirs formed during the second period were mainly distributed near the hydrocarbon generation sags. Vertically, it was characterized by a single phase in the upper layers and two phases in the lower layers of the Paleogene.

Keywords Dongpu Depression · Hydrocarbon charge history · Hydrocarbon generation and expulsion history · Fluid inclusion · Petrography · Fluorescence microscopy

1 Introduction

Hydrocarbon charge history is an important issue in the study of pool-forming. Determination of the hydrocarbon charge history is helpful to correctly understand the oil and gas reservoir formation and distribution and has important practical value for guiding petroleum exploration. Traditional hydrocarbon charge history analytical methods include the hydrocarbon generation and expulsion history analytical method, the trap development history method and the reservoir saturation pressure method. Since 1990s, many new methods such as fluid inclusion studies, reservoir bitumen analysis and diagenetic mineral dating have been widely used. In recent years, the fluid inclusion method has been widely used and has achieved good results (Zhao et al. 2013; Guo et al. 2012; Yang et al. 2014; Xiao et al. 2012, 2016; Jiang et al. 2015a, b, c; Liu et al. 2007a, b, 2013; Gui et al. 2015; Wu et al. 2013; Wang et al. 2015a, b). Fluid inclusions can provide valuable information on the reservoir pressure and temperature at the time of the fluid migration and entrapment as well as on the compositions of the fluids involved in diagenesis and may thus provide important insight into the mineral diagenesis and fluid dynamics within

✉ You-Lu Jiang
jiangyl@upc.edu.cn

[1] School of Geosciences, China University of Petroleum, Qingdao 266580, Shandong, China

[2] CNOOC Research Institute, Beijing 100028, China

[3] SINOPEC Zhongyuan Oil Company, Puyang 457001, Henan, China

Edited by Jie Hao

sedimentary basins (Dolníček et al. 2012; Shan et al. 2015; Wang et al. 2015a, b; Guo et al. 2014; Lü et al. 2015; Li 2016). Fluid inclusion entrapment temperature in conjunction with burial and thermal history plots can be used to determine the petroleum charge history (Parnell 2010; Liu et al. 2011; Pang et al. 2015). The key of fluid inclusion study—one of the most important methods—is the accurate division of hydrocarbon fluid inclusion formation stages. In hydrocarbon-rich depressions with multiple sets of hydrocarbon source rocks, oil and gas that generated in the same period but different structural positions could have different temperatures, different maturity and different fluorescent colors. Therefore, the fluorescent color and homogenization temperature cannot be used as the absolutely effective basis for dividing the phases of hydrocarbon inclusions (Tao 2006). Due to the irreversibility of diagenesis, and the simultaneity of the inclusions and their host minerals formation, it is more reliable to determine the hydrocarbon charge history according to the order of formation of the host minerals (Liu et al. 2007a, b; Tao 2006). But reservoirs of the same period but at different depths may be in different diagenetic evolution stages. So, fluid inclusions of different host minerals may be formed in the same period. Therefore, the fluid inclusion forming periods cannot be simply divided according to the sequence of host minerals in the diagenetic evolution stage. Because of the complexity of the formation and evolution of oil and gas reservoirs, a single method is often limited in the hydrocarbon charge history analysis, and it should be combined with a variety of methods, to ensure the reliability of the results. In this study, we integrated a variety of research methods to determine the hydrocarbon charge history of the Paleogene reservoir in the north of the Dongpu Depression. The hydrocarbon charge stages were divided qualitatively based on the analysis of generation and expulsion stages of major hydrocarbon source rocks and fluorescence microscopic features (Jarmołowicz-Szulc et al. 2012); the hydrocarbon fluid inclusion forming periods were divided according to the diagenetic evolution sequence of host minerals combined with reservoir burial evolution history. Then, the hydrocarbon charge time was determined according to the inclusions' homogenization temperature in conjunction with burial and thermal history plots.

2 Geological setting

The Dongpu Depression is located to the southwest of the Bohai Bay Basin with an area of about 5300 km^2. Faults are developed in the depression (Chen et al. 2007; Jiang et al. 2015a), and controlled by this, a tectonic framework of "two sags, one uplift and one slope" was formed (Fig. 1). The depression underwent the rift stage in the Paleogene and the depression stage in the Neogene and the

Quaternary. From the Kongdian period to the Es$_4$ period which was the early rifting stage, the former cratonic basin disintegrated and formed a half-graben basin prototype. During the Es$_3$ period, namely the strong rifting stage, the fault activity was strong, and the basic tectonic framework was established. The advanced fault depression stage was from the Es$_2$ period to the Dongying period, during which the fault activity weakened, and uplift occurred in the late Dongying period, and there was great erosion (Lu et al. 2007). During the Neogene and Quaternary which was the depression stage, the tectonic activity was weak and most of the faults stopped being active (Jiang et al. 2015b). There is a series of hydrocarbon generation subdepressions in the depression, two of which are in the study area, the Liutun-Haitongji sub-depression and the Pucheng-Qianliyuan sub-depression. Source rocks in the depression include the coal measures of the Carboniferous-Permian and the mud shale of the third and first member of the Shahejie Formation. The main hydrocarbon source rocks are the shale of the middle-lower sections of the third member of the Shahejie Formation. The reservoir strata are the sand layers of the Shahejie Formation. The several sets of mudstones and gypsum in the Shahejie Formation form high-quality caprocks. They were superposed and formed many sets of source rock–reservoir–caprock assemblage. More than 80% of the oil and gas reserves found in the depression are distributed in the north area (Fig. 1). Thus, the reservoir formation process of the northern part is also largely representative of the whole depression.

3 Methodologies and samples

3.1 Hydrocarbon generation and expulsion history modeling

The hydrocarbon generation and expulsion history of the source rocks in the Dongpu Depression were simulated using the basin modeling method (Makeen et al. 2016). The simulation parameters include stratigraphic ages, formation depth/thickness, erosion thickness, lithology, boundary conditions and source rock properties. The stratigraphic ages, formation depth/thickness, boundary conditions (heat flow, paleowater depth and sediment water interface temperature) and source rock properties (thickness, distribution, TOC and HI value) used the third resource evaluation results of Dongpu Depression. The erosion thickness was obtained using the vitrinite reflectance and sonic log methods (Lu et al. 2007). The porosity–depth curves of sandstones and mudstones are fitted according to the measured porosity and the porosity calculated from sonic logs, which have been used in the compaction correction of the simulation. The simulation results were calibrated and

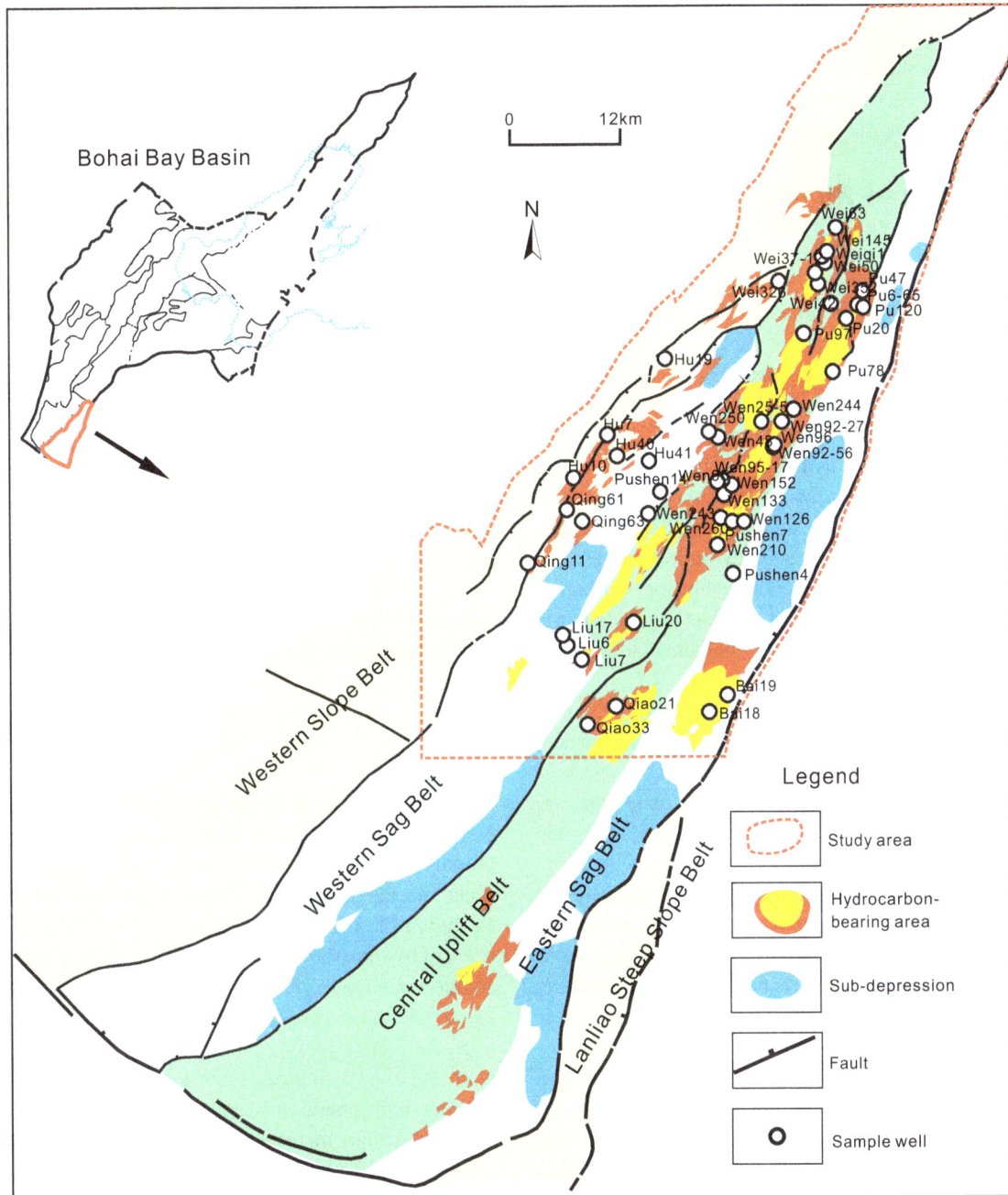

Fig. 1 Tectonic units division and hydrocarbon distribution of the Dongpu Depression, Bohai Bay Basin, showing the sample location

corrected using the measured temperature, pressure and vitrinite reflectance data.

3.2 Fluorescence microscopy

Fluorescence microscopy has been widely used in the field of biological sciences (Lin et al. 2015; Muhammad and Asifullah 2016). Since as early as the nineteenth century, it has been applied in petroleum geology research and has played an important role in the theory of petroleum origin

and the search for oil and gas reservoirs (Lang et al. 2008). Fluorescence microscopy uses ultraviolet or blue light as a light source to stimulate the asphalt material in the rocks to produce visible fluorescence and can be used to observe the distribution of the petroleum asphalt directly. Different components in the petroleum such as saturated hydrocarbons, aromatic hydrocarbons, non-hydrocarbon and asphaltene will show different fluorescence under the ultraviolet or blue light excitation. Different luminous pitch contains different petroleum components and will display different fluorescence (Lang et al. 2008). Oil and gas reservoirs formed in different periods which have experienced different evolution processes have different components. Therefore, the hydrocarbon charging stages can be analyzed by observing the fluorescent pitch in the reservoir.

3.3 The fluid inclusion petrography analytical method

Fluid inclusion petrography studies the relationship between hydrocarbon inclusions and the host minerals to establish the relationship between the hydrocarbon migration, accumulation and the diagenetic evolution time. The qualitative research of the fluid inclusion petrography is mainly based on the occurrence, fluorescence characteristics and the phase state of hydrocarbon inclusions.

The key of the fluid inclusion study is the accurate division of hydrocarbon fluid inclusion formation stages. The determination of the diagenetic evolution sequence is the premise of the division of the inclusion formation stages, and the relative position of the host minerals in the diagenetic evolution sequence is the fundamental basis for the correct classification of the inclusions.

The homogenization temperature of the fluid inclusions represents the temperature when the inclusion is formed, which can determine the time of oil and gas filling combined with the buried heating history. Oil inclusions are generally not captured at natural gas saturation, but when restored to a single-phase state it is in saturation condition, so the homogenization temperature of the oil inclusion is generally lower than the trapping temperature, while the homogenization temperature of the associated water inclusions is generally close to the capture temperature. Therefore, when analyzing the oil and gas reservoir forming period, we usually use the brine inclusions.

In this study, the diagenetic evolution sequence was determined by thin-section identification; then, inclusion formation stages were divided according to the host minerals' formation sequence combined with the inclusion petrography analysis. Homogenization temperature tests were carried out for the observed hydrocarbon inclusions and their associated water inclusions. The time of formation of hydrocarbon inclusions of each phase was determined by the analysis of the homogenization temperature in conjunction with burial and thermal history, which is namely the hydrocarbon charging time.

3.4 Instruments and samples

In this study, a Zeiss AXIO Imager D1m digital polarized fluorescence microscope was used in the fluorescence microscopy, cast thin-section observations and inclusion thin-section observations, and a Linkam THMS600 gas-

flow heating/freezing system was used in the homogenization temperature test.

The core sampling was carried out in the oil and gas intervals of the Shahejie Group of 49 wells (Fig. 1). A total of 136 inclusion thin sections were made for the fluorescence microscopic observation and the fluid inclusion studies. A total of 23 cast thin sections were made for the diagenesis research.

4 Results and discussion

4.1 Hydrocarbon generation and expulsion histories

The results of simulation of the evolution of major source rocks in study area showed that in the early sedimentary period of the second section of the Shahejie group (Es$_2$), all source rocks were at an immature stage and no hydrocarbon was expelled. From the Es$_1$ period to the early Dongying period, source rocks were in the low mature stage and there was some hydrocarbon generated but only a little expelled. In the middle and late period of Dongying group (about 31–27 Ma), most of the source rocks were at the mature to highly mature stage, at the peak hydrocarbon generating and expulsion period. Some hydrocarbon was expelled at the early uplift stage (27–23 Ma). Then, the evolution of source rocks stagnated until the early Minghuazhen period in the Neogene. From the late Minghuazhen period until now (about 5–0 Ma), the burial depth and maturity of source rocks reached and exceeded that before the uplift, and the evolution of source rocks continued. This was the second stage of hydrocarbon generation and expulsion, in which the source rocks were at the high maturity stage in the upper layers and the over mature stage in the lower layers. However, the source rocks which entered into the second stage were relatively limited, mainly distributed in the deep hydrocarbon generation sags.

In conclusion, there were two hydrocarbon generation and expulsion stages in the study area. The first stage was from the middle and the late Dongying deposition period to the early uplift stage in the late Paleogene (about 31–23 Ma), and the second was from the late Minghuazhen period of Neogene to Quaternary (about 7–0 Ma). According to the hydrocarbon expulsion quantity in each stage, the first stage was the main hydrocarbon expulsion stage (Fig. 2).

4.2 Fluorescence microscopy characteristics

4.2.1 Fluorescence characteristics of different types of asphalts

A lot of carbonaceous asphalts (without fluorescent), bituminous asphalts, colloidal asphalts and oleaginous

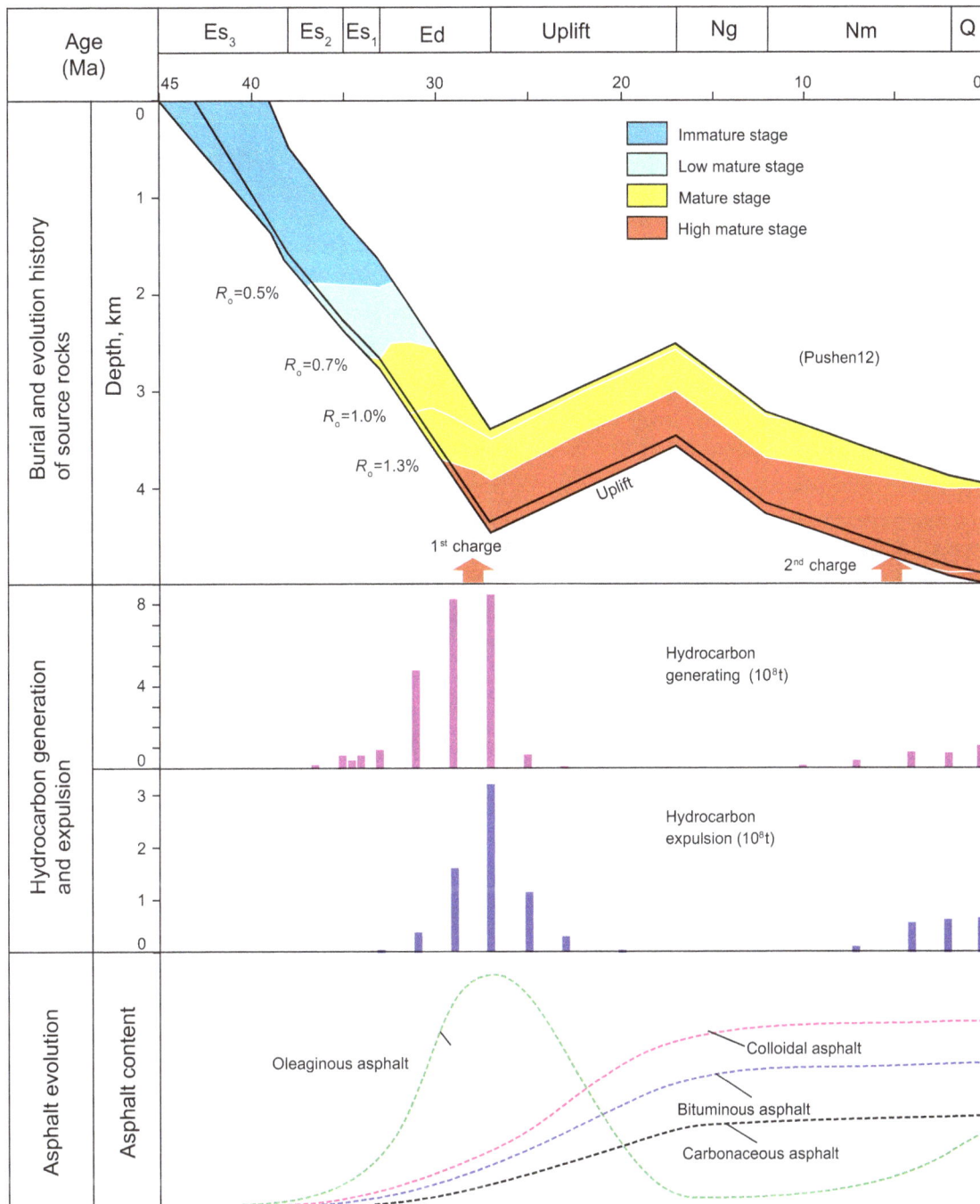

Fig. 2 Generation and expulsion history of Es$_3$ hydrocarbon source rocks and bitumen evolution history in the Dongpu Depression

asphalts were seen in the samples of the reservoirs in the study area (Fig. 3; Table 1).

Carbonaceous asphalts are the production of petroleum alteration (Shokrlu and Babadagli 2013). These are black under both the fluorescence and transmitted light. The occurrences can be divided into three types: (1) distributed in the secondary pore spaces between the quartz particles, blocky with straight clear edges (Fig. 3a, b). (2) Distributed around the particles, ring banded or irregular shape,

equivalent to the cements (Fig. 3c, d). (3) Distributed in fractures, banding (Fig. 3e, f).

The bituminous asphalts were formed due to the condensation of the resin and other heavy components of petroleum and were the residues of previous reservoirs (Qin and Guo 2002; Shalaby et al. 2012). Their main ingredients are resin, non-hydrocarbon and asphaltene, which are not soluble in petroleum ether. Bituminous asphalts are common in the samples of study area, and their

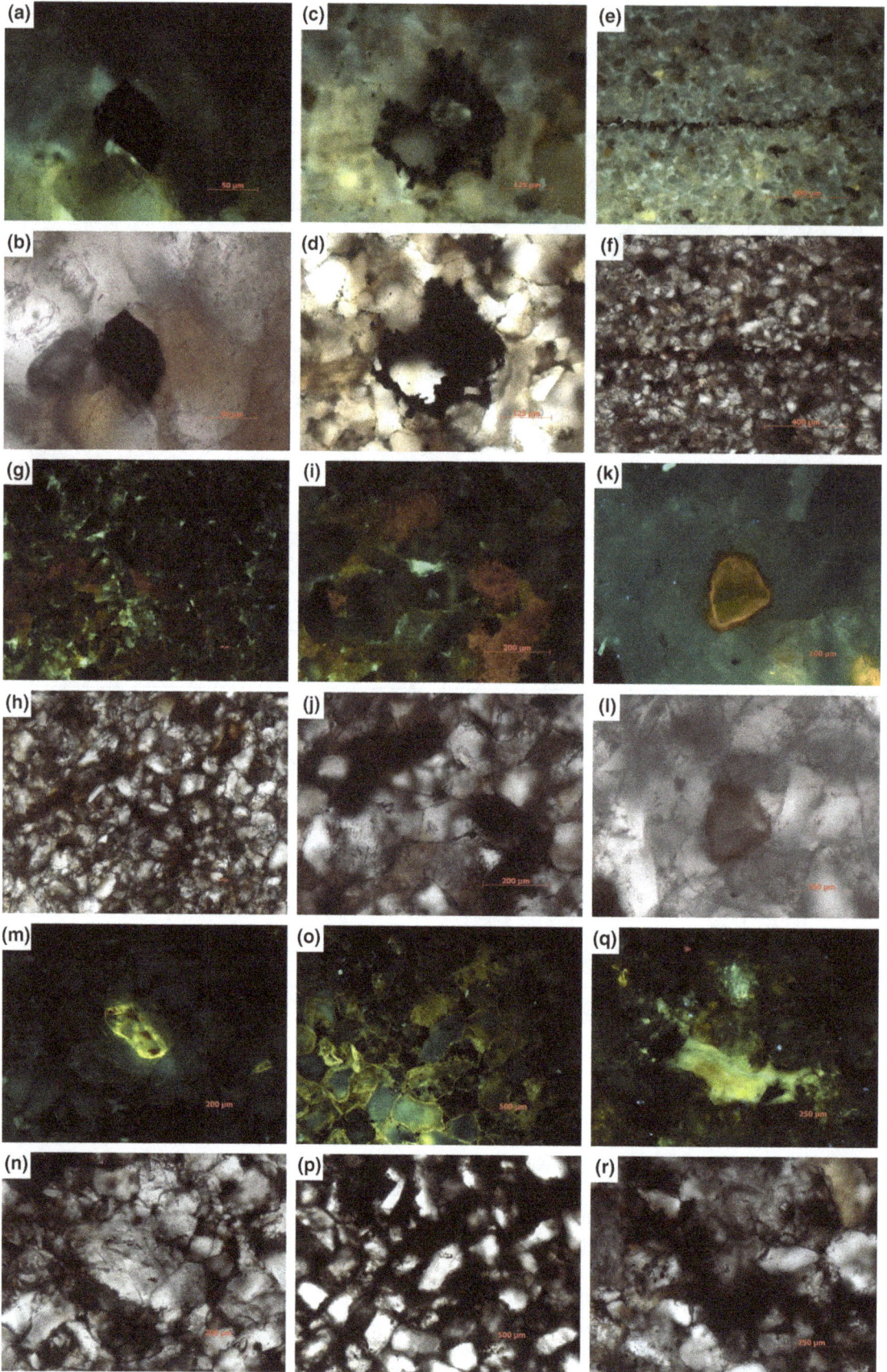

◄**Fig. 3** Photographs of asphalt from the northern Dongpu Depression under fluorescence and transmitted light. **a–f** Carbonaceous asphalt, **g–j** bituminous asphalt; **k–l** colloidal asphalt; **m–r** oleaginous asphalt; **a, b** Qiao21, 2525.4 m; **c, d** Wei145, 2785.32 m; **e, f** Wei47, 2863.78 m; **g, h** Hu40, 2660.3 m; **i, j** Qiao21, 2525.4 m; **k, l** Qiao59, 4556.8 m; **m, n** Liu6, 3882.8 m; **o, p** Pu120, 3264 m; **q, r** Pu80, 3676.05 m

main features and occurrence can be divided into two categories: (1) distributed in the intergranular pores, with weak intensity yellow fluorescence, layered, and disseminated along both sides of the pores, gray or gray brown under transmitted light (Fig. 3g, h). (2) Clustered or blocky distributed in the intergranular pores, yellow fluorescence, dark brown under transmitted light, disseminated to the surrounding cements (Fig. 3i, j).

The colloidal asphalts were distributed on the surface of the quartz grains like films, yellow brown fluorescent, disseminated to the interior of the particles in microfractures (Fig. 3k, l).

The oleaginous asphalts were mainly composed of light petroleum components, which were widely distributed in the reservoirs. Their occurrence can be divided into three types: (1) distributed along the calcite surface and in the cleavage cracks, yellow to yellow green fluorescence and colorless under transmitted light (Fig. 3m, n). (2) Distributed along the surface of mineral particles like thin films, yellow to yellow green fluorescence, disseminated to the interior of mineral particles (Fig. 3o, p). (3) Adsorbed in the intergranular spaces, mostly distributed in the parts where argillaceous matrix or clay mineral content is high, filling and disseminated to the matrix and matrix contraction joints, yellow green fluorescence (Fig. 3q, r).

4.2.2 Cause analysis and phase partition of asphalts

There are many reasons for the formation of asphalts in reservoirs. They are generally considered to be from liquid petroleum. There are 3 common reasons, which are thermal evolution, cold metamorphism and gas-washing (Luo et al. 2009). In addition to the nature of the oil, the environmental factors such as reservoir temperature, volume and pressure are very important to promote the occurrence of asphaltene deposition, and among which, pressure is the most important factor. A decrease in pressure can decrease the solubility of asphaltene in crude oil and thus cause deposition of asphalt (Qin and Guo 2002).

When the temperature of crude oil is over 150 °C, it will become unstable, macromolecular hydrocarbons and other heterocyclic compounds will be gradually transformed into low molecular compounds (condensate and gas hydrocarbon) and asphalts (Guo et al. 2008). The representative geothermal gradient in the Dongpu Depression is 3.4 °C/

Table 1 Distribution characteristics of reservoir bitumen in northern Dongpu Depression

Area	Wells	Formation	Depth, m	Asphalt type	Phases
Huzhuangji–Qingzuji area	Hu40	Es_3^z	2660.3	Mainly carbonaceous and colloidal asphalts	I
	Hu7	Es_3^z	1893.8	Mainly carbonaceous, occasionally oleaginous asphalts	I, II
Liuzhuang area	Liu22	Es_2^x	3628.4	Mainly colloidal, bituminous and carbonaceous asphalts	I
	Liu25	Es_3^s	4140.5	Carbonaceous, colloidal and oleaginous asphalts	I, II
Weicheng Horst zone	Wei145	Es_3^x	2785.32	Mainly carbonaceous asphalts	I
	Wei37-12	Es_2^x	2072.42	Carbonaceous, bituminous and oleaginous asphalts	I, II
	Wei63	Es_2^s	1855.35	Carbonaceous, colloidal, bituminous and oleaginous asphalts	I, II
	Weiqi1	Es_3^s	2560.9	Carbonaceous, bituminous and oleaginous asphalts	I, II
Qiaokou area	Qiao20	Es_3^s	3002.3	Carbonaceous, colloidal and oleaginous asphalts	I, II
	Qiao21	Es_2^x	2525.4	Mainly colloidal, bituminous carbonaceous	I
	Qiao33	Es_3^z	3164.8	Mainly oleaginous, then carbonaceous and bituminous asphalts	I, II
Baimiao area	Bai18	Es_3^s	3326.5	Mainly colloidal and carbonaceous asphalts	I
	Bai19	Es_3^x	3891.7	Mainly bituminous, colloidal and oleaginous, occasionally carbonaceous asphalts	I, II
	Bai20	Es_2^s	2655.2	Mainly carbonaceous and colloidal asphalts	I
Pucheng area	Pu80	Es_3^z	3673.9	Oleaginous, colloidal and carbonaceous asphalts	I, II
	Pu84	Es_3^z	3675.8	Carbonaceous, bituminous and oleaginous asphalts	I, II
	Pu6-65	Es_3^s	3187.6	Carbonaceous, colloidal and bituminous asphalts	I
Wenliu area	Wen101-16	Es_2^x	2304.9	Colloidal and carbonaceous asphalts	I
	Wen210	Es_3^z	3920.25	Carbonaceous, bituminous and colloidal asphalts	I
	Wen244	Es_3^z	3482.93	Carbonaceous, bituminous and colloidal asphalts	I

100 m (Liu et al. 2007a, b). The corresponding depth to reach the temperature of crude oil cracking is about 4400 m. But most of the carbonaceous asphalts observed in the study area did not reach this depth and were not of pyrolytic origin. However, asphalts in the deep reservoir could be formed by oil cracking.

The Dongpu Depression experienced a substantial rise in the Paleogene period (27–17 Ma), the formation suffered an intense erosion, and denudation thickness was up to more than 2000 m, with an average of about 1000 m (Lu et al. 2007). Strong tectonic uplift decreased the oil's burial depth and temperature and pressure became lower. Decrease in pressure caused the decrease in the solubility of asphaltene in crude oil and caused an increase in resins and bitumen and decrease in light components in the petroleum, which formed the carbonaceous asphalts, bituminous asphalts and colloidal asphalts (Fig. 2). Because of the stably distributed salt rocks in the first section of the Shahejie group and as the oil-bearing strata are mainly located under the salt rocks, the sealing and preservation condition is good, so the asphalt should not be formed by oxidation.

The above analysis indicates that carbonaceous asphalts, bituminous asphalts and colloidal asphalts were formed due to the strong uplift and erosion, which reflected the early oil and gas injection phase, while the oil and gas injected in the late period were not subjected to alteration and showed mostly oleaginous asphalts.

4.3 Diagenetic evolution sequence

From the detailed microscopic identification of thin sections, the diagenesis of reservoir sandstones of the Shahejie Formation in the study area mainly includes compaction, cementation, metasomatism and dissolution (Fig. 4).

According to the "Standard for the division of diagenetic stages of clastic rocks" (Ying et al. 2004), using paleotemperature, vitrinite reflectance (R_o) and authigenic clay mineral assemblages, the diagenetic stages of sandstones of the Shahejie Formation in the northern Dongpu Depression were divided into the early diagenetic stage B, middle diagenetic phase A and middle diagenetic phase B. The schematic diagram of diagenetic stages division of the reservoirs of the Shahejie Formation in the northern Dongpu Depression was compiled (Fig. 5). The characteristics of each diagenetic stages are as follows.

The early diagenetic stage B: depth from 2000 to 2500 m, temperature <95 °C, R_o < 0.5%, equivalent to the semi-mature period of the organic matter evolution, and the smectite layer of I/S mix-layer was in the second quick transformation zone (from 50% to 20%). In this stage, the diagenesis was dominated by compaction and early carbonate cementation. The cementation included calcite cementation, secondary outgrowth of quartz and the precipitation of dolomite.

The middle diagenetic phase A1: depth from 2500 to 3300 m, temperature from 95 to 135 °C, R_o from 0.5% to 0.7%, in the low mature stage of organic matter evolution. In this stage, the main diagenesis included dolomite cementation, ferroan dolomite cementation and clastic particles replacement by calcite, calcite replacement by ferroan calcite and the dissolution of quartz and feldspar particles in the later stage.

The middle diagenetic phase A2: depth from 3300 to 4200 m, temperature from 135 to 160 °C, vitrinite reflectance R_o 0.7% to 1.3%, in the mature stage of organic matter. This stage was characterized by a large amount of dolomite and ferroan dolomite cementation, and the dissolution of quartz and feldspar.

The middle diagenetic stage B: burial depth of 4200 to 5000 m, temperature of >160 °C, vitrinite reflectance R_o > 1.3%, the main features of this stage are the secondary outgrowth of quartz and the metasomatism of ferroan dolomite.

Microfractures are important places to capture fluid inclusions. Understanding the cause of the microfractures and their formation sequence in the diagenetic stage is the basis for determining the phase of hydrocarbon inclusions in the microfractures. Microfractures containing hydrocarbon inclusions in the study area include the internal cracks of quartz particles (Fig. 6e, f), and the cracks through quartz grains (Fig. 6i–l). The former did not cut through the quartz grains and the latter cut through the whole quartz grains. Research suggests that the internal cracks of quartz particles were formed due to the pressure of the overlying strata exceeding the critical fracture pressure of the debris particles (Li et al. 2014). These fractures had been gradually healed due to late diagenesis. In the samples, the internal cracks of quartz appeared at depths below 2500 m, which is equivalent to the middle diagenetic phase A1 and the later diagenetic stages. The cracks through quartz grains appeared in the whole section, not controlled by the depth. This indicates that the cracks through quartz grains should not be the result of diagenesis and presumably be formed due to the stress field changes caused by the tectonic uplift during late Paleogene, which captured the fluid inclusions during the subsidence in the Quaternary period.

4.4 Characteristics of fluid inclusion petrography

4.4.1 Fluid inclusion petrography

Generally speaking, petroleum inclusions formed in the low mature phase of hydrocarbon source rock evolution had a high density, with a high content of heavy hydrocarbon and asphaltene. And their fluorescence is mainly brown. As the maturity increases, or the migration and

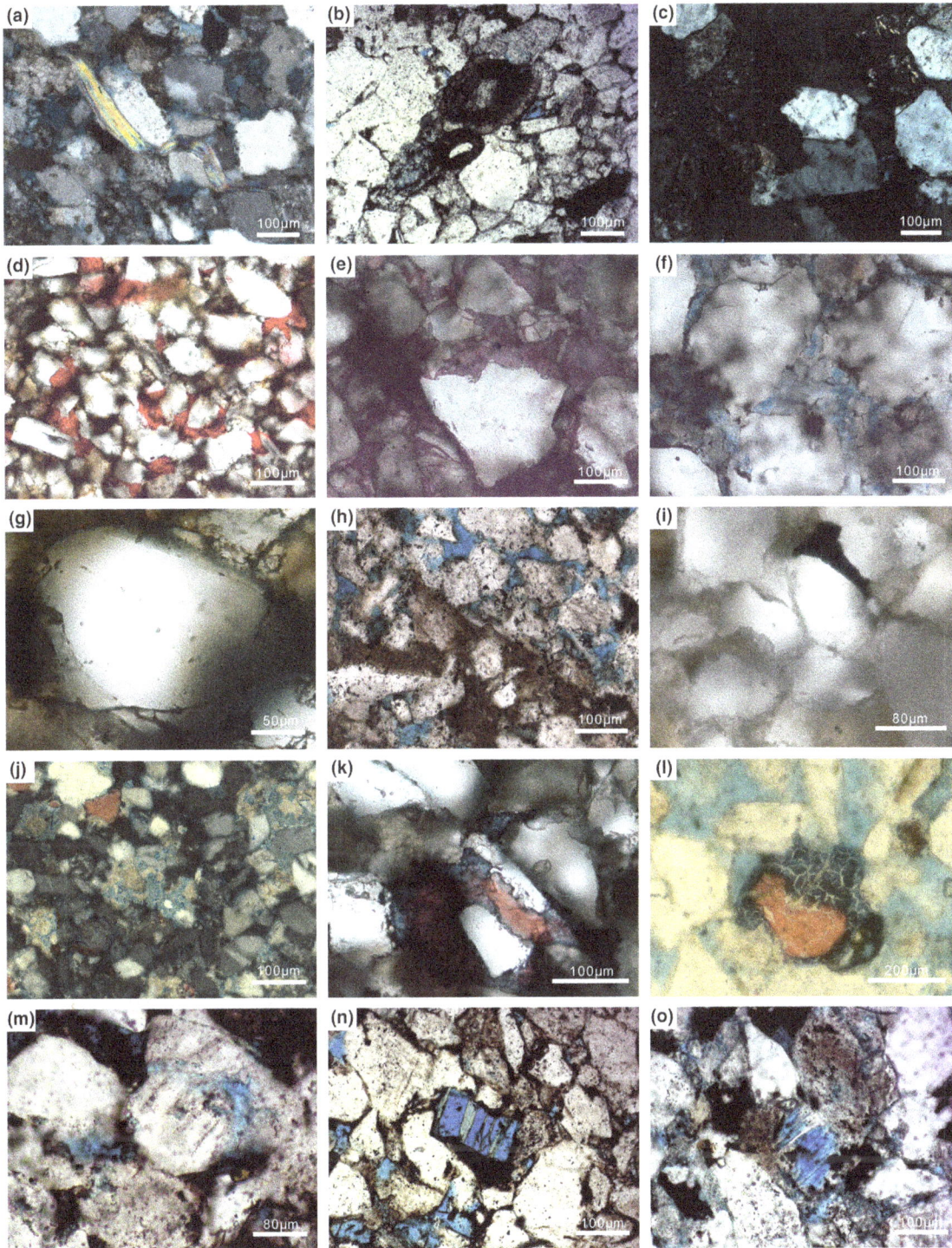

Fig. 4 Diagenesis of the Shahejie Formation in the northern Dongpu Depression. **a** Mica is extruded and deformed, Wen126, 2922.6 m; **b** Oolite deformation, Pushen7, 3696.7 m; **c** lined and concavo-convex intergranular contacts, Wen92-27, 2825.6 m; **d** calcite cementation, Wei327, 2511 m; **e** ferroan calcite cementation, Wei57, 2712.7 m; **f** dolomite and ankerite cement, Qing61, 2880.8 m; **g** quartz overgrowth, Liu6, 3882.8 m; **h** argillaceous cementation, Wen126, 2922.6 m; **i** asphalt cementation, Wen25-5, 2284.9 m; **j** ferroan dolomite replaces dolomite cements, Hu40, 2664.64 m; **k** ferroan dolomite replaces calcite cements, Hu64, 3410 m; **l** calcite replaces particles, ferroan calcite replaces calcite, Hu40, 2664.64 m; **m** quartz was dissolved, Wen243, 4273.5 m; **n** feldspar was dissolved along the cleavage, Pushen7, 3696.7 m; **o** feldspar was dissolved, Liu7, 3525.3 m

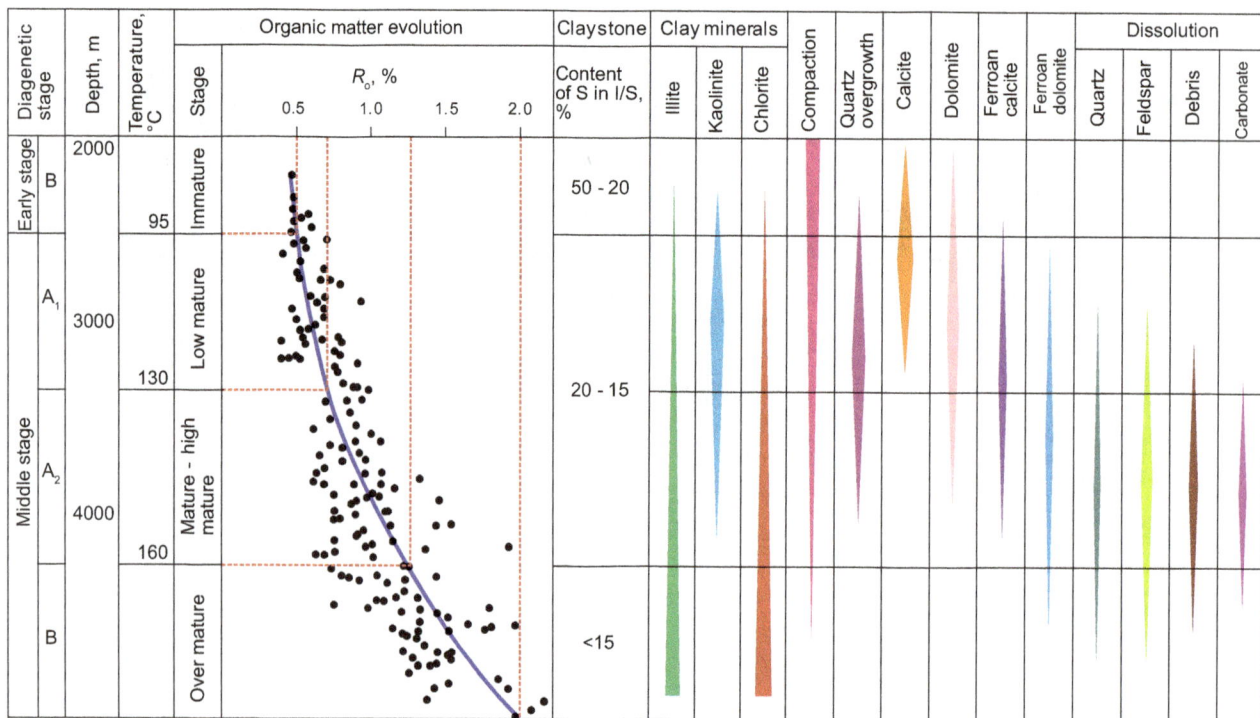

Fig. 5 Diagenetic stages of the Shahejie Formation in the northern Dongpu Depression

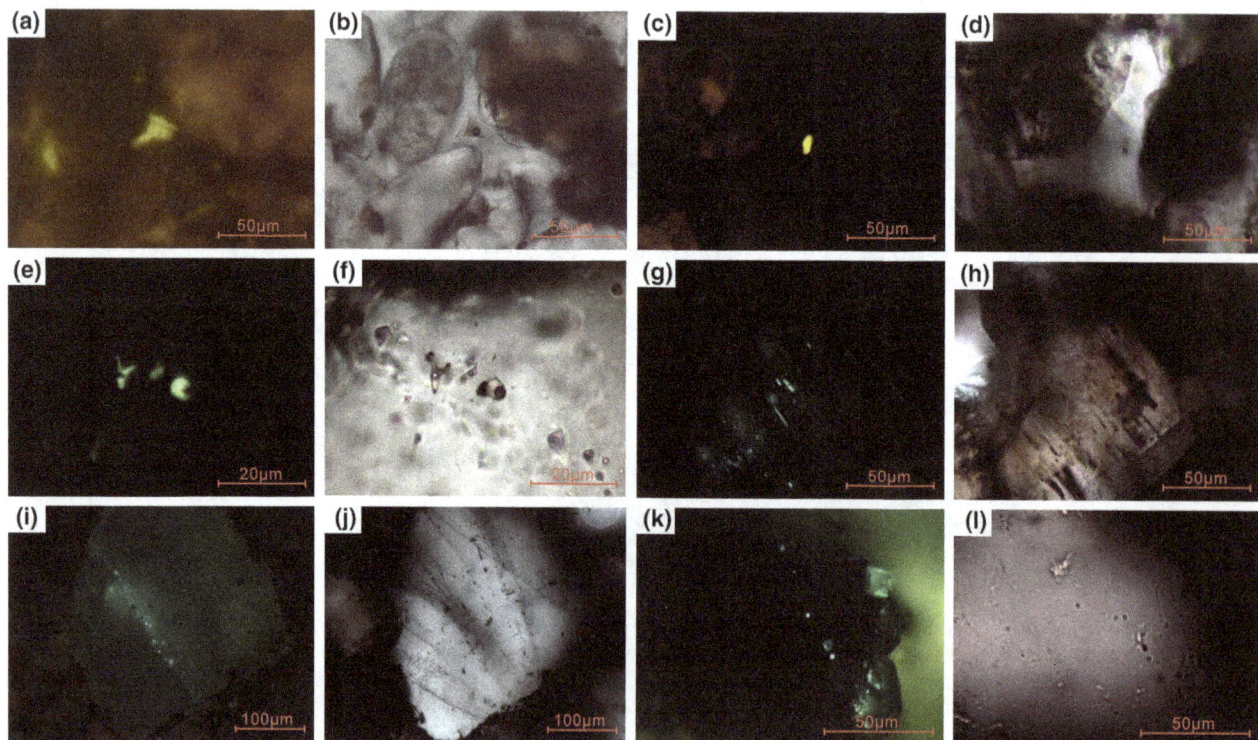

Fig. 6 Inclusion photographs under fluorescence and transmitted light of northern Dongpu Depression. **a**, **b** Inclusions occurred in calcite cements, Wen25-5, 2282.8 m; **c**, **d** inclusions occurred in ferroan calcite cements, Hu40, 2660.3 m; **e**, **f** inclusions occurred in fractures within quartz grains, Pu47, 3039.49 m; **g**, **h** inclusions occurred in dissolution fractures of feldspar, Wen88-1, 3603.38 m; **i**, **j** inclusions occurred in fractures through the quartz grains, Hu10, 2233.9 m; **k**, **l** inclusions occurred in fractures through the quartz grains, Weiqi1, 2658.05 m

differentiation increased, crude oil density became low, and fluorescence colors of the petroleum inclusions captured in this stage are mainly white, yellow or yellow green. Petroleum inclusions formed in the highly mature phase have fluorescence colors mainly milky blue or blue (Liu et al. 2007a, b). Fluorescent colors of hydrocarbon inclusions observed in the study area include yellow, yellowish white, yellow green and blue white, which indicates that the oil and gas in the study area were mainly charged at mature and highly mature stages (Fig. 6).

Hydrocarbon inclusions in the study area are mainly gas–liquid two-phase inclusions, mainly distributed in the carbonate cements (including calcite dolomite and ferroan dolomite), feldspar dissolution pores, internal cracks in quartz grains and late cracks through quartz grains (Fig. 6).

4.4.2 Phase division of fluid inclusions

According to the analysis of diagenetic evolution sequence, the calcite and dolomite were early carbonate cements, formed in the early diagenetic stage B, when the reservoir depth was up to 2000 m. The ferroan dolomite was late carbonate cement formed in the middle diagenetic A phase, when the reservoir was buried below 2500 m. The internal

cracks of quartz grains were formed in the middle diagenetic phase A1, when buried deep below 2500 m. The feldspar dissolution pores and fractures were the result of feldspar dissolution, which happened in the late period of middle diagenetic phase A1, when buried below 3000 m. The cracks through quartz grains were formed after the uplift in the late Paleogene.

The hydrocarbon inclusions in different minerals or fractures distributed in different depths: The inclusion in the carbonate cements distributed at depths of 2000–2700 m, among which the early carbonate cements distributed at 2000–2500 m and the late carbonate cements in 2500–2700 m. The inclusions in the internal cracks of quartz grains distributed at depths of 2500–4000 m, and inclusions in the feldspar dissolution pores and fractures of the quartz grains are distributed at 3100–4000 m. Inclusions in the cracks through quartz grains are distributed in all depths.

Samples of different depths have different burial histories. The geological periods corresponding to the diagenetic stages of the host minerals are estimated in the reservoir burial history, from the time of formation of different host minerals of hydrocarbon inclusions (Fig. 7). The results show that the inclusions in the carbonate

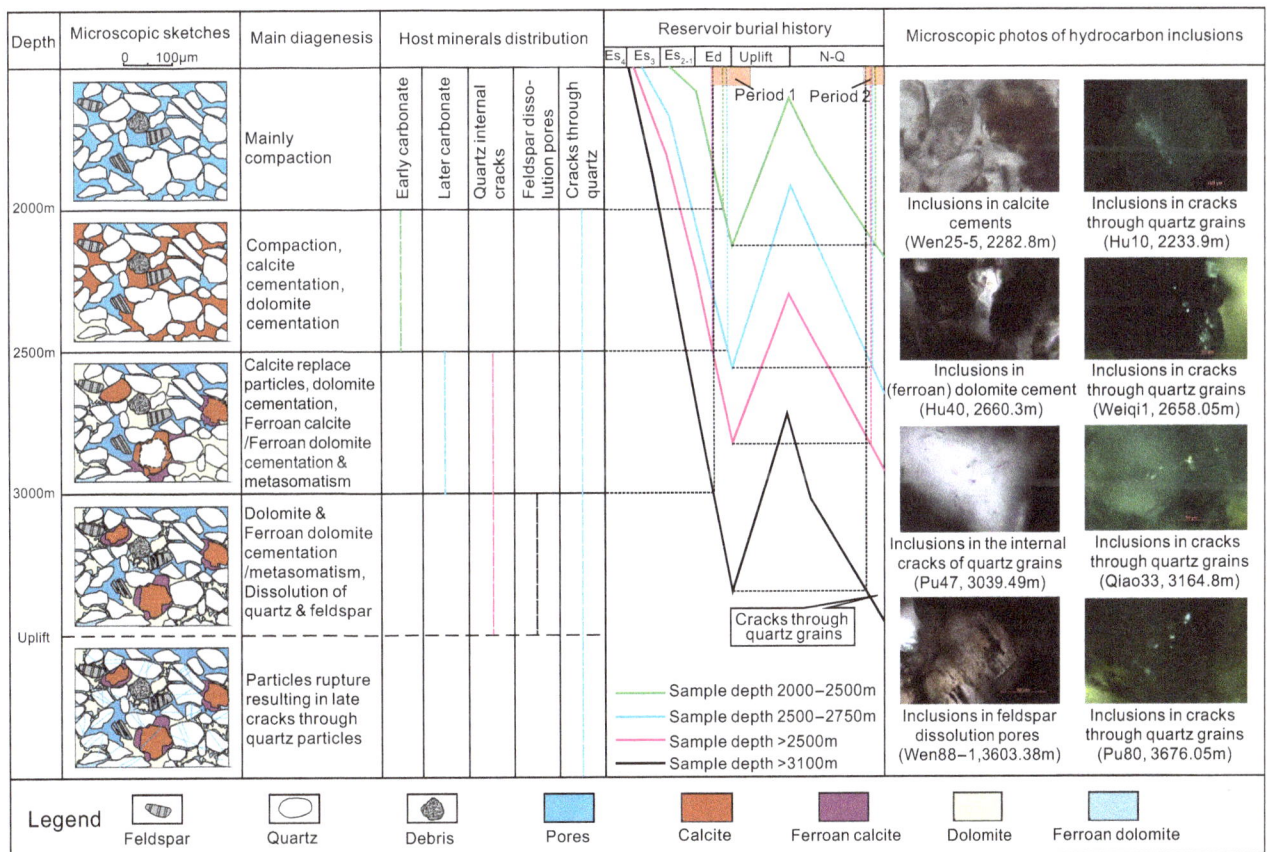

Fig. 7 Using diagenetic stages and burial history to determine the time of formation of hydrocarbon inclusions in the North Dongpu Depression

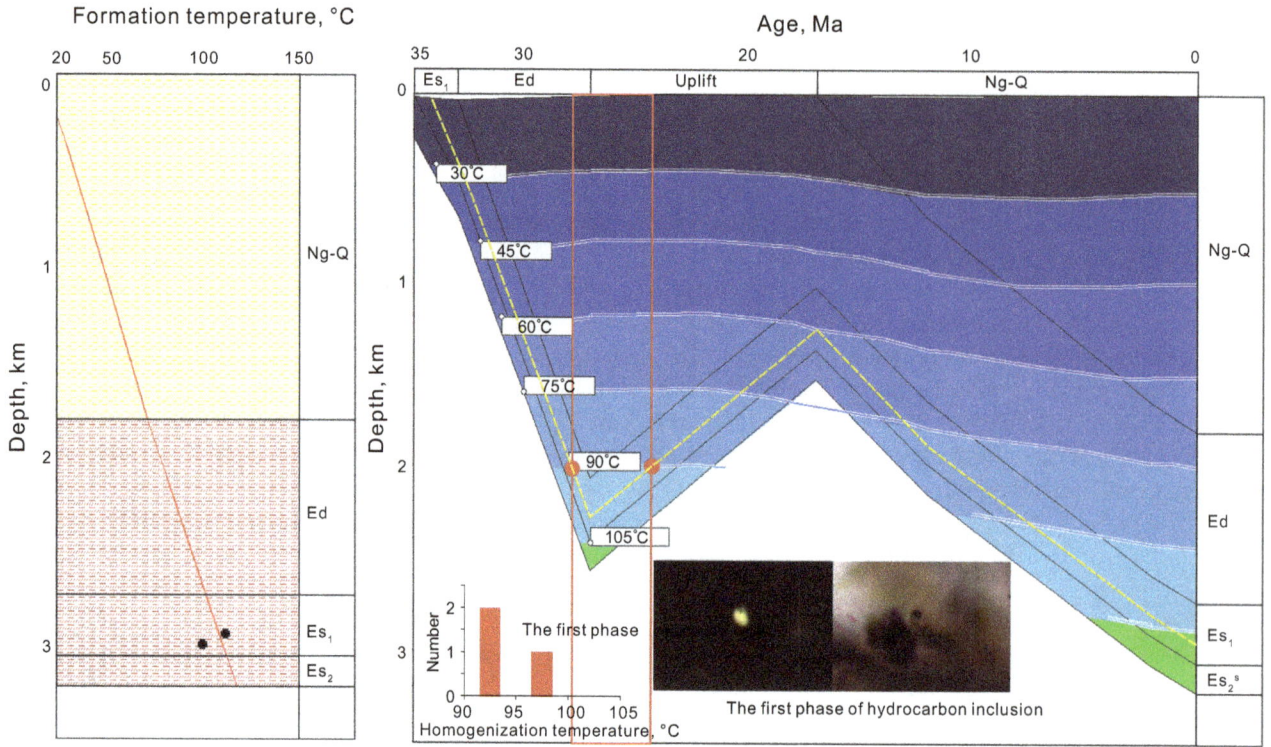

(a) Liu17, 2959.15m

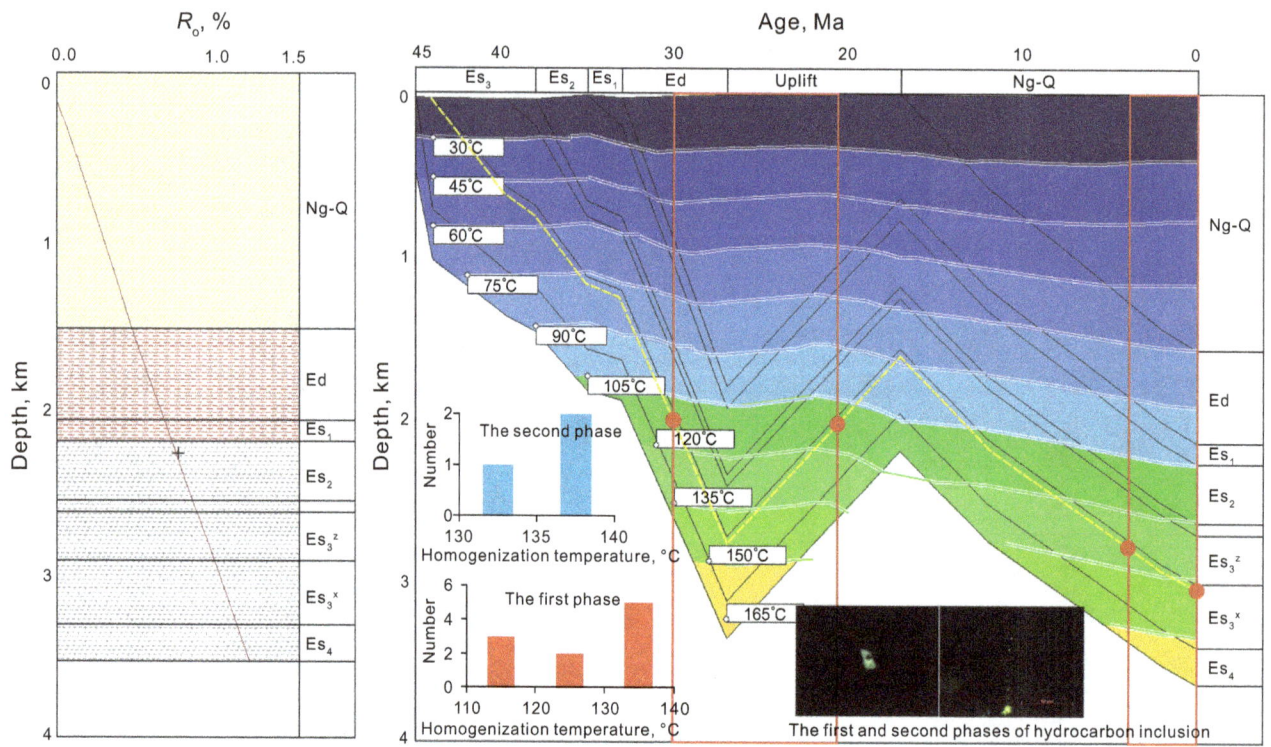

(b) Qiao 33, 3442.3m

Fig. 8 Hydrocarbon accumulation time determined by homogenization temperature combined with geothermal and burial history

Table 2 Characteristics of fluid inclusions and hydrocarbon accumulation time in different areas in the northern Dongpu Depression, Bohai Bay Basin

Area	Tectonic belt	Well	Formation	Depth, m	Phases	Associated brine inclusion homogenization temperature, °C	Hydrocarbon charge time, Ma
Wenliu–Liuzhuang area	Eastern sag belt	Pushen4	Es_3^s	3703.2	I, II	105–130, 125–130	30–27, 5–0
	Western sag belt	Pushen14	Es_3^s	3983.9	I, II	140–155, 135–145	28.8–23.8, 5–0
		Wen243	Es_3^z	4273.5	I, II	140–155, 150–160	30.5–28.8, 5.5–0
	Wenxi fault zone	Wen250	Es_3^z	3605.8	I	120–135	30.5–28.8
	Wendong graben zone	Wen260	Es_3^z	3577.7	I	105–135	30.9–27.8
		Wen133	Es_2	2903.3	I	110–125	31.6–29.8
		Wen152	Es_3^s	3194	I	105–120	29.2–28
	Wenzhong Horst zone	Wen48	Es_2	2548.59	I	100–120	29–23.5
		Wen95	Es_3^z	2753.9	I	85–110	29.2–24
		Wen95-17	Es_3^z	2876.75	I	90–110	29.4–23.2
	Wendong rolling anticlinal zone	Wen244	Es_3^z	3477.6	I	120–130	29.8–28.8
		Wen25–5	Es_2	2282.8	I	85–100	29.5–28
		Wen92-27	Es_3^s	2825.6	I	95–110	28.8–24.8
		Wen92-56	Es_3^s	2748.45	I	90–110	28.3–25.1
		Wen96	Es_2	2524.55	I	80–100	28.7–23.6
	Wendong reverse roof ridge zone	Pushen7	Es_3^s	4156.9	I	115–135	32.2–29.9
		Pushen7	Es_3^z	3696.7	I, II	80–135, 120–140	32.5–27, 7.5–0
		Wen126	Es_2	2922.6	I, II	100–120, 95–110	28.4–23.3, 6–0
		Wen210	Es_3^z	3920.25	I	120–135	30.5–29.2
	Liuzhuang area	Liu17	Es_1	2959.15	I	90–100	28–24.5
		Liu20	Es_2	3554.2	I	100–130	30.6–28
		Liu6	Es_3^s	3882.8	I, II	120–140, 125–140	29.5–27, 7–2.5
		Liu7	Es_2	3525.3	I, II	110–1252, 10–135	29.8–28.4, 5.4–0
Pucheng–Weicheng area	Pucheng reverse roof ridge zone	Pu120	Es_3^s	3054.3	I	110–120	29.3–28
		Pu120	Es_3^z	3264	II	120–135	6–0
		Pu20	Es_2	2756.5	I, II	90–130, 100–115	30.5–27, 5–0
		Pu47	Es_3^s	3104.53	I	110–115	29.5–27.8
		Pu6-65	Es_3^s	3108.25	I	120–130	28.4–24
		Pu78	Es_3^s	3279.25	I, II	105–125, 310–140	30–28.6, 6.5–0
	Weicheng Horst zone	Pu97	Es_3^s	3120	I, II	120–140, 105–120	29–27, 5–0
		Wei145	Es_3^s	2785.32	I	135–150	28.1–25
		Wei352	Es_2	2252.6	I, II	80–105, 70–110	29.5–27.5, 3.5–0
		Wei37-12	Es_2	2072	I, II	80–100, 75–90	29.2–27.5, 4.5–0
		Wei42	Es_3^z	3446	I, II	125–135, 115–130	29.5–28, 4.5–0
		Wei50	Es_3^s	2737	I, II	120–130, 95–100	27.8–25.7, 4.2–0
		Wei63	Es_2	1860	I, II	80–90, 56–80	28–25, 2.3–0
		Weiqi1	Es_3^s	2660	I, II	115–130, 100–110	29.4–27.8, 3.6–0
Western slope belt	Mazhai	Wei326	Es_3^s	3344.8	I, II	115–135, 105–120	31–29, 8.5–2
	Xinzhuang	Hu19	Es_3^z	1820	I	90–120	28.8–24.5
	Huzhuangji–Qingzuji Area	Hu7	Es_3^z	1892.6	I, II	80–95	28.8–24.8
		Hu40	Es_3^z	2660.3	I	90–105	30.5–28.5
		Hu41	Es_3^z	3888.4	I	150–170	29.2–23.1
		Hu10	Es_3^s	2233.9	I, II	90–100, 105–130	30.7–29, 3.6–0
		Qing61	Es_3^s	2880.8	I	95–130	30–22.2
		Qing63	Es_3^s	3144	I, II	130–135, 120–130	28.2–26.3, 6–0
		Qing11	Es_3^s	2382	I, II	90–110, 90–120	29.2–23.7, 7–0

Table 2 continued

Area	Tectonic belt	Well	Formation	Depth, m	Phases	Associated brine inclusion homogenization temperature, °C	Hydrocarbon charge time, Ma
Qiaokou–Baimiao area	Baimiao	Bai18	Es_3^s	3326.5	I	100–140	31.5–28
		Bai19	Es_3^s	3891.7	I, II	100–120, 100–130	31.5–29, 7–0
	Qiaokou	Qiao21	Es_2	2525	I	95–110	28–25.4
		Qiao33	Es_3^z	3164	I, II	110–140, 130–140	30–27, 4–0

cements, internal cracks of quartz grains and feldspar dissolution pores were formed in the late Dongying deposition period, and they were formed in the same period. The hydrocarbon inclusions occurred in the cracks through quartz grains were formed in the reburial period after the tectonic uplift, the second phase of hydrocarbon inclusions. Therefore, the hydrocarbon inclusions can be divided into two phases, which reflects the two hydrocarbon accumulation processes, the main characteristics of each stage of the hydrocarbon inclusions are as follows.

The first phase of hydrocarbon inclusions is distributed in carbonate cements, internal cracks of quartz grains and feldspar dissolution pores which are gas–liquid two-phase hydrocarbon inclusions with sliver shape or irregular shape, isolated or sporadic distribution, variable size, yellow, yellowish white or yellowish green fluorescence, and brown, light brown or colorless under transmitted light. They correspond with the early oil and gas filling (Fig. 6a–h).

The second phase of hydrocarbon inclusions occurs in cracks through quartz grains in a beaded distribution, with ellipse shapes, usually small, yellow green to bluish green in fluorescence, colorless under transmitted light. They correspond to the late highly mature stage of oil and gas filling (Fig. 6i–l).

4.5 Inclusion homogenization temperature characteristics and determination of hydrocarbon accumulation timing

The time of formation of hydrocarbon inclusions of different phases was determined by the combined analysis of homogenization temperature and reservoir burial heating history (Fig. 8). The homogenization temperature of fluid inclusions was tested, and the hydrocarbon charge period was determined in different tectonic units and different layers in this study (Table 2).

Based on the comprehensive analysis of hydrocarbon generation and expulsion histories of major source rocks, fluorescence microscopic characteristics and reservoir fluid inclusions, the pool-forming period was divided into two phases: The first phase is from late Dongying deposition period to the early uplift stage in late Oligocene and the

second is from late Minghuazhen deposition period in Pliocene to Quaternary. The first period is the main accumulation period.

By comparing the pool-forming time in different tectonic belts and different oil and gas bearing series, it can be seen that the first stage of hydrocarbon accumulation occurs widely, distributed in the whole region, and the second stage of the hydrocarbon reservoirs was mainly distributed near the sub-depressions, including the Puwei sag belt, Huqing area, Qiaokou and Baimiao area, Wendong area, the south of Wenxi area, the west of Liuzhuang area, Qianliyuan sub-sag and Haitongji sub-depression. The central uplift belt and the western slope zone of northern Xinzhuang, Mazhai, and the distant-free hydrocarbon accumulation area only have the first stage of pool-forming. In the vertical direction, the first stage is widely distributed in a large depth span, mainly distributed in the depth range of 2000–3500 m. From the view of layers, there was a single stage in the upper layers and two stages in lower layers. Specifically, the second section of the Shahejie group and the upper and middle layers of the third section of the Shahejie group were mainly in the first stage, while the lower part of the third section of the Shahejie group was in two stages (Fig. 9).

The difference of the hydrocarbon charge history in different parts and different layers is closely related to fault activity. The fault system is highly developed in the Dongpu Depression and is the main channel for oil and gas migration. The activity rate of the main secondary oil source faults was calculated using the fault activity rate method (Table 3). The main oil source faults in the study area became active in the early stage of Es_3. The activity rate was the highest in the late Es_3 and Es_2 and decreased gradually in the Es_1 and Dongying period. The rate of fault activity could not be calculated during the uplift period because there is no sediment, but the tectonic movement in this period is very intense, and it is inferred that most faults are active. Most of the faults have not cut through the Guantao Formation, so that most of the fault activity stopped in the Early Neogene.

In the first stage of oil and gas accumulation, the source of oil and gas was abundant, and the fault activity was

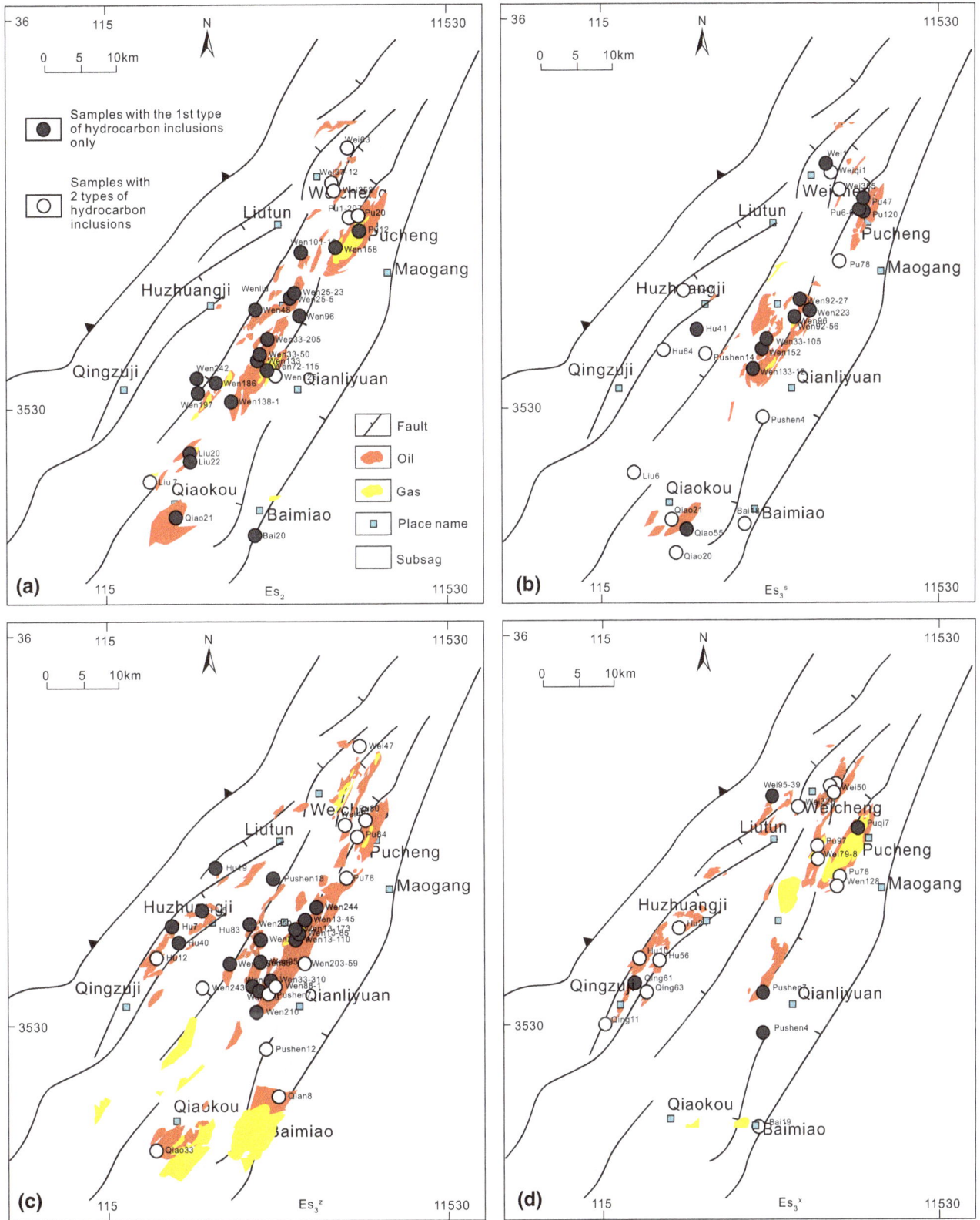

Fig. 9 Distribution of hydrocarbon inclusions in different stages in the major petroleum bearing formations of the northern Dongpu Depression, Bohai Bay Basin

Table 3 Activity rate and time of the main source faults of the Dongpu Depression, Bohai Bay Basin

Fault name	Fault active rate, m/Ma						Fault active time
	Ed	Es_1	Es_2	Es_3^s	Es_3^z	Es_3^x	
Huanghe	6	51	88	114	76	–	Es_3^z–Ed
Wendong 2	68	–	–	–	–	–	Ed
Wendong 3	21	53	13	–	–	–	Es_2–Ed
Wenxi 1	19	56	139	237	–	–	Es_3^s–Ed
Wenxi 2	40	32	12	–	–	–	Es_2–Ed
Xulou	11	32	55	37	–	–	Es_3^s–Ed
Changyuan	146	123	292	885	52	152	Es_3^s–Ed
Mazhai	–	–	–	–	7	92	Es_3^s–Es_3^z

strong, so the oil and gas migrated to different layers and depths to accumulate. During the period of Dongying movement, an uplift occurred throughout the depression, the fault activity increased, and had great influence on the distribution and reorganization of oil and gas reservoirs, which resulted in a wide range of oil and gas distribution in the plane and vertical direction. In the second phase of hydrocarbon charging period, most of the faults were not active, and due to the plugging of salt rocks, oil and gas could not migrate along faults at a large-scale, so most accumulated near the sources. This results in the oil and gas formed in the second phase being mainly distributed in the vicinity of the sags in the horizontal plane and mainly concentrated in the main hydrocarbon generation layers vertically.

5 Conclusions

1. The main hydrocarbon source rocks of the Dongpu Depression mainly had two expulsion periods. The fluorescence microscopic features also indicated two stages of hydrocarbon accumulation. The carbonaceous asphalts, bituminous asphalts and colloidal asphalts reflect an early oil and gas injection phase, whereas the oleaginous asphalt reflects a second oil and gas injection phase.

2. According to the diagenetic evolution sequence, fluid inclusion petrography and the reservoir burial history analysis, hydrocarbon inclusions were divided into two formation phases. Phase I occurs in carbonate cements, internal cracks of quartz grains and dissolution pores of feldspars, corresponding to the early hydrocarbon filling. Phase II occurs in late cracks through quartz grains, corresponding to the late oil and gas filling.

3. Based on the comprehensive analysis of the hydrocarbon generation and expulsion histories of the major source rocks, fluorescence microscopic characteristics

and reservoir fluid inclusions, the pool-forming in northern Dongpu Depression is found to be in two phases. The first phase was from the late Dongying depositional period to the early uplift period in the late Oligocene, while the second was from the late Minghuazhen period of the Pliocene to the Quaternary, with the first phase being the main hydrocarbon charge phase. The first phase of hydrocarbon reservoir charging was distributed across the whole region, whereas the second phase of the hydrocarbon reservoir accumulation was mainly distributed near the sub-sags and mainly in the hydrocarbon generation layers. In the vertical direction, it was characterized by a single phase in the upper layers and two phases in the lower layers within the Shahejie group.

Acknowledgements This work was supported by the Important National Science & Technology Specific Projects (Grant No. 2011ZX05006-003/004).

References

Chen SP, Qi JF, Wang DR, et al. Fault systems and transfer structures in Dongpu Sag. Acta Pet Sin. 2007;28(1):43–9.

Dolníček Z, Kropáč K, Janíčková K, et al. Diagenetic source of fluids causing the hydrothermal alteration of teschenites in the Silesian Unit, Outer Western Carpathians, Czech Republic: petroleum-bearing vein mineralization from the Stříbrník site. Mar Pet Geol. 2012;37(1):27–40.

Gui Y, Liu K, Liu S, et al. Hydrocarbon charge history of Yingdong Oilfield, Western Qaidam Basin. Earth Sci J China Univ Geosci. 2015;40(5):890–9.

Guo LG, Tian H, Jin YB, et al. Reaction mechanism, medium influencing factors and identification and evaluation of oil-cracking gas. Geochimica. 2008;37(5):499–511 **(in Chinese)**.

Guo S, Tan LJ, Lin CY, et al. Hydrocarbon accumulation characteristics of beachbar sandstones in the southern slope of the Dongying Sag, Jiyang Depression, Bohai Bay Basin, China. Pet Sci. 2014;11(2):220–33.

Guo XW, Liu KY, He S, et al. Petroleum generation and charge history of the northern Dongying Depression, Bohai Bay Basin, China: insight from integrated fluid inclusion analysis and basin modelling. Mar Pet Geol. 2012;32(1):21–35.

Jarmołowicz-Szulc K, Łukasz K, Marynowski L. Fluid circulation and formation of minerals and bitumens in the sedimentary rocks of the Outer Carpathians—based on studies on the quartz–calcite–organic matter association. Mar Pet Geol. 2012;32(1):138–58.

Jiang FJ, Pang XQ, Yua S, et al. Charging history of Paleogene deep gas in the Qibei sag, Bohai Bay Basin, China. Mar Pet Geol. 2015a;2015(67):617–34.

Jiang YL, Liu H, Song GQ, et al. Relationship between geological structures and hydrocarbon enrichment of different depressions in the Bohai Bay Basin. Acta Geol Sin. 2015b;89(6):1998–2011.

Jiang YL, Liu P, Song GQ, et al. Late Cenozoic faulting activities and their influence upon hydrocarbon accumulations in the Neogene in Bohai Bay Basin. Oil Gas Geol. 2015c;2015(4):525–33 **(in Chinese)**.

Lang DS, Jiang DH, Yue X, et al. Fluorescence microscopy, light hydrocarbon analysis technology and their application in oil exploration and development. Beijing: Petroleum Industry Press; 2008. p. 15–48 (**in Chinese**).

Li HT. Accumulation process and pattern of oolitic shoal gas pools in the platform: a case from member 3 of Lower Triassic Feixianguan Formation in the Heba area, northeastern Sichuan Basin. Pet Explor Dev. 2016;43(5):18–23.

Li N, Chen YJ, Deng XH, et al. Fluid inclusion geochemistry and ore genesis of the Longmendian Mo deposit in the East Qinling Orogen: implication for migmatitic-hydrothermal Mo-mineralization. Ore Geol Rev. 2014;63(1):520–31.

Lin ZM, Zhu XP, You F, et al. Nuclei fluorescence microscopic observation on early embryonic development of mitogynogenetic diploid induced by hydrostatic pressure treatment in olive flounder (Paralichthys olivaceus). Theriogenology. 2015;83(8):1310–20.

Liu DH, Lu HZ, Xiao XM. Fluid inclusions and their application in hydrocarbon exploration and development. Guangzhou: Guangdong Science and Technology Press; 2007. p. 29–44 (**in Chinese**).

Liu KY, Julien B, Zhang BS, Zhang N, et al. Hydrocarbon charge history of the Tazhong Ordovician reservoirs, Tarim Basin as revealed from an integrated fluid inclusion study. Pet Explor Dev. 2013;40(02):171–80.

Liu L, Ren ZL, Cui YB, et al. Distribution of present-day geothermal field in the Dongpu Sag. Chin J Geol. 2007;42(4):787–94 (**in Chinese**).

Liu Y, Zhong NN, Tian YJ, et al. The oldest oil accumulation in China: meso-proterozoic Xiamaling Formation bituminous sandstone reservoirs. Pet Explor Dev. 2011;38(4):503–12.

Lu XS, Jiang YL, Chang ZH, et al. Calculation of the erosion thickness of Dongying Formation in Dongpu Depression and its significance. Geol Sci Technol Inf. 2007;26(2):8–12 (**in Chinese**).

Luo XP, Cao J, Shen ZM. Geochemical characteristics and genesis of reservoir bitumen of Xujiahe Formation in the Upper Triassic of Western Sichuan Depression. J Mineral Pet. 2009;29(1):93–8 (**in Chinese**).

Lü ZX, Ye SJ, Yang X, et al. Quantification and timing of porosity evolution in tight sand gas reservoirs: an example from the Middle Jurassic Shaximiao Formation, western Sichuan, China. Pet Sci. 2015;12(2):207–17.

Makeen YM, Abdullah WH, Pearson MJ, et al. Thermal maturity history and petroleum generation modelling for the Lower Cretaceous Abu Gabra Formation in the Fula Sub-basin, Muglad Basin, Sudan. Mar Pet Geol. 2016;2016(75):310–24.

Muhammad T, Asifullah K. Protein subcellular localization of fluorescence microscopy images: employing new statistical and Texton based image features and SVM based ensemble classification. Inf Sci. 2016;2016(345):65–80.

Pang XQ, Jia CZ, Wang WY. Petroleum geology features and research developments of hydrocarbon accumulation in deep petroliferous basins. Pet Sci. 2015;12(1):1–53.

Parnell J. Potential of palaeofluid analysis for understanding oil charge history. Geofluids. 2010;10(1–2):73–82.

Qin KZ, Guo SH. Petroleum asphaltene. Beijing: Petroleum Industry Press. 2002; p. 2–16 (**in Chinese**).

Shalaby MR, Hakimi MH, Wan HA. Geochemical characterization of solid bitumen (migrabitumen) in the Jurassic sandstone reservoir of the Tut Field, Shushan Basin, northern Western Desert of Egypt. Int J Coal Geol. 2012;100(3):26–39.

Shan XQ, Zhang BM, Zhang J, et al. Paleofluid restoration and its application in studies of reservoir forming: a case study of the Ordovician in Tarim Basin, NW China. Pet Explor Dev. 2015;42(3):301–10.

Shokrlu YH, Babadagli T. In-situ upgrading of heavy oil/bitumen during steam injection by use of metal nanoparticles: a study on in situ catalysis and catalyst transportation. SPE Reserv Eval Eng. 2013;16(3):333–44.

Tao SZ. Sequence of diagenetic authigenic mineral: the basis of timing the inclusions formation in sedimentary rocks. Pet Explor Dev. 2006;33(2):154–60 (**in Chinese**).

Wang B, Feng Y, Zhao YQ, et al. Determination of hydrocarbon charging history by diagenetic sequence and fluid inclusions: a case study of the Kongquehe area in the Tarim Basin. Acta Geologica Sinica (English Edition). 2015a;89(3):876–86.

Wang TT, Yang SY, Duan SS, et al. Multi-stage primary and secondary hydrocarbon migration and accumulation in lacustrine Jurassic petroleum systems in the northern Qaidam Basin, NW China. Mar Pet Geol. 2015b;62:90–101.

Wu N, Cai ZX, Yang HJ, Wang ZQ, et al. Hydrocarbon charging of the Ordovician reservoirs in Tahe-Lunnan area, China. Sci China (Earth Sci). 2013;56(5):763–72.

Xiao H, Zhao JZ, Yang HJ, et al. Fluid inclusion and micro-FTIR evidence for hydrocarbon charging fluid evolution of the Ordovician reservoir of Halahatang Depression, the Tarim Basin. Earth Sci J China Univ Geosc. 2012;37:163–73.

Xiao ZY, Li MJ, Huang SY, et al. Source, oil charging history and filling pathways of the Ordovician carbonate reservoir in the Halahatang Oilfield, Tarim Basin, NW China. Mar Pet Geol. 2016;73:59–71.

Yang P, Xie Y, Wang ZJ, et al. Fluid activity and hydrocarbon accumulation period of Sinian Dengying Formation in northern Guizhou, South China. Pet Explor Dev. 2014;41(3):346–57.

Ying FX, Luo P, He D B. Diagenesis and diagenetic numerical simulation of clastic-rock reservoirs in petroleum basins of China. Beijing: Petroleum Industry Press. 2004; p. 47–86 (**in Chinese**).

Zhao XZ, Jin Q, Jin FM, et al. Origin and accumulation of high-maturity oil and gas in deep parts of the Baxian Depression, Bohai Bay Basin, China. Pet Sci. 2013;10(3):303–13.

Permissions

All chapters in this book were first published in PS, by Springer International Publishing AG.; hereby published with permission under the Creative Commons Attribution License or equivalent. Every chapter published in this book has been scrutinized by our experts. Their significance has been extensively debated. The topics covered herein carry significant findings which will fuel the growth of the discipline. They may even be implemented as practical applications or may be referred to as a beginning point for another development.

The contributors of this book come from diverse backgrounds, making this book a truly international effort. This book will bring forth new frontiers with its revolutionizing research information and detailed analysis of the nascent developments around the world.

We would like to thank all the contributing authors for lending their expertise to make the book truly unique. They have played a crucial role in the development of this book. Without their invaluable contributions this book wouldn't have been possible. They have made vital efforts to compile up to date information on the varied aspects of this subject to make this book a valuable addition to the collection of many professionals and students.

This book was conceptualized with the vision of imparting up-to-date information and advanced data in this field. To ensure the same, a matchless editorial board was set up. Every individual on the board went through rigorous rounds of assessment to prove their worth. After which they invested a large part of their time researching and compiling the most relevant data for our readers.

The editorial board has been involved in producing this book since its inception. They have spent rigorous hours researching and exploring the diverse topics which have resulted in the successful publishing of this book. They have passed on their knowledge of decades through this book. To expedite this challenging task, the publisher supported the team at every step. A small team of assistant editors was also appointed to further simplify the editing procedure and attain best results for the readers.

Apart from the editorial board, the designing team has also invested a significant amount of their time in understanding the subject and creating the most relevant covers. They scrutinized every image to scout for the most suitable representation of the subject and create an appropriate cover for the book.

The publishing team has been an ardent support to the editorial, designing and production team. Their endless efforts to recruit the best for this project, has resulted in the accomplishment of this book. They are a veteran in the field of academics and their pool of knowledge is as vast as their experience in printing. Their expertise and guidance has proved useful at every step. Their uncompromising quality standards have made this book an exceptional effort. Their encouragement from time to time has been an inspiration for everyone.

The publisher and the editorial board hope that this book will prove to be a valuable piece of knowledge for researchers, students, practitioners and scholars across the globe.

List of Contributors

Chun-Mao Chen
State Key Laboratory of Heavy Oil Processing, China University of Petroleum, Beijing 102249, China
Department of Molecular Biosciences and Bioengineering, University of Hawaii at Manoa, Honolulu, HI 96822, USA

Jin-Ling Wang, Qing-Hong Wang and Jing Wang
State Key Laboratory of Heavy Oil Processing, China University of Petroleum, Beijing 102249, China

Qing X. Li
Department of Molecular Biosciences and Bioengineering, University of Hawaii at Manoa, Honolulu, HI 96822, USA

Jung Bong Kim
Department of Agro-Food Resources, National Institute of Agricultural Sciences, Rural Development Administration, Jeonju 55365, Republic of Korea

Brandon A. Yoza
Hawaii Natural Energy Institute, University of Hawaii at Manoa, Honolulu, HI 96822, USA

Zhen-Zhen Jia, Cheng-Yan Lin, Li-Hua Ren and Chun-Mei Dong
School of Geosciences, China University of Petroleum, Qingdao 266580, Shandong, China
Key Laboratory of Reservoir Geology in Shandong Province, Qingdao 266580, Shandong, China

Shao-Lin Qiu and Lai-Bin Zhang
China University of Petroleum (Beijing), Beijing 102249, China

Mu Liu
CNPC Research Institute of Safety and Environment Technology, Beijing 102206, China

Yi Jin, Xu Tang, Cui-Yang Feng, Jian-Liang Wang and Bao-Sheng Zhang
School of Business Administration, China University of Petroleum, Beijing 102249, China

Zheng-Xiang Lü
College of Energy Resources, Chengdu University of Technology, Chengdu 610059, Sichuan, China
State Key Laboratory of Oil-Gas Reservoirs Geology and Exploitation, Chengdu University of Technology, Chengdu 610059, Sichuan, China

Shun-Li Zhang, Chao Yin, Hai-Long Meng, Xiu-Zhang Song and Jian Zhang
College of Energy Resources, Chengdu University of Technology, Chengdu 610059, Sichuan, China

Guangwei Ren
Petroleum and Geosystems Engineering Department, University of Texas at Austin, Austin, TX, USA
Total E&P R&T USA, Houston, TX, USA

Quoc P. Nguyen
Petroleum and Geosystems Engineering Department, University of Texas at Austin, Austin, TX, USA

Dian-Fa Du, Yan-Wu Zhao and Xi Xia
School of Petroleum Engineering, China University of Petroleum, Qingdao 266580, Shandong, China

Yan-Yan Wang
Sinopec Petroleum Exploration and Production Research Institute, Sinopec, Beijing 100083, China

Pu-Sen Sui
Shengli Production Plant, Shengli Oilfield Branch Company, Sinopec, Dongying 257051, Shandong, China

Mehrdad Soleimani
Faculty of Mining, Petroleum and Geophysics, Shahrood University of Technology, Shahrood, Iran

Jin Su, Yu Wang, Xiao-Mei Wang, Kun He, Hui-Tong Wang, Hua-Jian Wang, Bin Zhang, Ling Huang, Na Weng and Li-Na Bi
Key Laboratory of Petroleum Geochemistry, CNPC, Beijing 100083, China
State Key Laboratory for Enhancing Oil Recovery, Research Institute of Petroleum Exploration and Development, PetroChina, Beijing 100083, China

Hai-Jun Yang
Research Institute of Tarim Oilfield Company, PetroChina, Korla 841000, Xinjiang, China

Zhi-Hua Xiao
PetroChina Coalbed Methane Company Limited, Beijing 100028, China

Wen-Tao Zhao
The Key Laboratory of Orogenic Belts and Crustal Evolution of Ministry of Education, School of Earth and Space Sciences, Peking University, Beijing 100871, China
China Huaneng Clean Energy Research Institute, Beijing 102209, China
PetroChina Research Institute of Petroleum Exploration and Development, Beijing 100083, China

Gui-Ting Hou
The Key Laboratory of Orogenic Belts and Crustal Evolution of Ministry of Education, School of Earth and Space Sciences, Peking University, Beijing 100871, China

Saad Mehmood
United Energy Pakistan, Bahria Complex-1, M.T. Khan Road, Karachi 74000, Sindh, Pakistan

Abeeb A. Awotunde
Department of Petroleum Engineering, King Fahd University of Petroleum and Minerals, Dhahran 31261, Saudi Arabia

Lun Zhao, Shu-Qin Wang, Man Luo, Cheng-Gang Wang, Hai-Li Cao and Ling He
Research Institute of Petroleum Exploration and Development, CNPC, Beijing 100083, China

Wen-Qi Zhao
Research Institute of Petroleum Exploration and Development, CNPC, Beijing 100083, China
School of Energy Resources, China University of Geosciences, Beijing 100083, China

D. A. Angus
School of Earth and Environment, University of Leeds, Leeds, UK
ESG Solutions, Kingston, Canada

Q. J. Fisher
School of Earth and Environment, University of Leeds, Leeds, UK

J. M. Segura
Formerly School of Earth and Environment, University of Leeds, Leeds, UK
Repsol, Madrid, Spain

J. P. Verdon and J.-M. Kendall
Department of Earth Sciences, University of Bristol, Bristol, UK

M. Dutko
Rockfield Software Ltd., Swansea, UK

A. J. L. Crook
Three Cliffs Geomechanics, Swansea, UK

You-Lu Jiang, Jing-Dong Liu and Hong-Jin Hu
School of Geosciences, China University of Petroleum, Qingdao 266580, Shandong, China

Lei Fang
CNOOC Research Institute, Beijing 100028, China

Tian-Wu Xu
SINOPEC Zhongyuan Oil Company, Puyang 457001, Henan, China

Index

A

Anaerobic Stage, 4-6

Analcite, 13-24, 26-27

B

Bio-stimulation, 1-6, 10

Bottom Water Reservoirs, 88-89, 98

C

Calcite, 13-24, 26-27, 48, 50, 54, 108, 201-205, 210

Carbon Nutrient, 3, 6, 10

Carbonate Rock, 1, 54, 106, 166

Co2-soluble Surfactant, 56-60, 62, 64-65, 67-68, 70-71, 81-84

Condensates, 116-117, 120-122, 124, 126-127

Coupled Fluid-flow, 179-180, 193

Cresting Model, 88, 98

Crude Oil, 1-2, 6, 8-9, 12, 36, 44, 117, 127, 201-202, 205

Cumulative Free-water Production, 88, 97

D

Discrete Fracture Model, 100-101

Dissolution, 13-14, 16-22, 24, 26-27, 45, 47-50, 52-54, 168, 170, 175, 202, 204-205, 208, 210

Distribution Regularity, 166, 175

Dolomite Reservoir, 116, 177

Dongpu Depression, 195-197, 199, 201-205, 207-211

E

Efficiency, 1, 11, 20, 40, 56, 58, 61, 63-68, 70-71, 73, 75-77, 80, 82-86, 112

Elastic Gridding, 100, 102-103, 107, 112

Energy Strategy, 39

Eodiagenetic Cements, 13

Expulsion History, 195-196, 199

F

Finite Element Modeling, 129, 140, 148-149

Fluid Inclusion, 45, 47, 54, 195-196, 198, 202, 210-211

Fluorescence Microscopy, 195, 197-198, 210

Foam, 56-68, 70-77, 79-87

Formation Water, 2, 80, 109, 116-117, 119-120, 122-126

Fracture Prediction, 129-131, 137-140, 145-149, 151

G

Generalized, 39-41, 43-44, 131, 138

Geomechanics, 114-115, 179, 193-194

Global Oil Supply, 39, 43-44

Grassroots Posts, 29-30, 35, 37

Gravity Segregation, 56-57, 63, 65, 67, 69-72, 82, 84-86

H

Hazard Identification, 31-32, 36

High-quality Reservoirs, 19, 45-46, 49-50

History Matching, 61, 100-103, 110, 112, 152-154, 157-165, 180

Hse, 29-33, 35-38

Hydrocarbon Charge History, 195-196, 208, 210

Hydrocarbon Generation, 46, 116, 195-196, 198-199, 210

I

Inflow Profile, 88-89, 95, 97, 99

Inverse Analysis, 152-154, 157, 163-165

M

Microbial Community, 1, 3-4, 7-12

Microbial Diversity, 1-3, 7-10

Mixed Sediments, 45-47, 50

Multi-cycle, 39-41, 43

N

Naturally Fractured Reservoir, 100, 115

O

Oil Market, 39-40, 44

Oil Production, 30-32, 34, 36, 39-41, 43-44, 88-89, 97-99, 106

Optimal Injection Strategy, 56-59

P

Perforation Completion, 88-89, 95

Petrochemical Enterprise, 29

Petrography, 13, 15, 17, 19, 24, 46, 178, 195, 198, 202, 210

Petrography Textures, 13, 19

Petroleum, 1, 8, 11-13, 26-30, 35, 39, 44, 46-47, 54, 56, 85-88, 98, 100, 116-117, 126-130, 147, 149, 152, 165-166, 177-178, 193, 195-197, 199, 201-202, 205, 209-211

Platform Facies, 166, 168, 170, 177
Pore Water, 13, 23-24, 26

Q
Quality Evolution, 13-14, 17

R
Reservoir Characterization, 100, 115, 154, 193
Reservoir Performance, 100
Reservoir Space Type, 166
Reservoir Type, 166, 174-175
Response Surface Methodology (RSM), 2
Rice Bran, 1-6, 10-12

S
Secondary Porosity, 13, 19-20, 26-27
Seismic Anisotropy, 179-182, 184, 190-194
Selective Dissolution, 13-14, 16, 18, 20-22, 26-27, 168
Sensitivity, 86, 115, 152-154, 157, 161-165, 180, 190-191
Siliciclastic-carbonate, 45-48, 50-55
Single Factor Optimization (SFO), 2
Sterilization, 2
Stress Path, 179-180, 182-183, 185-186, 190-192, 194

Sulfate-cips, 116, 124-126
Surface Tension, 2-6, 8
Sweep, 56-58, 61-63, 66-68, 70-71, 73, 75, 80, 82-87

T
Tarim Basin, 55, 116-117, 120, 127-128, 147, 149, 211
Template, 29-30
Template Development, 29-30
Tight-sand Reservoirs, 129, 149
Training Matrices, 29-33, 35-38
Tsr, 116-117, 122, 124-128
Two-factor Method, 129, 138, 140, 144, 147, 149

U
Upscaling, 85, 114-115, 152-154, 156-165

V
Variable Density, 88-89, 95-96, 99

W
Wavelets, 152, 159, 163-165
Weng Model, 39-41, 43-44

Y
Yanchang Formation, 27, 129, 132, 146, 148-151

www.ingramcontent.com/pod-product-compliance
Lightning Source LLC
Chambersburg PA
CBHW082041190326
41458CB00010B/3426